AF353111

Technologies of Coatings and Surface Hardening for Tool Industry II

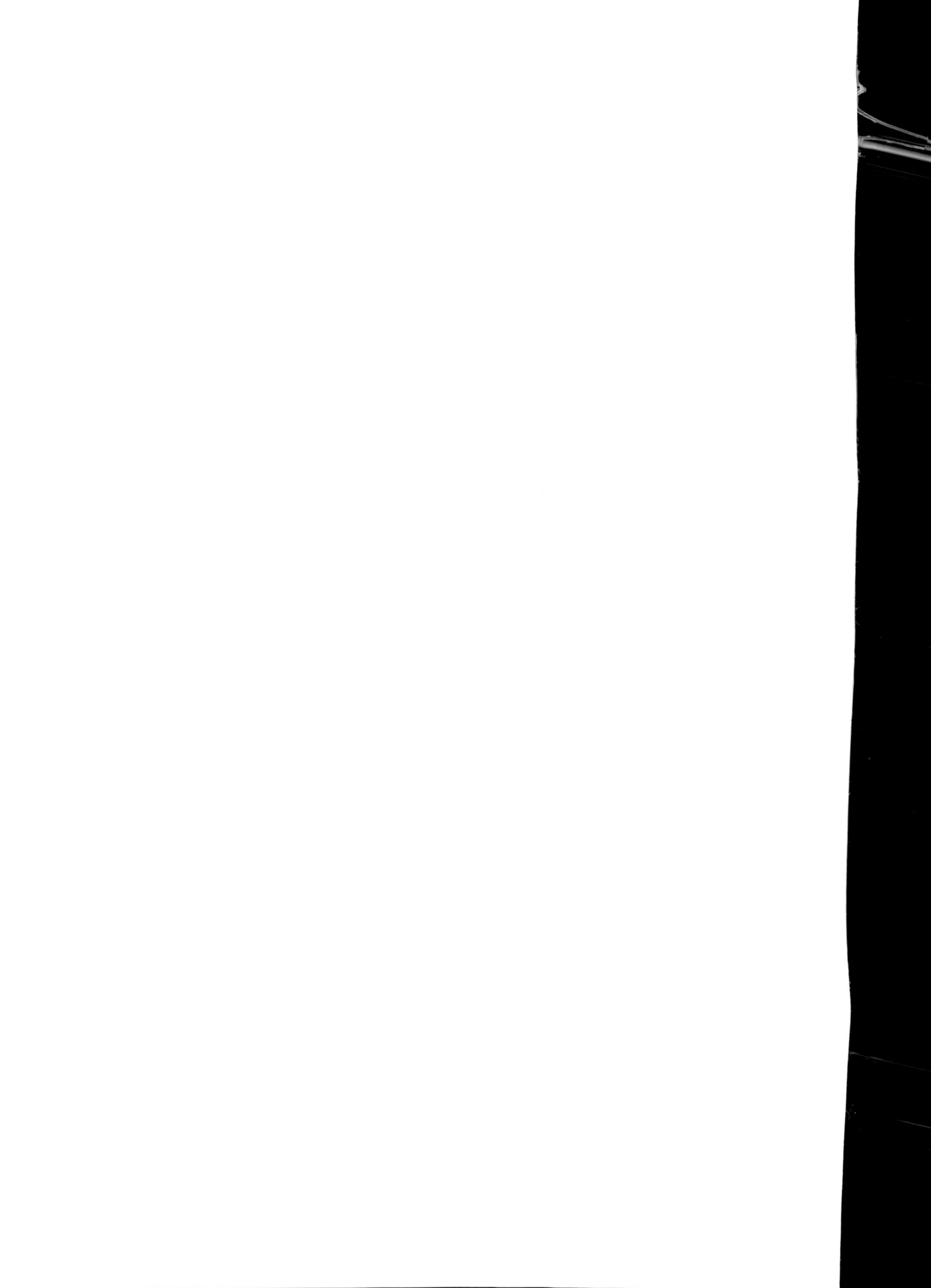

Technologies of Coatings and Surface Hardening for Tool Industry II

Editor

Sergey N. Grigoriev

Basel • Beijing • Wuhan • Barcelona • Belgrade • Novi Sad • Cluj • Manchester

Editor
Sergey N. Grigoriev
Moscow State University of
Technology STANKIN
Moscow, Russia

Editorial Office
MDPI
St. Alban-Anlage 66
4052 Basel, Switzerland

This is a reprint of articles from the Special Issue published online in the open access journal *Coatings* (ISSN 2079-6412) (available at: https://www.mdpi.com/journal/coatings/special_issues/Coat_Surf_Tool_II).

For citation purposes, cite each article independently as indicated on the article page online and as indicated below:

Lastname, A.A.; Lastname, B.B. Article Title. *Journal Name* **Year**, *Volume Number*, Page Range.

ISBN 978-3-0365-8888-9 (Hbk)
ISBN 978-3-0365-8889-6 (PDF)
doi.org/10.3390/books978-3-0365-8889-6

Cover image courtesy of Moscow State University of Technology STANKIN, Moscow, Russian Federation

Contents

About the Editor

Sergey N. Grigoriev

Sergey N. Grigoriev, Doctor of Engineering Science, Professor, head of the Department of High-Efficiency Processing Technologies of Moscow State University of Technology STANKIN completed his higher education in 1982. The scientific team of the Department has received awards from the most prestigious international scientific competitions and exhibitions under the supervision of Dr. of Eng. Sci., Prof. S. N. Grigoriev. The brightest achievements present the advances in surface engineering, coating, and hardening technologies. The researchers and academic staff of the Department are multiple winners of grants from the President of the Russian Federation for the leading scientific schools in the field of Engineering and Technical Sciences. The research and academic staff of the Department develop projects in the fields of cutting tools, the efficiency of the most advanced and new technologies, adaptive, multi-component, and nanocomposite structures, and diagnostics of the technology processes. Prof. S.N. Grigoriev initiated and participated in various invited lectures at conferences. He serves as editor in the most authoritative international scientific journals all over the world, the author of over 180 patents, and more than 500 international scientific articles in the journals indexed in the Web of Science and Scopus with H-index 52 and 53, correspondingly.

Preface

The innovative Coatings and Surface Hardening technologies developed in recent years allow us to obtain practically any physical–mechanical or crystal–chemical complex property of the metalworking tool and other responsible products' surface layer. Today, the scientific approach to improving the operational characteristics of the product surface layers made of traditional industrial materials is a highly costly and long-lasting process. Different technological techniques, such as coatings, mainly physical and chemical methods, surface hardening and alloying (chemical–thermal treatment, implantation), a combination of the listed methods, and other solutions are used. The high efficiency of this approach may be explained by the fact that under the diversified operating conditions of the product, in all cases, the most loaded is its surface layer. First, its properties precisely define the product's workability during exploitation. There is no all-purpose coating and surface hardening method; everything is very particular for each type of operating condition.

The metalworking tool as a primary research object was not chosen accidentally. Production experience shows that even with the most advanced machine tools, it is not possible to achieve high technical and economical rates of part machining with a tool's low operation life rate. Diverse conditions of the tool operation bring on diverse injuries and failures of the technological system. Namely, the cutting tool is subjected to the most extreme mechanical and thermal loads in operation. The tool wear is significantly higher than the wear rate of other machine tool parts and units. Therefore, the operational ability of the technological system depends on the metalworking tool.

The leading scientific schools in the field of coating and surface hardening technologies, such as

- Moscow State University of Technology STANKIN,
- Korea Institute of Industrial Technology,
- Xi'an Jiaotong University,
- National Taipei University of Technology,
- National University of Science and Technology MISiS,

and others are the leading scientific and educational centers in the field of science and technology for the purposes of the machinery industry and other science-intensive manufacturing sectors, introduce traditional and innovative approaches in the industry, and conduct advanced research activities. The non-interrupted development of projects related to the critical stages of modern technology progress in Coatings and Surface Hardening for the tool industry and microelectronics needs demonstrates the most current and innovative published scientific results in the most outstanding high-rank scientific journals. The years of experience contribute to developing innovative products and technologies in developing multi-layer, nanocomposite, adaptive, and nano-thin coatings and technological features of their deposition and diagnostics. The most remarkable results of scientific activities are commissioned in engineering enterprises for manufacturing reliable and high-performance products. The most recent achievements of the invited colleagues from the leading research teams in Technologies of Coatings and Surface Hardening for cutting tools and microelectronics are presented in the Special Issue of Coatings.

Sergey N. Grigoriev

Editor

 *coatings*

Editorial

Technologies of Coatings and Surface Hardening: Industrial Applications

Sergey N. Grigoriev

Department of High-Efficiency Processing Technologies, Moscow State University of Technology STANKIN, Vadkovsky per. 1, 127055 Moscow, Russia; s.grigoriev@stankin.ru; Tel.: +7-499-972-94-92

Citation: Grigoriev, S.N. Technologies of Coatings and Surface Hardening: Industrial Applications. *Coatings* **2023**, *13*, 511. https://doi.org/10.3390/coatings13030511

Received: 17 February 2023
Accepted: 23 February 2023
Published: 25 February 2023

1. Introduction and Scope

The most advanced and recently developed coating and surface-hardening technologies make it possible to obtain almost the full range of physical–mechanical and crystal–chemical properties of the metalworking tool surface and electronic component surface for a wide range of applications to enlarge product operational life for working under the most extreme mechanical and thermal loads [1–3]. The scientific attitude to improving the surface layer parameters of the product made of traditional industrial materials or new advanced composites and nanocomposites requires budget material and time investments [4,5]. It hinders the industrial introduction of innovations and slows down the progress of the transition to the technological level associated with the concept of nanoindustry [6–8]. Different technological techniques, such as physical and chemical methods of coating deposition [9,10], surface hardening, and alloying, including chemical–thermal treatment and implantation technique [11,12], a combination of the above and other solutions, are widely proposed on the market. A wide range of energy sources such as vacuum-arc and laser beam in various working media (vacuum, gas–vapor medium, liquid solutions, etc.) can be used for product surface modification [2,13–15].

The product's operational life can be improved via introduced innovations in the coating and surface layer structure deposed to such materials as high-speed tool steel [16], Al_2O_3 + TiC ceramic [17,18], Si and hard alloy substrate [19–25], plastic substrate [26]. The innovative multilayer coatings demonstrate the destruction of the coating from the contact load passing through one layer and developing along the interface between the layers, significantly increasing the product's durability [19–25]. At the same time, nanocomposite coatings containing the amorphous phase with the introduced nano-crystallines of 3–5 nm hamper the development of the cracks and redirect them [23,24]. It is shown that annealing up to 1000 °C can transform an amorphous homogeneous coating structure into a nanocomposite one [18]. The interlayer introduced in the coating also improves its durability by improving coating adhesion [17,19,23]. Some of the coatings exhibit a self-healing effect under mechanical and thermal loads—with reaching a contact temperature of 800–1000 °C, some of the coating components may oxidize and form amorphous phases that can provide superlubricity in the processing zone and increase the operational life of the cutting tool [16,21,23]. Thin oxide films of a few nanometers effectively hamper moisture and protect microelectronics from mechanical and thermal loads [26]. The quality of the surface where the coating is supposed to be deposed plays a critical role in the product's durability [17] when special measures can reduce the adhesion of the workpiece material and favor the descent of material flow on the tool face [24]. The questions of reliable coating deposition diagnostics remain open and require special measures [27].

The surface layer plays the primary role in the potential for durability of the products operating under loads. Therefore, the coating and surface hardening technologies should be chosen individually to prolong the product's service life following the known contact loads.

This Special Issue is devoted to the latest advances in the coatings and surface hardening technologies for cutting tools and microelectronic components that can ensure a valuable increase in the product durability to ensure its reliable operation. The main emphasis lies in the results of the research and engineering works that have proven successful in laboratory or manufacturing conditions.

2. Contributions

Twelve research articles and communication were published in the presented Special Issue. The discussed subjects cover the most advanced achievements in the field of technologies of coatings and surface hardening for the tool industry. Among them, there are questions of

- Properties including high-temperature and triboproperties of the advanced multi-[19–25] and mono-layer [16,17], nanocomposite [23,24], nanostructured [18], and nanothin coatings [26];
- Tool surface conditions and micro-texturing on machining performance of hardened steel [17] and heat-resistant alloy [24], correspondingly;
- Self-healing adaptive coatings in milling titanium alloys [23], protective coatings for microelectronic needs [26], diagnostics of the coating deposition process [27];
- Coating technologies such as physical vapor deposition (PVD) [17,23,24,27], including a radio frequency magnetron sputtering [16], high-power impulse magnetron sputtering [18], a closed-field unbalanced magnetron sputtering [19,20], filtered cathodic vacuum arc deposition [21,22,25], as well as atomic layer deposition [26];
- Full-scale tests in conditions close to industrial ones in milling hardened steel 100CrMn6 [17] and titanium alloys [23], turning steel 1045 [22], nickel–chromium alloy [21,25], and iron–nickel alloy [24], annealing tests [19,20], open/short tests for electronic components [26];
- Development of a mathematical basis for future research [23,24].

The influence mechanism of temperature on the tribological properties of hexagonal boron nitride coatings deposed on the high-speed tool steel by radio frequency magnetron sputtering method (PVD) was investigated in [16]. Thermotribological behavior was evaluated in friction pair against ZrO_2 ball in the range of temperatures from 500 to 800 °C and compared with the steel sample. It is experimentally shown that the coefficient of friction of h-BN coating and steel sample decreases to 600 °C and grandly decreases to 800 °C when the coefficient of friction of h-BN coating decreases from 0.135 to 0.02, and it decreased from 0.3 to 0.14 for the uncoated sample. Thus, h-BN coating exhibits high-temperature superlubricity at 800 °C. The wear mechanism of h-BN coatings is tribooxidation; α-Fe_2O_3, Fe_3O_4, and γ-Fe_2O_3 were detected on the disc's worn surface, and α-Fe_2O_3, γ-FeOOH, γ-Fe_2O_3 and TiO_2 were detected on the ball worn surface at 800 °C, which is the main factor for the antifriction behavior at high temperature. The superlubricity mechanism is attributed to the γ-Fe_2O_3/h-BN formation due to tribochemistry.

An increase in the tool's average resistance by 1.4 times compared to the base-coated tool was shown in milling hardened steels of the 100CrMn type by Al_2O_3 + TiC ceramic inserts with the (TiZr)N, (TiAl)N, TiN coatings deposed by PVD method [17]. However, the detected tool resistance values significantly varied (by 30%). Additional surface lapping and polishing increase the wear resistance by two times, and the coefficient of resistance variation decreases by more than two times (14%). The additionally lapped and polished Al_2O_3 + TiC ceramic inserts with (TiZr)N coating demonstrate an increase in the average resistance by 1.7 times compared to the ceramic inserts without additional pre-processing. The increase in average resistance is 1.2-foldcompared to ceramic inserts after diamond grinding and (TiZr)N coating. Additional pre-processing of the inserts decreases surface roughness parameter Ra from 0.33 to 0.034 μm, reduces the coefficient of friction during high-temperature heating at 800 °C from 0.6 to 0.2 at the beginning of friction, demonstrates more stable behavior in the friction path, and results in a decrease in the range of change in the average resistance by ~1.9 times. The variation in the dispersion of resistance decreases up to 15%, which is two times less than for (TiZr)N-coated ceramic inserts.

The Mo-Zr-Si-B coatings of the dense and homogeneous structure were deposed by high-power impulse magnetron sputtering on industrial Al_2O_3 + TiC ceramic inserts at frequencies of 10, 50, and 200 Hz [18]. The minimal growth rate of 7.5 nm/min was observed at 10 Hz for the highest maximum peak current. The growth rate increased to 17.5 and 67.5 nm/min when the frequency was increased to 50 and 200 Hz. The maximum hardness of the Mo-Zr-Si-B coatings was 23 GPa and obtained at 200 Hz, when the coefficient of friction was 0.81. The coating deposed at 50 Hz was characterized by higher wear resistance, cyclic–dynamic impact loading, and high-temperature oxidation resistance at 1300 and 1500 °C. The coating deposed at 50 Hz remained amorphous during heating in the transmission electron microscope column at 20–1000 °C. The segregation and growth of $MoSi_2$ phase grains were observed in the temperature range of 600–1000 °C during vacuum annealing. The hardness and Young's modulus of the coatings that were deposited at frequencies of 10 and 200 Hz decreased after vacuum annealing. The hardness and Young's modulus increased from 15 and 250 GPa to 37 and 380 GPa, respectively, when heating from normal temperature to 1000 °C due to the crystallization and transformation to a nanocomposite structure.

The CrAlN coatings with three interlayers (CrN, CrZrN, and CrN/CrZrSiN) were deposited by a closed-field unbalanced magnetron sputtering system (PVD) on Si and hard alloy (WC-6 wt.% Co) substrates [19]. The coefficient of friction of the CrAlN coating with the CrN/CrZrSiN interlayer was ~0.3, which is less than for the CrN and CrZrN interlayer, where the coefficient of friction was 0.35 and 0.45, respectively. The adhesion strength of the CrAlN coating with the CrN and CrN/CrZrSiN interlayer was 69.3 N and 69.2 N, which is ~3–4 times higher than for those with CrZrN interlayer and without interlayer. The hardness of the CrAlN coating with the CrN/CrZrSiN interlayer was approximately 28 GPa up to 1000 °C after annealing test in the air for 30 min, which is 1.6 times higher than for the samples with CrN, CrZrN interlayer and the sample without interlayer. Up to 500 °C, all considered coatings' surface roughness and coefficient of friction were similar. The oxidation by the residual oxygen explains the variation in the coating surface roughness. Thus, introducing an interlayer could improve the mechanical properties and thermal stability of the CrAlN coatings. These properties increase the operational life of the hard alloy tool in milling in the conditions of the thermal loads.

The CrZrN/CrZrSiN multilayer coatings with different bilayer periods were deposited by a closed-field unbalanced magnetron sputtering (PVD) on Si and hard alloy (WC-6 wt.% Co) substrates to improve their mechanical properties and thermal stability [20]. The thickness of the bilayer periods varied from 1.35 to 0.45 μm. The hardness and elastic modulus of the CrZrN/CrZrSiN multilayer coatings gradually increased with the decrease in the bilayer period from 1.35 to 0.54 μm when coating with 0.45 μm bilayer period thickness showed a trend to decrease in the hardness and elastic modulus. The coefficient of friction showed the same trend: the lowest coefficient of friction of 0.24 was achieved for the CrZrN/CrZrSiN multilayer coating with a bilayer period of 0.54 μm. The adhesion strength increased with the decrease in the bilayer period. The highest values of 75 and 79 N were observed for a bilayer period of 0.54 and 0.45 μm. The chipping or delamination was not detected for coatings with a bilayer thickness of 0.67–0.45 μm. The hardness of the developed coatings was approximately above 27 GPa up to 800 °C, demonstrating excellent thermal stability. The CrZrN/CrZrSiN multilayer coatings with a bilayer period showed improved hardness, friction coefficient, adhesion, and thermal stability compared to monolithic one that corresponds to the requirements of the high-performance cutting-tool application.

It was shown in [21] that the substrate Zr-ZrN-(Zr,Mo,Al)N coating deposited by PVD to hard alloy consisted of two cubic nitride phases, (Zr,Mo,Al)N (ZrN-based solid solution) and (Mo,Zr,Al)N (MoN-based solid solution). Cr-CrN-(Cr,Mo,Al)N and Ti-TiN-(Ti,Mo,Al)N coatings were characterized by a single cubic nitride phase, (Ti,Mo,Al)N (TiN-based solid solution) or (Cr,Mo,Al)N (CrN-based solid solution). The tool with the Zr-ZrN-(Zr,Mo,Al)N coating had the highest wear intensity in turning Ni-Cr alloy at the cutting speed of 45 m/min, when the wear intensity was similar for three coatings at the cutting speed of 60 m/min. The tool with the Zr-ZrN-(Zr,Mo,Al)N coating demonstrated the lowest wear intensity at cutting speed of 75 and 90 m/min, when the average flank wear for the Zr-ZrN-(Zr,Mo,Al)N coated tool was ~30% lower compared to Cr-CrN-(Cr,Mo,Al)N and Ti-TiN-(Ti,Mo,Al)N coatings. A reliably identified transition layer of ZrO_2 at the interface between the Zr-ZrN-(Zr,Mo,Al)N coating and the machined workpiece was observed at the cutting speed of 90 m/min. A transition layer of the same thickness was also detected for the Cr-CrN-(Cr,Mo,Al)N coating. Such a transition layer was absent for the Ti-TiN-(Ti,Mo,Al)N coating. That may indicate the process of tribooxidation under the influence of probable increased mechanical and thermal loads (turning Ni-based alloys) and create conditions for increased lubricity [28]. There is supposed to be a dependence between the protective transition oxide layer that exhibits thermochemical inertness and phase stability up to ~1200 °C for ZrO_2 (zirconium dioxide is polymorphic—when heated up to ~1200 °C, it changes from monoclinic phase m-ZrO_2 to tetragonal one t-ZrO_2 by a martensitic mechanism with volumetric changes up to ~5%–9%; upon further heating to 2300 °C, it diffusional passes into the cubic phase c-ZrO_2) and more than ~1000 °C for chromium trioxide CrO_3 (different oxidation states of chromium are achieved by calcination, $CrO_3 \rightarrow Cr_2O_3 \rightarrow CrO$) and the depth of nickel diffusion [29–31]. The Zr-ZrN-(Zr,Mo,Al)N coating demonstrated Ni-diffusion up to the depth of 360 nm where the transition layer was detected. The Cr-CrN-(Cr,Mo,Al)N coating demonstrated that the diffusion of nickel into the coating was less than 260 nm.

It was shown that Cr,Mo-(Cr,Mo,)N-(Cr,Mo,Al)N coating with the introduction of 20 at.% Mo deposed on hard alloy tool by PVD increases its wear resistance compared to (Cr,Al)N coating [22]. The developed Cr,Mo-(Cr,Mo,)N-(Cr,Mo,Al)N coating had detected a nanolayer structure with a modulation period $\lambda = 50$ nm. The (Cr,Al)N coating exhibits more active wear crater growth on the rake face in turning steel 1045 than the (Cr,Mo,Al)N system nanolayer coating. The detected pattern of nanolayer coating wear reveals the following fracture mechanisms:

- Penetration of particles of the workpiece between nanolayers, resulting in interlayer delamination;
- Plastic deformation of nanolayers of the coating under the influence of the moving flow of the machined steel;
- Fracture of fragments of the coating's nanolayers with their further removal by the cut material.

The Fe-diffusion into the coating up to the depth up to 200 nm was observed when the Cr- and Mo-diffusion did not exceed the depths of 250 nm. Possible impact on wear resistance of the considered coatings and the operational life of the tool can have formed oxides of iron Fe_xO_y detected at the "coating—machined steel" interface that exhibit hardness superior to the workpiece material.

The thermodynamic model of wear of the hard alloy cutting tool coated by PVD for face milling two types of titanium alloy (Ti64 and Ti811) is proposed in [23]. It allows determining the ways to reduce the wear intensity and presents the dissipative function of the tool material shape change and the conditions for improving the wear resistance of the cutting tool using the phenomenon of adaptation (self-organization) under friction. Experimental studies on the wear resistance of cutting tools coated with innovative multilayer nanostructured coatings show an increase in the average operational life of the tool by 1.5–2 times. The following coatings with a thickness of less than 10 μm were under study:

- (CrAlSi)N + diamond-like carbon (DLC);
- TiB_2 (nanocomposite coating);
- nACRo (nanocomposite coating based on chrome and aluminum carbonitrides with microhardness of 42 GPa, coefficient of friction is 0.45);
- nACo3 (nanocomposite coating based on aluminum nitride and titanium with microhardness of 45 GPa, coefficient of friction is 0.35);
- nACRo + TiB_2;
- nACo3 + TiB_2;
- nACRo + TiB_2 + epilama;

- $nACo_3$ + epilama;
- uncoated + epilama.

The grain size of nanocomposite coatings was in the range of 3–5 nm. A decrease in temperature–force loading in the cutting zone was 15%–25% and can be explained by the formation of secondary structures on the friction surfaces of aluminum and titanium oxides that were detected by X-ray spectral analysis. As it is known, oxides provide heat-reflecting (shielding) and lubricating properties and exhibit thermochemical inertness and stability up to 800 °C (properties of γ-Al_2O_3 obtained at 450–600 °C persist up to 800 °C when the formation of a new crystalline phase σ-Al_2O_3 begins; it forms thermodynamically stable amorphous α-Al_2O_3 at 850–1000 °C [32,33]; thermal stability of amorphous anatase is up to 500–800 °C and even higher for crystalline rutile). It should be noted that the thermal stability of oxides is determined by Tamman and Hüttig temperatures. There is a noticeable mobility of the atoms of the crystal lattice at 0.5 Tm and surface mobility of the atoms at 0.3 Tm, where Tm is the melting point. Accordingly, the higher melting temperature provides higher thermal stability. A decrease in the coefficient of friction (adhesive component) by 13%–17% was observed in a temperature range of 550–950 °C. Contact processes reveal a phenomenon of adaptation (self-organization) of friction surfaces at milling by a coated cutting tool, promoting the formation of oxide films of amorphous structure that exhibit protective and lubricating properties.

The problem of increasing the efficiency of iron–nickel heat-resistant alloy turning with a hard alloy tool is aimed to be solved by tool surface microtexturing and multilayer nanocomposite PVD-coating in [24]. Microtexturing of the rake face of carbide insert in the form of strips were provided by nanosecond laser ablation and in the form of mesh structure by an indenter. The obtained microtexture by laser beam revealed fragile phases' formation from molten material that were resolidified on the surface of the texture when easy-to-melt components of the material are sublimated in the environment. At the same time, ejections of material from the molten pool were observed, characterized by the explosive behavior of the liquid from the impact of laser pulses. The indenter allowed obtaining meshes by plastic deformation with improved surface properties with a force of 20 N. The square-shaped prints were with a width of 30 μm and a maximum depth of 7 μm. The multilayer $nATCRo_3$ on (TiCrAlSi)N basis that includes (CrTi)N and (AlTi)N layers alternation and (AlTiCr)N/SiN nanocomposite layer formed by (AlTiCr)N crystals of 5 nm evenly distributed in amorphous Si_3N_4. It is shown that a carbide insert with a microtexture filled with a MoS_2-based lubricant decreases the workpiece material adhesion. The tests of cutting tool operational life under various turning modes with cutting depth from 0.3 to 0.5 mm and feed from 0.1 to 0.15 mm/rev shows temperatures increased up to 240–330 °C in the processing zone. The temperature measurements provided a basis for the mathematical modeling of the cutting thermal power parameters. The function of predictive evaluation is proposed and justified using the criterion of relative efficiency. Predictive estimates of the turning efficiency established that the calculated values converged highly with the experiments. The microtextured tool's operational life was increased by 1.3–1.5 times.

The properties of Ti-TiN-(Ti,Y,Al)N multilayer composite PVD-coating with the high content (approximately 40 at.%) of yttrium in its wear-resistant layer were studied comparing Ti-TiN-(Ti,Cr,Al)N coated and uncoated hard alloy tools in turning nickel–chromium alloy [25]. The hardness was HV 2758 ± 78, and the elastic modulus was 356 ± 24 GPa. The coating was characterized by cubic phases c-(Ti,Y,Al)N and c-(Y,Ti,Al)N. The wear resistance on the rake face was improved by 250%–270% for the tools with both coatings under consideration. The wear rates were similar when the operational life of the carbide tool with (Ti,Cr,Al)N coating was increased by 10%–15%. The active oxidation of the Ti-TiN-(Ti,Y,Al)N coating in contact with the workpiece material with Y_2O_3 formation was detected. Thus, the dominant wear mechanism was oxidation. Despite the known chemical activity of Y towards an oxygen-containing environment when heating and hydrolysis (actively forms $Y(OH)_3$ that decreases to YO(OH) in the presence of water under pressure at 90–350 °C), the rake face of the tool with the Ti-TiN-(Ti,Y,Al)N coating demonstrated undamaged fragments in the transition layer and the wear-resistant layer after 16 min of turning. The oxide layers formed on the surface of the coating improve the tribological conditions in the cutting zone and reduce the tool wear rate.

The Al_2O_3 protective thin film layer with a thickness of 14–15 nm deposed by atomic layer deposition technology to protect electronic components improves their service life in harsh environments [26]. The excavator integrated circuit is prone to generating high heat during operation as a packaging technology in a flip-chip land grid array package type, whereby the stress caused by high temperature can destroy the interface between plastic substrates, solder, and metal mats and reduce their service life. The protective thin film significantly reduces the component damage caused by thermal stress and shows the same lowest operation voltage at 0.66 V before and after the 1000 h unbiased highly accelerated temperature and humidity stress test. The atomic layer can effectively block moisture and oxygen from entering electronic devices to avoid their rapid deterioration, improve mechanical protection and reduce the damage caused by moving collisions, which has potential applications in organic light-emitting diodes, micro light-emitting diodes, and semiconductor packages for resisting moisture. The atomic layer deposition can be used to fabricate luminescent material layers, biomedicine sensor components, and water-resistant thin films for use in energy engineering and other applications related to coating technology.

The diagnostic methods concerning the flow of technology processing in a vacuum for solid bodies' surface modification are shown in [27]. The analysis of energy distribution curves shows that the average ion energy decreases with a decrease in the discharge current, and the ratio of the average ion energy to the discharge voltage is approximately 0.68–0.71. The magnetic induction method of diagnostics presented in the study allows controlling the parameters of glow discharge plasma, determining the moment when the unit enters the operating mode and the moment of completion of technology transitions and correcting the process according to the changes in electromagnetic pulses. The presented electromagnetic radiation signal spectrum analyzer is based on a multichannel inductive sensor. It makes it possible to survey broadband plasma electromagnetic radiation signals, obtain spectra at the sensor output, and resolve the signal spectrum of glow discharge plasma electromagnetic radiation into its constituent elements.

3. Conclusions and Outlook

Various technologies of coating and surface hardening of manufacturing tools, including the study of tribotechnical properties of multilayer nanostructured coatings, the study of the polishing impact on coating performance, surface texturing effect, the study of coating low friction behavior, the study of the interlayer effect on properties and thermal stability of a CrAlN coating, the study of structure and properties of Mo-Zr-Si-B coatings, the study of Zr-ZrN-(Zr,Mo,Al)N, Cr-CrN-(Cr,Mo,Al)N and Ti-TiN-(Ti,Mo,Al)N multilayer coatings, the study of adaptive and self-organized nanocomposite coatings based on chrome and aluminum carbonitrides and aluminum nitride and titanium, the study of Ti-TiN-(Ti,Y,Al)N and Ti-TiN-(Ti,Cr,Al)N multilayer composite coatings, Al_2O_3 nanofilm, and diagnostics of plasma vapor deposition processes, have been presented in the Special Issue. It covers current evolution of achievements in coating techniques for providing the most advanced structures offering superior properties and technologies of deposition for traditional and the most-advanced coating basis. However, there are still many steps that should be taken to transfer a few of the most outstanding results and proposals into applications in the tool industry. The Guest Editor of the Special Issue hopes that the presented studies will complete the previously published advances in the first Special Issue devoted to the same topic and contribute to the transfer of the tool industry to the next technological paradigm.

Guest Editor highly appreciates the high requirements of Coatings for the quality of published papers' presentation, their up-to-date scientifically valuable theoretical and practical content provided by the Authors, and the kind efforts of the team of Reviewers, Editors, and Assistants in contributing to the Special Issue.

Acknowledgments: Many thanks to the Coatings Editorial Office and personally to Miya Liu, Section Managing Editor, for her kind help and assistance in publishing the most outstanding research results.

Conflicts of Interest: The authors declare no conflict of interest.

References

1. Volosova, M.; Grigoriev, S.; Metel, A.; Shein, A. The role of thin-film vacuum-plasma coatings and their influence on the efficiency of ceramic cutting inserts. *Coatings* **2018**, *8*, 287. [CrossRef]
2. Vereschaka, A.; Tabakov, V.; Grigoriev, S.; Sitnikov, N.; Milovich, F.; Andreev, N.; Sotova, C.; Kutina, N. Investigation of the influence of the thickness of nanolayers in wear-resistant layers of Ti-TiN-(Ti,Cr,Al)N coating on destruction in the cutting and wear of carbide cutting tools. *Surf. Coat. Technol.* **2020**, *385*, 125402. [CrossRef]
3. Grigoriev, S.; Vereschaka, A.; Zelenkov, V.; Sitnikov, N.; Bublikov, J.; Milovich, F.; Andreev, N.; Mustafaev, E. Specific features of the structure and properties of arc-PVD coatings depending on the spatial arrangement of the sample in the chamber. *Vacuum* **2022**, *200*, 111047. [CrossRef]
4. Li, B.; Xu, Y.; Rao, G.; Wang, Q.; Zheng, J.; Zhu, R.; Chen, Y. Tribological properties and cutting performance of AlTiN coatings with various geometric structures. *Coatings* **2023**, *13*, 402. [CrossRef]
5. Heo, S.-B.; Kim, W.R.; Kim, J.-H.; Choe, S.-H.; Kim, D.; Lim, J.-H.; Park, I.-W. Effects of copper content on the microstructural, mechanical and tribological properties of TiAlSiN–Cu superhard nanocomposite coatings. *Coatings* **2022**, *12*, 1995. [CrossRef]
6. Glaziev, S.Y. The discovery of regularities of change of technological orders in the central economics and mathematics institute of the soviet academy of sciences. *Econ. Math. Methods* **2018**, *54*, 17–30. [CrossRef]
7. Korotayev, A.V.; Tsirel, S.V. A spectral analysis of world GDP dynamics: Kondratiev waves, Kuznets swings, Juglar and Kitchin cycles in global economic development, and the 2008–2009 economic crisis. *Struct. Dyn.* **2010**, *4*, 3–57. [CrossRef]
8. Perez, C. Technological revolutions and techno-economic paradigms. *Camb. J. Econ.* **2010**, *34*, 185–202. [CrossRef]
9. Grigoriev, S.; Vereschaka, A.; Zelenkov, V.; Sitnikov, N.; Bublikov, J.; Milovich, F.; Andreev, N.; Sotova, C. Investigation of the influence of the features of the deposition process on the structural features of microparticles in PVD coatings. *Vacuum* **2022**, *202*, 111144. [CrossRef]
10. Kwok, F.M.; Sun, Z.; Yip, W.S.; Kwok, K.Y.D.; To, S. Effects of coating parameters of hot filament chemical vapour deposition on tool wear in micro-drilling of high-frequency printed circuit board. *Processes* **2022**, *10*, 1466. [CrossRef]
11. Guan, J.; Gao, C.; Xu, Z.; Yang, L.; Huang, S. Lubrication Mechanisms of a nanocutting fluid with carbon nanotubes and sulfurized isobutylene (CNTs@T321) composites as additives. *Lubricants* **2022**, *10*, 189. [CrossRef]
12. Morozow, D.; Siemiątkowski, Z.; Gevorkyan, E.; Rucki, M.; Matijošius, J.; Kilikevičius, A.; Caban, J.; Krzysiak, Z. Effect of yttrium and rhenium ion implantation on the performance of nitride ceramic cutting tools. *Materials* **2020**, *13*, 4687. [CrossRef] [PubMed]
13. Liu, W.; Chu, Q.; Zeng, J.; He, R.; Wu, H.; Wu, Z.; Wu, S. PVD-CrAlN and TiAlN coated Si3N4 ceramic cutting tools −1. Microstructure, turning performance and wear mechanism. *Ceram. Int.* **2017**, *43*, 8999–9004. [CrossRef]
14. Vopát, T.; Sahul, M.; Haršáni, M.; Vortel, O.; Zlámal, T. The tool life and coating-substrate adhesion of AlCrSiN-coated carbide cutting tools prepared by LARC with respect to the edge preparation and surface finishing. *Micromachines* **2020**, *11*, 166. [CrossRef] [PubMed]
15. Doctor, A.Q.; Mazhar, A.A. *Khan Research Labs*; Mazhar, A.A., Khan, M.A., Mirza, J.A., Khan, A.Q., Eds.; Islamabad, Pakistan, 2001; pp. 470–472, 520p.
16. Zeng, Q. High temperature low friction behavior of h-BN coatings against ZrO_2. *Coatings* **2022**, *12*, 1772. [CrossRef]
17. Volosova, M.A.; Stebulyanin, M.M.; Gurin, V.D.; Melnik, Y.A. Influence of surface layer condition of Al_2O_3 +TiC ceramic inserts on quality of deposited coatings and reliability during hardened steel milling. *Coatings* **2022**, *12*, 1801. [CrossRef]
18. Kiryukhantsev-Korneev, P.V.; Sytchenko, A.D.; Loginov, P.A.; Orekhov, A.S.; Levashov, E.A. Frequency effect on the structure and properties of Mo-Zr-Si-B coatings deposited by HIPIMS using a composite SHS target. *Coatings* **2022**, *12*, 1570. [CrossRef]
19. Kim, H.-K.; Kim, S.-M.; Lee, S.-Y. Influence of interlayer materials on the mechanical properties and thermal stability of a CrAlN coating on a tungsten carbide substrate. *Coatings* **2022**, *12*, 1134. [CrossRef]
20. Kim, H.-K.; Kim, S.-M.; Lee, S.-Y. Mechanical properties and thermal stability of CrZrN/CrZrSiN multilayer coatings with different bilayer periods. *Coatings* **2022**, *12*, 1025. [CrossRef]
21. Grigoriev, S.; Yanushevich, O.; Krikheli, N.; Vereschaka, A.; Milovich, F.; Andreev, N.; Seleznev, A.; Shein, A.; Kramar, O.; Kramar, S.; et al. Investigation of the nature of the interaction of Me-MeN-(Me,Mo,Al)N coatings (where Me = Zr, Ti, or Cr) with a contact medium based on the Ni-Cr system. *Coatings* **2022**, *12*, 819. [CrossRef]
22. Vereschaka, A.; Seleznev, A.; Gaponov, V. Wear resistance, patterns of wear and plastic properties of Cr,Mo-(Cr,Mo,)N-(Cr,Mo,Al)N composite coating with a nanolayer structure. *Coatings* **2022**, *12*, 758. [CrossRef]
23. Migranov, M.S.; Shehtman, S.R.; Sukhova, N.A.; Mitrofanov, A.P.; Gusev, A.S.; Migranov, A.M.; Repin, D.S. Study of tribotechnical properties of multilayer nanostructured coatings and contact processes during milling of titanium alloys. *Coatings* **2023**, *13*, 171. [CrossRef]
24. Stebulyanin, M.; Ostrikov, E.; Migranov, M.; Fedorov, S. Improving the efficiency of metalworking by the cutting tool rake surface texturing and using the wear predictive evaluation method on the case of turning an iron–nickel alloy. *Coatings* **2022**, *12*, 1906. [CrossRef]

25. Grigoriev, S.; Vereschaka, A.; Milovich, F.; Sitnikov, N.; Bublikov, J.; Seleznev, A.; Sotova, C.; Rykunov, A. Investigation of the Properties of Multilayer Nanostructured Coating Based on the (Ti,Y,Al)N System with High Content of Yttrium. *Coatings* **2023**, *13*, 335. [CrossRef]
26. Chen, P.-C.; Chang, S.-M.; Kuo, H.-C.; Chang, F.-C.; Li, Y.-A.; Ting, C.-C. Reliability enhancement of 14 nm HPC ASIC using Al_2O_3 thin film coated with room-temperature atomic layer deposition. *Coatings* **2022**, *12*, 1308. [CrossRef]
27. Grigoriev, S.; Dosko, S.; Vereschaka, A.; Zelenkov, V.; Sotova, C. Diagnostic techniques for electrical discharge plasma used in PVD coating processes. *Coatings* **2023**, *13*, 147. [CrossRef]
28. Grigoriev, S.N.; Volosova, M.A.; Fedorov, S.V.; Okunkova, A.A.; Pivkin, P.M.; Peretyagin, P.Y.; Ershov, A. Development of DLC-coated solid SiAlON/TiN ceramic end mills for nickel alloy machining: Problems and prospects. *Coatings* **2021**, *11*, 532. [CrossRef]
29. Ikim, M.I.; Spiridonova, E.Y.; Belysheva, T.V.; Gromov, V.F.; Gerasimov, G.N.; Trakhtenberg, L.I. Structural properties of metal oxide nanocomposites: Effect of preparation method. *Russ. J. Phys. Chem. B* **2016**, *10*, 543–546. [CrossRef]
30. Surzhikov, A.P.; Ghyngazov, S.A.; Frangulyan, T.S.; Vasil'ev, I.P. Thermal transformations in ultrafine plasmochemical zirconium dioxide powders. *J. Therm. Anal. Calorim.* **2015**, *119*, 1603–1609. [CrossRef]
31. Grigoriev, S.N.; Volosova, M.A.; Okunkova, A.A.; Fedorov, S.V.; Hamdy, K.; Podrabinnik, P.A.; Pivkin, P.M.; Kozochkin, M.P.; Porvatov, A.N. Electrical discharge machining of oxide nanocomposite: Nanomodification of surface and subsurface layers. *J. Manuf. Mater. Process.* **2020**, *4*, 96. [CrossRef]
32. Li, L.; Pu, S.; Liu, Y.; Zhao, L.; Ma, J.; Li, J. High-purity disperse alpha-Al_2O_3 nanoparticles synthesized by high-energy ball milling. *Adv. Powder Technol.* **2018**, *29*, 2194–2203. [CrossRef]
33. Morozova, L.V. Mechanochemical Activation of precursor powders for the preparation of dense Al_2O_3-ZrO_2 <Y_2O_3> nanoceramics. *Inorg. Mater.* **2019**, *55*, 295–301.

Article

Investigation of the Nature of the Interaction of Me-MeN-(Me,Mo,Al)N Coatings (Where Me = Zr, Ti, or Cr) with a Contact Medium Based on the Ni-Cr System

Sergey Grigoriev [1], Oleg Yanushevich [2], Natella Krikheli [2], Alexey Vereschaka [1,*], Filipp Milovich [3,4], Nikolay Andreev [3], Anton Seleznev [1], Alexander Shein [1], Olga Kramar [2], Sergey Kramar [2] and Pavel Peretyagin [1,2]

[1] Spark Plasma Sintering Research Laboratory, Moscow State University of Technology, STANKIN, 127055 Moscow, Russia; s.grigoriev@stankin.ru (S.G.); a.seleznev@stankin.ru (A.S.); ppy@digitalfabrika.ru (A.S.); p.peretyagin@stankin.ru (P.P.)

[2] Scientific Department, A.I. Evdokimov Moscow State University of Medicine and Dentistry, 127473 Moscow, Russia; olegyanushevich@mail.ru (O.Y.); nataly0088@mail.ru (N.K.); dr.ovkramar@gmail.com (O.K.); kramarsv@mail.ru (S.K.)

[3] National University of Science & Technology (MISIS), 119049 Moscow, Russia; filippmilovich@mail.ru (F.M.); andreevn.misa@gmail.com (N.A.)

[4] Dianov Fiber Optics Research Center, Prokhorov General Physics Institute of the Russian Academy of Sciences, 119333 Moscow, Russia

* Correspondence: dr.a.veres@yandex.ru; Tel.: +7-9169100413

Citation: Grigoriev, S.; Yanushevich, O.; Krikheli, N.; Vereschaka, A.; Milovich, F.; Andreev, N.; Seleznev, A.; Shein, A.; Kramar, O.; Kramar, S.; et al. Investigation of the Nature of the Interaction of Me-MeN-(Me,Mo,Al)N Coatings (Where Me = Zr, Ti, or Cr) with a Contact Medium Based on the Ni-Cr System. *Coatings* 2022, *12*, 819. https://doi.org/10.3390/coatings12060819

Academic Editor: Michał Kulka

Received: 28 April 2022
Accepted: 8 June 2022
Published: 10 June 2022

Abstract: This paper discusses the results of a study focused on the nature of the interaction of Me-MeN-(Me,Mo,Al)N coatings (where Me = zirconium (Zr), titanium (Ti), or chromium (Cr)) with a contact medium based on the Ni-Cr system. The studies were carried out during the turning of nickel–chromium alloy at different cutting speeds. The hardness of the coatings was found, and their nanostructure and phase composition were studied. The experiments were conducted using transmission electron microscopy (TEM), X-ray diffraction (XRD), and selected area electron diffraction (SAED). According to the studies, at elevated cutting speeds, the highest wear resistance is demonstrated by the tools with the ZrN-based coating, while at lower cutting speeds, the tools with the TiN- and CrN-based coatings had higher wear resistance. At elevated cutting speeds, the experiments detected the active formation of oxides in the ZrN-based coating and less active formation of oxides in the CrN-based coating. No formation of oxides was detected in the TiN-based coating. The patterns of cracking in the coatings were also studied.

Keywords: oxidation; coatings; nickel alloy; wear; diffusion

1. Introduction

Surface-modifying systems in general, and modifying coatings in particular, are increasingly used in various spheres [1–3]. Along with such traditional spheres of application of coatings as the manufacturing of metal-cutting and stamping tools, new areas of surface modification are emerging, including in medicine (in particular, coatings for implants operating in the human body environment) and in the manufacturing of friction pairs, including those functioning in aggressive media and at elevated temperatures [4–7]. There is an interaction between a coating, on the one hand, and a contact medium (material being machined, human body, or counterbody in a friction pair), on the other hand. In general, the mentioned interaction is a complex and integral phenomenon, which includes, in particular, such factors as adhesion, impact action, interdiffusion, and chemical interaction [8–12]. Since, as a rule, the contact processes take place in an oxygen-containing medium, the processes of oxidation of both the coating and the contact medium can also occur. As an example of such interaction, there is a consideration of the turning of nickel alloy using tools with coatings of various compositions. When the material being machined (in this case, the Ni-Cr alloy) and the coating of Me-MeN-(Me,Mo,Al)N (where Me = Zr, Ti, or Cr) interact at high temperatures, most of the forms of interaction mentioned above take place. In particular, the interdiffusion of elements of the coating and the material being machined, adhesive interaction, and related fatigue processes can occur [13–16]. The processes of oxidation of both the coating and the material being machined can also take place. Thus, a complex interaction can be expected, not only between the initial phases of the material being machined and the coating, but also between their oxides [17–22]. The key trend of modern manufacturing is that increased cutting speeds lead to rising temperatures in the cutting zone [23–25]. In turn, with an increase in temperature, the diffusion and oxidation processes noticeably intensify [26–28]. Accordingly, modern coatings for cutting tools should not only effectively resist adhesive and abrasive wear, but they should also be characterized by high heat resistance and resistance to diffusion and oxidation wear [29–31].

The system of (Mo,Al)N is one of the possible coating compositions for operation under the specified conditions [32–35]. Coatings with the described composition are characterized by high hardness (to 38.4 GPa [34]) and relatively low compressive residual stresses [33], with an active oxidation beginning at a considerably high level of temperature [33,35].

It should be noted that the surface films of the compounds of MoO_3 and Al_2O_3, formed during the process of oxidation of the coating, can play a positive role in the cutting process due to their good properties preventing further oxidation and reducing the coefficient of friction [35]. Further improvement in the performance properties of the (Mo,Al)N coating can be achieved due to the introduction in its composition of additional elements, in particular, Ti, Zr, or Cr [36–40]. The introduction of Ti decreases the coefficient of friction [36], increases hardness and wear resistance [36,41–45], and simultaneously provides good resistance to cracking [43]. The introduction of Cr in the composition of the (Mo,Al)N coating also increases its hardness (to 41.2 GPa [46]) and wear resistance [38,46]. The good tribological properties of the coating are also noted in [47]. The introduction of Cr into the composition of the (Mo,Al)N coating leads to the formation of two basic phases—CrN and Mo_2N [39,47–50]. A similar effect is also detected upon the introduction of Zr in the composition of the (Mo,Al)N system [40], when two basic phases—ZrN and Mo_2N—are also formed [40].

The introduction of molybdenum (Mo) considerably increases the tribological properties and wear resistance of the two- and three-component coatings of TiN, (Al,Ti)N, CrN, ZrN [51], and (Cr,Al)N [52,53] and the multicomponent coating of (Cr,Zr,Nb,Al)N [52,54]. The Mo-containing coatings can combine high hardness and wear resistance with sufficient plasticity [54], and they also have good barrier properties in terms of diffusion [52–54]. The earlier studies, focused on the coatings of (Cr,Al)N [55,56], (Ti,Al)N [55], and (Zr,Al)N [56], detected their very good performance properties. It can be expected that the introduction of Mo into the composition of the mentioned coatings can further improve their tribological properties and their resistance to wear and brittle fracture, which, in turn, can lead to an improvement of the cutting properties of the tools in metal machining.

Based on the foregoing, the main goal of this study has been formulated, which consists of the comparative analysis of the following coatings: Zr-ZrN-(Zr,Mo,Al)N (hereafter referred to as Coating M1), Ti-TiN-(Ti,Mo,Al)N (hereafter referred to as Coating M2), and Cr-CrN-(Cr,Mo,Al)N (hereafter referred to as Coating M3).

Earlier papers considered the tribological properties of the described coatings [53,57,58] and the patterns of wear and fracture during the turning of 1045 steel using tools with the described coatings [57,59–61]. The general pattern of wear on metal-cutting tools during the tuning of nickel alloys was also considered [62–64]. The task of this paper is to investigate the patterns of cracking and the oxidation and diffusion processes in the coatings of Zr-ZrN-(Zr,Mo,Al)N (hereafter referred to as Coating M1), Ti-TiN-(Ti,Mo,Al)N (hereafter referred to as M2), and Cr-CrN-(Cr,Mo,Al)N (hereafter referred to as M3) under the conditions of high-temperature plastic contact with the Ni-Cr system during the cutting process.

2. Materials and Methods

The coatings under study have a three-layer architecture [65–70], which includes an adhesive layer of pure metal (Zr, Ti, or Cr, respectively, 20–50 nm thick), a transition layer of two-component nitride (ZrN, TiN, or CrN, respectively, about 1 μm thick), and a wear-resistant layer, which, in turn, has a nanolayer structure ((Zr,Mo,Al)N, (Ti,Mo,Al)N, or (Cr,Mo,Al)N, respectively, about 3 μm thick) with identical structure parameters (total thickness, thicknesses of functional layers, and nanolayer period λ).

The composition of the coatings was determined based on an analysis of the studies available and earlier experimental results [53–64]. In particular, the content of molybdenum (Mo) and aluminum (Al) in all the coatings under consideration was 40 at.% and about 10 at.%, respectively. The content of Ti, Zr, or Cr was about 50 at.%, respectively.

On the one hand, the specified proportion provides the best combination of hardness and plasticity with high heat resistance. On the other hand, the considerably high content of Mo makes it possible to predict the active formation of an oxide film of MoO_3.

Filtered cathodic vacuum arc deposition (FCVAD) technology was used to deposit on samples the coatings under study [67,68,71,72]. The coating deposition took place on the experimental unit of VIT-2 (VIT–IDTI RAS) [67,68,72]. During the process of coating deposition, cathodes of the following compositions were used: Mo 99.98 at.%, Zr 99.97 at.%, Ti 99.98 at.%, and Cr 99.97 at.%. A vacuum arc evaporator with controlled accelerated motion of a cathode spot of Arc-PVD (CAA-PVD) was applied [72]. The cathode of Al 99.95 at.% was installed on an evaporator of the FCVAD system [67,68]. The standard procedure for preparation of the samples included their washing using special solutions and ultrasonic simulation, followed by drying in flows of hot pre-purified air. No additional special treatment of the surface for deposition (for example, polishing) was carried out. The samples were installed in the tooling, which, in turn, was placed on the turntable, providing planetary rotation during the deposition process [68,73,74]. The rotation rate of the turntable was n = 0.7 rpm, which provided the formation of the nanolayer structure of the coatings. Before the deposition of the coatings, the samples were subjected to ion cleaning in gas (argon) plasma. The coating deposition process was carried out under the following parameters: arc current was 160 A for the Al-cathode, 75 A for the Ti- and Zr-cathodes, 125 A for the Mo-cathodes, and 73 A for the Cr-cathode. During the deposition of the coatings, the nitrogen pressure was 0.42 Pa. Substrate bias voltage was −600 DC.

The coatings were deposited on carbide (WC+15% TiC+6% Co) cutting inserts of SNUN ISO 1832:2012. The hardness of the coatings was measured using a micro-indentometer hardness tester (CSM Instruments, Needham, MA, USA), using the Oliver–Pharr method [75], with a stress of 10 mN. The fracture threshold stress value was found according to ASTM C1624-05 [76] using a Nanovea M1 scratch-test tester (Micro Scratch, Nanovea, Irvine, CA, USA).

For the microstructural studies of samples of carbide substrates with coatings, a scanning electron microscope (SEM) FEI Quanta 600 FEG (Materials & Structural Analysis Division, Hillsboro, OR, USA) was used. The studies of nanostructure involved a high-resolution transmission electron microscope (TEM) JEM 2100, manufactured by JEOL Company, Tokyo, Japan. When presenting the results of the EDX analysis of the elemental composition of the coating, the content of the main elements was indicated without taking into account nitrogen (the coating contains 48–52 at % nitrogen).

The samples (lamellas) of the material being machined were prepared using the Strata focused ion beam (FIB) 205 (FEI, Houston, TX, USA). Diffraction spectra were obtained on a DRON 4 automated X-ray diffractometer (LNPO "Burevestnik" (Leningrad Scientific and Production Enterprise, St. Petersburg, Russia) using monochromatic CuKα radiation. The survey was carried out in symmetrical geometry (Bragg–Brentano geometry). The obtained spectra were processed using a package of special software developed at the Physical Materials Science Department of the National University of Science and Technology (MISiS). A PDF database was used to identify the phases.

The influence of the coatings under consideration on the wear resistance of tools was assessed during the turning of the heat-resistant titanium alloy NiCr20TiAl (machinability group S under ISO 513:2004-07).

A CU 500 MRD (ZMM-BULGARIA HOLDING, Sofia, Bulgaria) lathe was used while testing the cutting properties of the tools with the coatings under study. The tests were carried out in the longitudinal turning of the heat-resistant nickel-based alloy WNr NiCr20TiAl (EN 10090:1998, HRC 32).

Cutters with cemented carbide SNUN ISO 1832:2012 [77] inserts were used as cutting tools. The following cutting geometry was used: $\gamma = -7°$, $\alpha = 7°$, $K = 45°$, $\lambda = 0°$, and $R = 0.8$ mm. During the turning of the nickel-based alloy WNrNiCr20TiAl, the following cutting modes were applied: $f = 0.11$ mm/rev; $a_p = 0.5$ mm; $v_c = 45, 60, 75$, and 90 m/min. Flank wear $VB_{max} = 0.3$ mm was assumed as a wear criterion for all cutting speeds.

3. Results and Discussion

3.1. Investigation of the Mechanical Properties of the Coatings and Wear Resistance of Coated Cutting Tools

The coatings under consideration had almost identical nanolayer periods λ: 40, 43, and 48 nm (for Coatings M1, M2, and M3, respectively) (Figure 1a–c).

The phases were identified using PDF standard diffraction cards with the reference codes: no. 03-065-2899, 00-038-1420, and 00-031-1493. Judging by the obtained data on the parameters of the crystal structures, these coatings represent cubic solid solutions of nitrides (type B1): (Zr,Mo,Al)N, (Ti,Mo,Al)N, and (Cr,Mo,Al) N, respectively. The XRD phase analysis of the coatings revealed the presence of WC and TiC together with substrate phases (Figure 1d–f). The cubic solid solutions had the same structural type, cF8/2, and differed only in the lattice parameters. Table 1 shows the cubic nitride phases of the coating with the calculated lattice parameters. When comparing the lattice parameters of the coating phases and nitride phases without Mo and Al, the following was obtained: for (Cr,Mo,Al)N, the lattice parameter increased relative to the CrN phase (a = 4.149), while for (Ti,Mo,Al)N and (Zr,Mo,Al)N, the lattice parameter decreased compared to the TiN (a = 4.241) and ZrN (a = 4.574) phases, respectively.

Table 1. Phase composition with lattice parameters.

Cubic Solid Solution	(Zr,Mo,Al)N	(Ti,Mo,Al)N	(Cr,Mo,Al)N
Structural type	cF8/2	cF8/2	cF8/2
a (Å)	4.565 (1)	4.236 (1)	4.186 (1)

Judging by the intensity of the diffraction lines in the (Zr,Mo,Al)N coating, the predominant grain orientation is observed—<111>, (Figure 1d). The texture is also clearly visible on SAED from this coating. On the (Ti,Mo,Al)N and (Cr,Mo,Al)N coatings, the grain texture is not so pronounced.

The XRD phase analysis of the coatings revealed the presence of a solid solution phase of nitrides ((Zr,Mo,Al)N, (Ti,Mo,Al)N, and (Cr,Mo,Al)N, respectively) together with substrate phases of WC and TiC (Figure 1d–f).

The highest hardness was detected for Coating M2 (30.70 ± 1.2 GPa); the hardness of Coating M1 was slightly lower (28.30 ± 0.70 GPa), and Coating M3 had the lowest hardness (26.60 ± 1.3 GPa) (Figure 1g). For all three coatings under consideration, the fracture threshold stress values in scratch testing were considerably high (36–40 N). In general, it can be argued that, in terms of the hardness and the fracture threshold stress value in scratch testing, the coatings under consideration have fairly close properties.

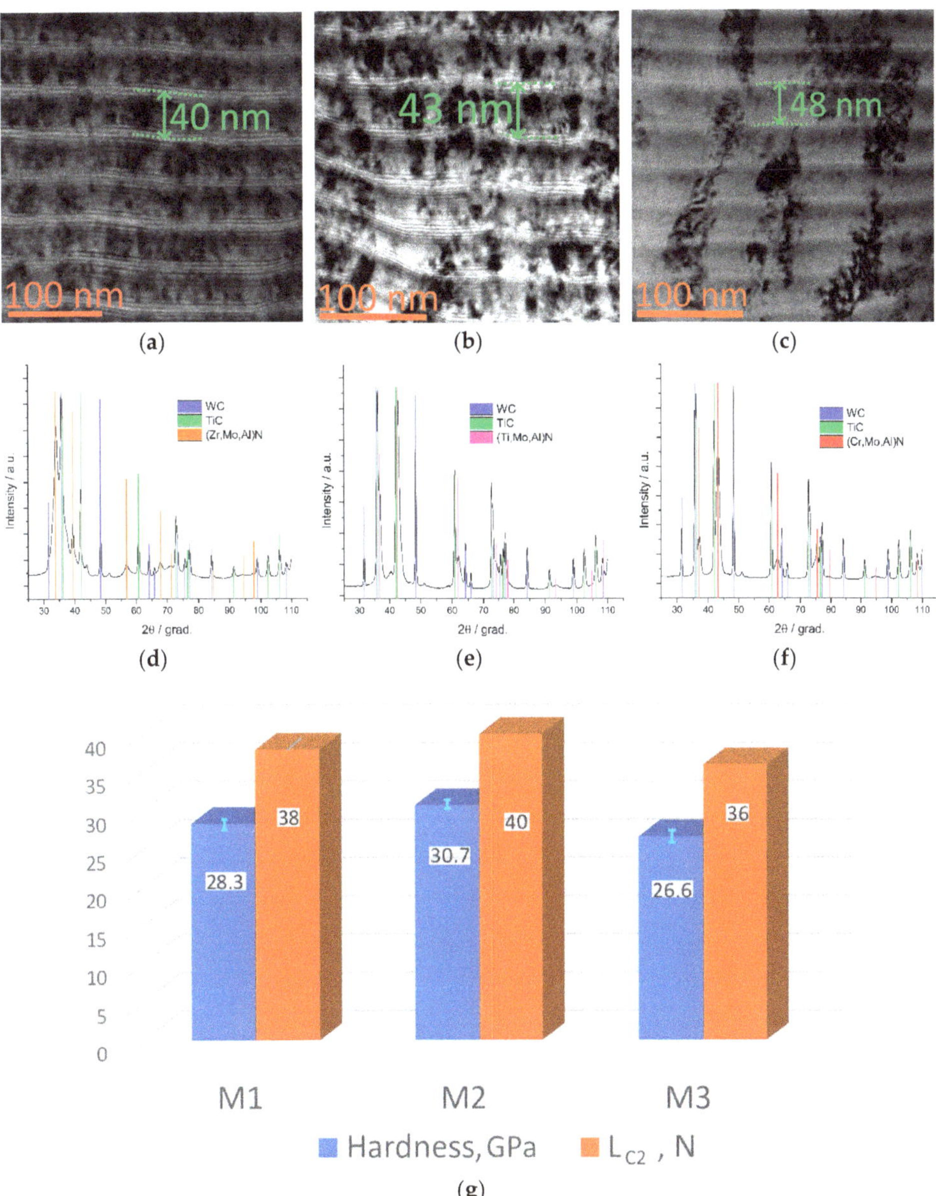

Figure 1. Parameters of the nanostructure of Coatings: (**a**) M1, (**b**) M2, (**c**) M3; XRD phase analysis of Coatings: (**d**) M1, (**e**) M2, (**f**) M3; (**g**) comparison of the values of hardness and the fracture threshold stress values in scratch testing.

The wear resistance of coated metal-cutting tools was studied at cutting speeds of 45, 60, 75, and 90 m/min (see Figure 2). It should be noted that, at the cutting speed of 45 m/min, the tool with Coating M1 demonstrated the highest wear intensity; then, when the cutting speed was 60 m/min, at the initial stage of cutting, the highest wear intensity was also detected for the tool with Coating M1. Closer to the moment when the wear criterion is reached, the wear intensity for all three samples became almost equal. With a further increase in cutting speed up to 75 and 90 m/min, there was a noticeable change in the trends of wear on the tools with different coatings. At the given cutting speeds, the lowest wear intensity was detected for the tool with Coating M1. The average value of flank wear for the tool with Coating M1 was approximately 30% less than for the tools with Coatings M2 and M3. The tools with Coating M2 and M3 demonstrated almost equal values of wear intensity at cutting speeds of 45–90 m/min. The tool with Coating M3 had an insufficient advantage in terms of wear intensity compared to the tool with Coating M2. It should be noted that the tool with Coating M2, characterized by the highest hardness and the highest fracture threshold stress value in scratch testing, demonstrated the lowest wear resistance in the given series of tests. Such an effect can occur, as with an increase in the cutting speed, the temperature in the cutting zone rises [23–25,78–80]. In turn, the increase in temperature leads to the intensification of the oxidation and diffusion processes. While at lower temperatures in the cutting zone, the prevailing wear mechanism is adhesive–fatigue and adhesive wear, with an increase in the cutting speed, the oxidation and diffusion factors become increasingly important [23–27,78]. At the same time, the tribologically active oxide films can be formed, which have a positive effect on the cutting conditions [18–22]. According to earlier studies, while at a temperature of 400 °C, the minimal value of the coefficient of friction was detected on the samples with Coatings M2 and M3; then, with an increase in temperature to 550 °C and higher, the lowest value of the coefficient of friction was detected for the sample with Coating M1 [63,65]. The results, obtained during the turning of 1045 steel using the tools with the coatings under consideration, did not detect the formation of any noticeable amount of oxide during the cutting process.

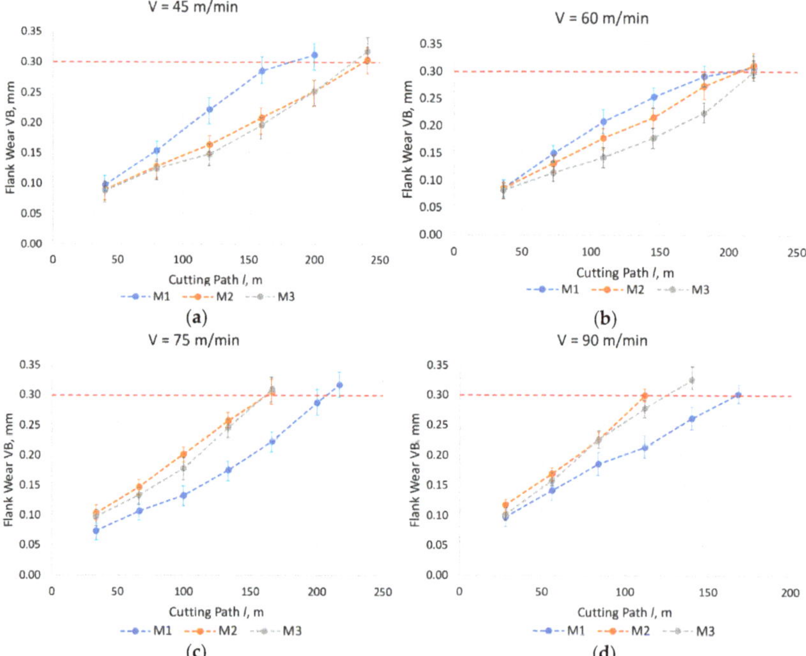

Figure 2. Results of the comparative tests focused on the wear resistance of the tools with the coatings under study in turning a nickel-based alloy at cutting speeds of (**a**) 45, (**b**) 60, (**c**) 75, and (**d**) 90 m/min. The red dotted line indicates the value of limit wear on the flank face $VB_{max} = 0.3$ mm.

The patterns of wear and fracture on the coatings under consideration after cutting were considered. Lamellas, cut from the area of the coating fracture boundary on the wear crater boundary farthest from the cutting edge, were selected for the study (see Figure 3).

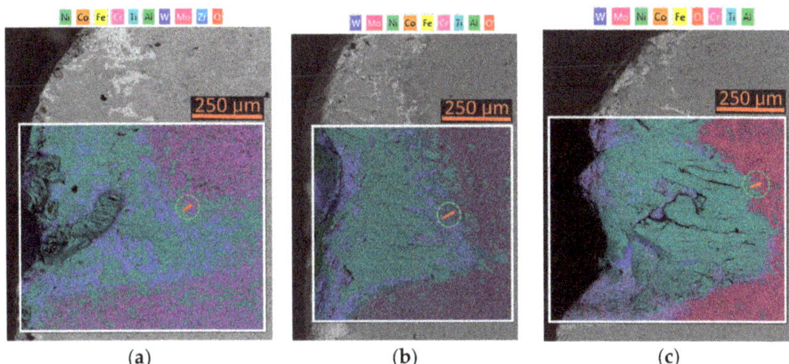

Figure 3. General view of the worn-out rake face of the tool; distribution map of the main elements and localization of the areas for cutting out the lamellas for further studies. Coatings: (**a**) M1, (**b**) M2, (**c**) M3.

3.2. Investigation of the Processes of Fracture, Diffusion, and Oxidation in Zr–ZrN–(Zr,Mo,Al)N (Coating M1)

There was a consideration of the pattern of wear in Coating M1. Typical areas were selected on a cut-out lamella for further studies (Figure 4). The structure of Coating M1 contained considerably extended delaminations between nanolayers, which at the edges turn into the cracks that cut through the nanolayers. Such delaminations can be associated both with a high level of compressive residual stresses and with insufficient strength of the cohesive bond between nanolayers [80–85]. While such delaminations undoubtedly weaken the overall structure of the coating, they can also play some positive role by reducing the values of internal stresses [80–85].

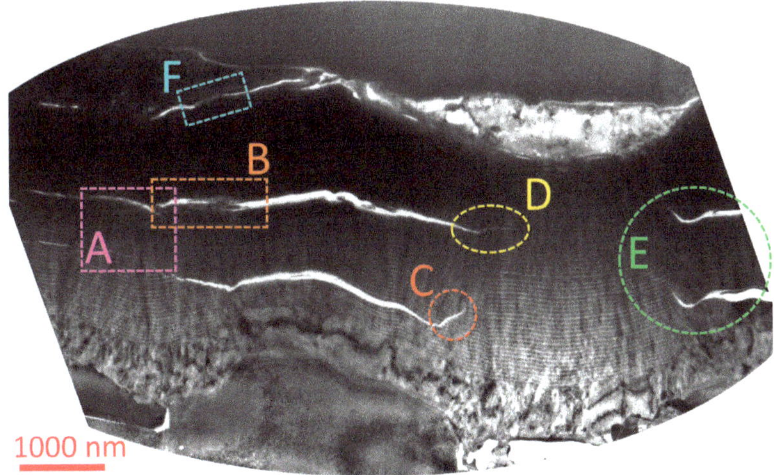

Figure 4. Investigation of the patterns of wear and fracture in Coating M1. General view of the lamella under study. Areas for further research are labeled A, B, C, D, E and F.

The pattern of cracking in Coating M1 was considered in detail (Figure 5). During the growth of delaminations, bond bridges between the nanolayers were retained (Figure 5a,b). The very nanolayers demonstrated sufficient strength, and no formation of through-transverse cracks was detected. Thus, a sequence of the nanolayers between two delaminations was essentially a curved multilayer elastic beam, resembling a semi-elliptical leaf spring, the sheets of which are tightly fastened to each other. During the cutting process, the center of the described "curved beam" was subjected to a variable load, which caused deflection. As a result, the edges of the given "curved beam," coinciding with the boundary of delamination, acted on the coating with a certain force. Such a cyclic action led to the formation of fatigue cracks, exhibited in Figure 5c–f. According to the preliminary analysis, the angle between the direction in which delamination grew and the initial direction in which a subsidiary crack propagated was 90–110° (see Figure 5a,c–e). It should be noted that it was the initial direction of the crack propagation which was considered, and that the described direction can change with further movement of the crack tip. Such close values of the angles may indicate that the conditions for the formation of cracks in different parts of the sample under consideration were fairly close, that is, the physical conditions leading to the formation of cracks. Another typical feature of the formation of cracks in Coating M1 was the discrete nature of their propagation trajectories detected in some areas. In particular, Figure 5f depicts a fully open crack transforming into a discrete one; when the crack propagation trajectory intersected the nanolayers that are lighter in contrast (accordingly, with a high content of aluminum), gaps (relaxations) were formed in the material structure (in particular, the area circled by the yellow dotted line in the inset in the upper right corner of Figure 5f). At the same time, there were no noticeable changes in the structure of the nanolayers that are darker in contrast (with a prevailing content of Zr and Mo). With the further propagation of the crack, the described gaps (relaxations) merged, and as a result, the crack moved (see the inset in the lower right corner of Figure 5f). A similar mechanism of crack propagation in the structure, combining harder and more plastic phases, was described in [80–86].

The area of contact between the flow of the material being machined (an adherent on the lamella) and the coating surface (Area F in Figure 4, exhibited in Figure 6a) was also investigated. There was a transition layer about 100 nm thick between the coating surface and the adherent of the material being machined. The elemental composition of the layer, as well as the phase composition (SAED) of the areas of the adherent adjacent to the coating and the coating layers adjacent to the adherent, were considered (the localization of the line of study of the elemental composition and the SAED analysis are exhibited in Figure 6b). The analysis of the distribution of elements detected a noticeable increase in the oxygen content in the region of the transition layer under consideration (Figure 6c). Accordingly, oxide phases could be detected in the specified layer. Given the high content of Zr and Mo in the layer under consideration, the oxides of Zr and Mo were primarily expected.

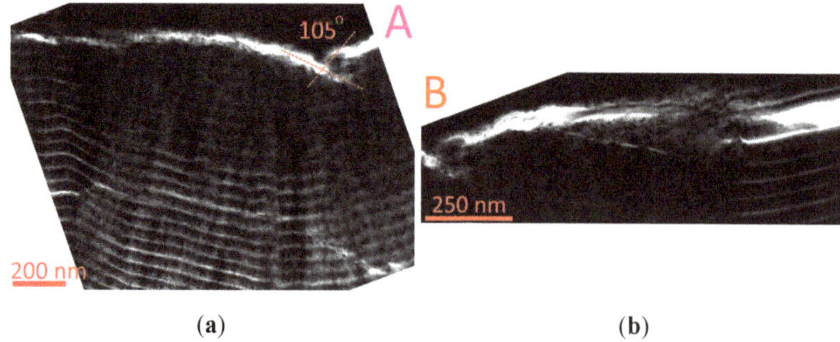

(a)   (b)

Figure 5. *Cont.*

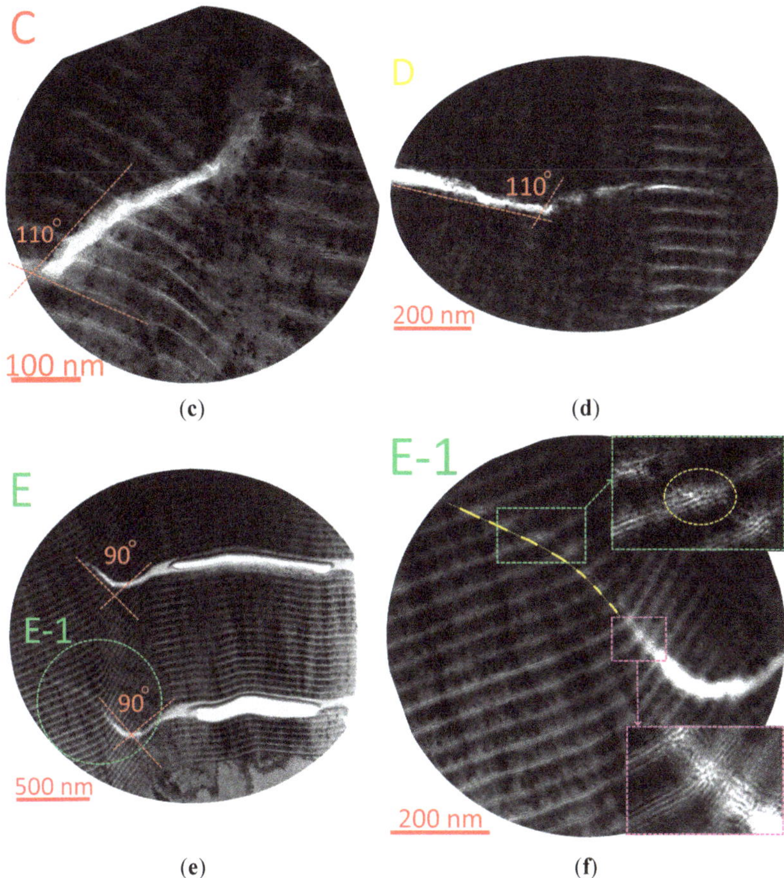

Figure 5. Investigation of the patterns of wear and fracture in Coating M1. (**a–e**) Investigation of the pattern of cracking and the specific features of the relationship between the direction of propagation of delamination and cracks cutting through the nanolayers; (**f**) mechanism of cracking in the nanolayer structure. The localization of areas A,B,C,D and E is shown in Figure 4.

The above assumptions were partially confirmed by the results of the SAED analysis (Figure 6d,e).

The identification of the adherent revealed the presence of metallic nickel (Figure 6e), and two phases of solid solution of nitrides were detected in the coating—(Zr,Mo,Al)N, with a slightly larger lattice parameter compared to the second detected phase, (Mo,Zr,Al)N (Figure 6d). The SAED ring distribution pattern also indicated a slight texturization of the grains of the nitride phases. In addition to the two nitride phases, a small amount of cubic zirconium oxide phase was detected in the coating at the interface with the adherent, which correlated well with the distribution of oxygen and zirconium at the interface. While the layer under consideration also contained regions with an increased content of Mo and Al (see the local peaks of Mo and Al in Figure 6c), no oxide phases of Mo and Al were detected. Such a result may indicate either the absence of oxides of molybdenum and aluminum in the layer under consideration or the presence of an amount of them too small for reliable identification with the methods used. Thus, it is clear that during the cutting process, a transition layer about 100 nm thick was formed between the flow of the material being machined and the surface of Coating M1. The described transition layer included zirconium oxide ZrO_2 and, possibly, small amounts of oxides of molybdenum and aluminum (based on the presence of the regions with an increased content of Mo and Al in the layer under consideration). No diffusion of nickel into the coating was detected. It is possible that the considered oxide layer inhibited the diffusion processes. At the same time, the studies detected the diffusion of zirconium into the material being machined to a depth not exceeding 150 nm, and the diffusion of molybdenum and aluminum to a similar depth. Since the studies also revealed the diffusion of oxygen into the material being machined, it can be assumed there was a diffusion of metals from the oxide layer or a direct diffusion of metal oxides. This issue requires additional study, as the available equipment does not allow reliable phase analysis in such small regions.

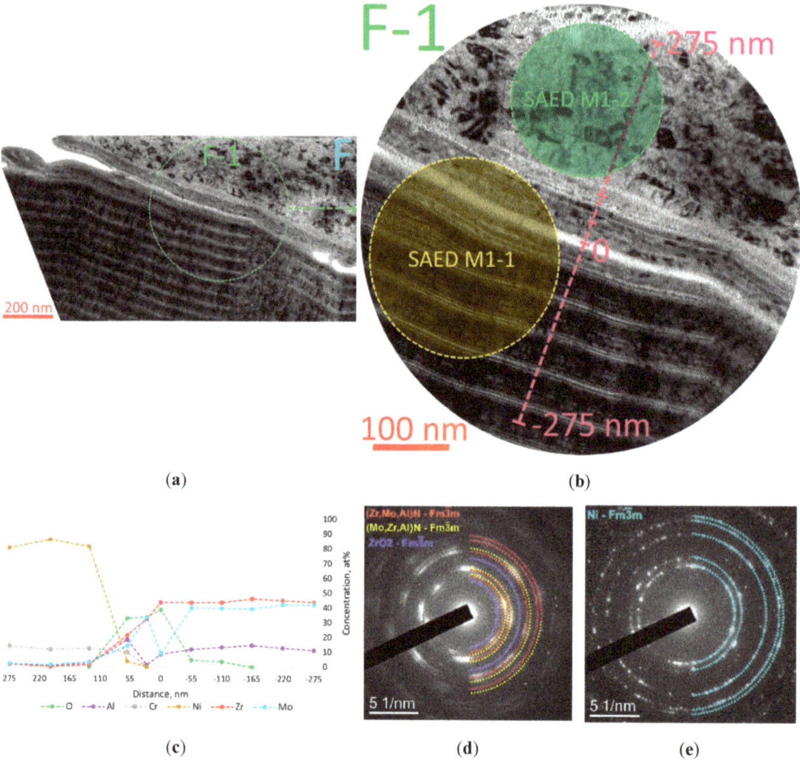

Figure 6. (**a**) General view of Area F under consideration; (**b**) "coating–adherent" interface with indication of the line for study of the elemental composition and areas of SAED; (**c**) results of the study of the elemental composition along the specified line; (**d**) SAED M1-1 from the coating area at the "coating–adherent" interface; (**e**) SAED M1-2 from the area of contact between the adherent and the coating. The localization of area F is shown in Figure 4.

Thus, the considered oxide layer performed barrier functions with respect to the diffusion of nickel into the coating due to the low tendency of nickel to form oxides and the actual absence of nickel in the layer under consideration. At the same time, zirconium, molybdenum, and aluminum diffused into the material being machined, as pure zirconium, molybdenum, and aluminum or their oxides were detected in the layer under consideration.

3.3. Investigation of the Processes of Fracture, Diffusion, and Oxidation in Ti–TiN–(Ti,Mo,Al)N (Coating M2)

There was a consideration of the pattern of wear in Coating M2. Figure 7a exhibits the general view of a cut-out lamella. The SAED analysis from the coating area adjacent to an adherent of the material being machined revealed a uniquely identified phase of cubic solid solution of c-(Ti,Mo,Al)N. No other phases, including oxide phases, were detected. The area of the adherent was identified as metallic nickel.

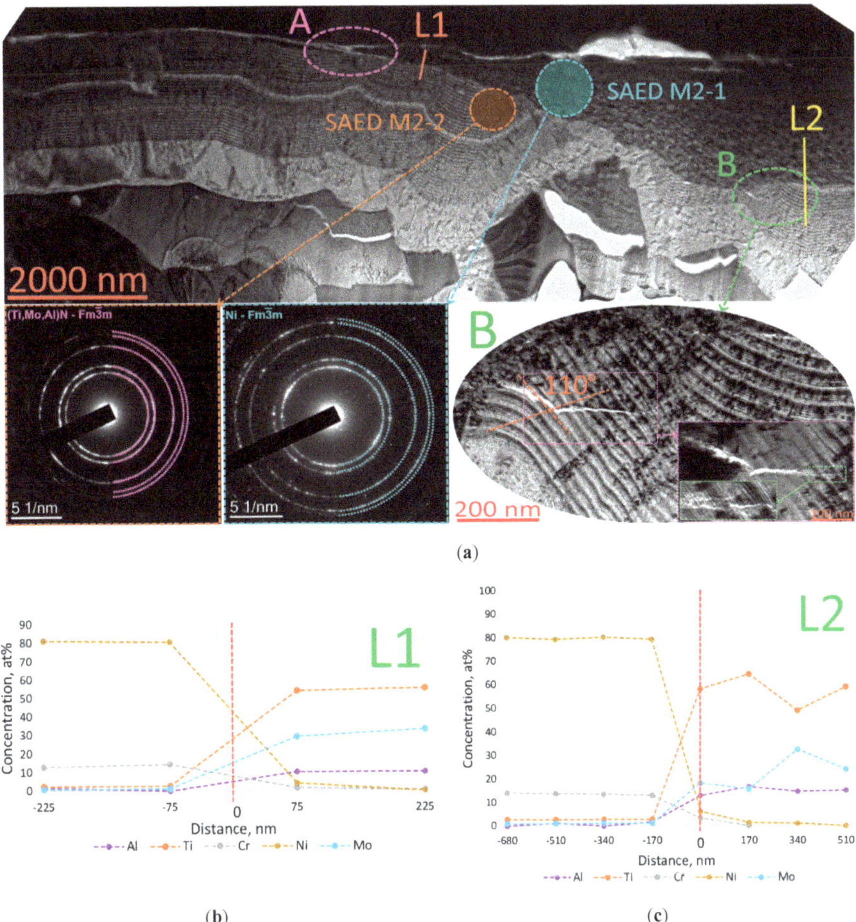

Figure 7. (**a**) General view of the lamella under consideration with the localization of areas for further study, lines of the elemental composition analysis, SAED areas, and results of the analysis of SAED obtained; (**b**) analysis of distribution of elements along line L1; (**c**) analysis of distribution of elements along line L2. The localization of area B and the lines of study of the elemental composition L1 and L2 are shown in (**a**).

According to the results of the analysis of elemental composition along lines L1 and L2 (depicted in Figure 7a, with the results of the analysis presented in Figure 7b,c), there were no signs of oxygen in the area of the adherent–coating interface. The data obtained correlated well with the SAED data.

Unlike Coating M1, the coating under consideration demonstrated no active formation of delaminations between the nanolayers. The analysis of the existing cracks revealed some common features of the cracking processes in Coatings M1 and M2. Figure 7a exhibits delamination and a crack extending from it and cutting through the nanolayers (Area B). It should be noted that the angle between the propagation trajectories of the delamination and the crack was 110°, which is almost identical to the trend considered in the analysis of Coating M1.

Since the absence of oxygen and, accordingly, the absence of oxides in the studied area between the coating surface and the adherent of the material being machined required additional confirmation, the local Area A was studied (the location of Area A is indicated in Figure 7a, and the results of the analysis are presented in Figure 8). The access of oxygen to the areas considered earlier could be hindered, and any formation of oxides could thus be prevented, so the area of the outer boundary of the adherent (Area A-1 in Figure 8b) was chosen for further studies. It is clear there was a possibility of oxygen access in the region, and, accordingly, the presence of oxide phases could be predicted. However, the analysis of the elemental composition of the considered areas along lines L3 and L4 (Figure 8c,d) revealed the absence of oxygen at the adherent–coating interface (at least, the absence of sufficient amounts of oxygen for its identification). Thus, it can be argued there was no active formation of oxides in the considered area of Coating M2 under the cutting conditions specified. In particular, the studies detected the diffusion of nickel and chromium from the material being machined into the coating to a depth of 60–100 nm, and the diffusion of elements of the coating (primarily titanium) into the material being machined to a depth of 60–170 nm.

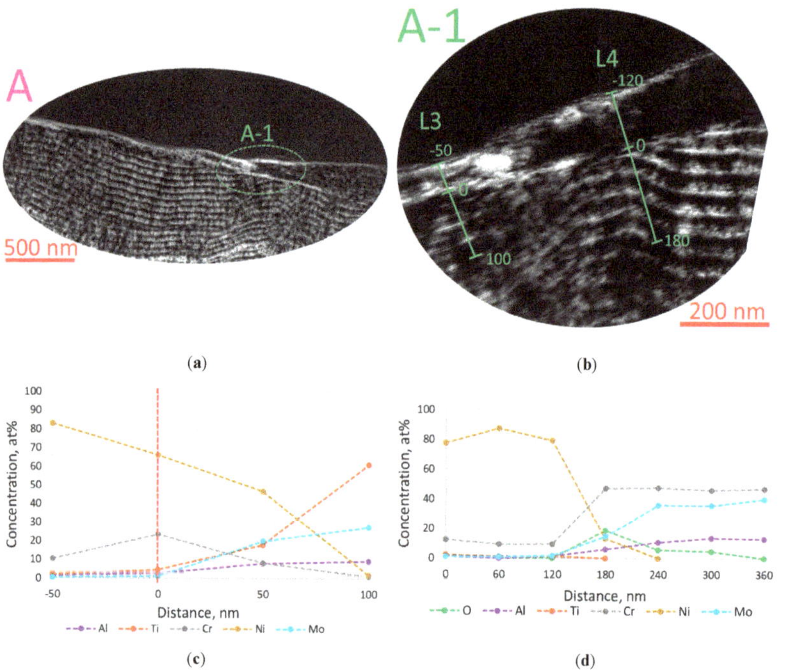

Figure 8. (**a**) General view of Area A; (**b**) localization of lines L3 and L4 for study of the elemental composition; (**c**) change in the elemental composition along line L3; (**d**) change in the elemental composition along line L4. The localization of area A is shown in Figure 7.

3.4. Investigation of the Processes of Fracture, Diffusion, and Oxidation in Cr–CrN–(Cr, Mo, Al)N (Coating M3)

Figure 9a depicts the general view of the lamella under study. Areas A and B were chosen for further investigation, since in these areas, the coating clearly contained a transition layer that resembled the oxide layer studied in Coating M1. The study of the elemental composition along lines L1 and L2 detected the presence of a certain amount of oxygen in the indicated layer, in both Area A and Area B (Figure 9d). The analysis of the HTEM images of the layer under study (Figure 9b, the inset on the right and Area A-1, Figure 9e) revealed the absence of crystalline grains in the region. This indicated a more likely amorphous structure, which may be typical for assumed oxides. However, the SAED analysis did not detect any oxide phases (Figure 9c). Such a contradiction may occur due to the extremely small dimensions of the areas under study and, accordingly, the impossibility of obtaining SAED data with the required accuracy. The coating phase from the near-interface region was identified as a cubic solid solution of c-(Cr,Mo,Al)N. The area of the adherent was identified as metallic nickel. Nickel diffusion into the coating was detected to a depth of 155–260 nm, which was the maximum depth for the three samples under consideration.

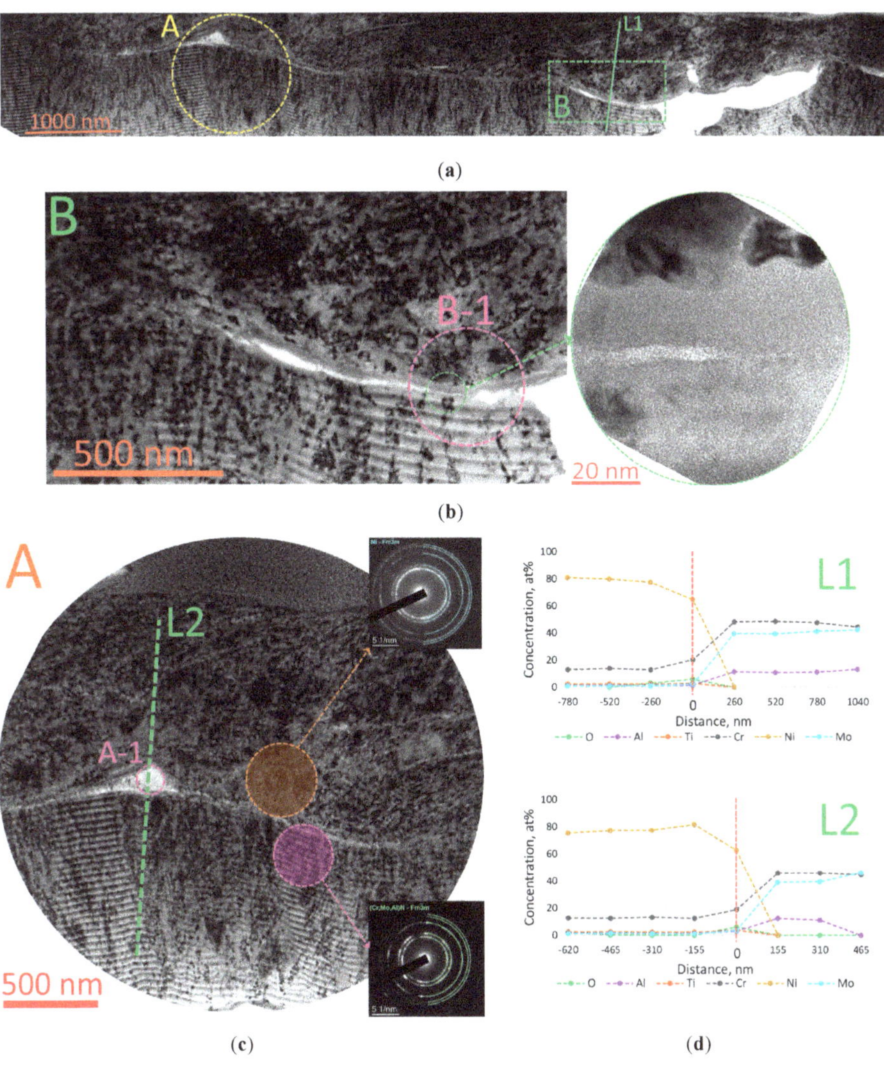

Figure 9. *Cont.*

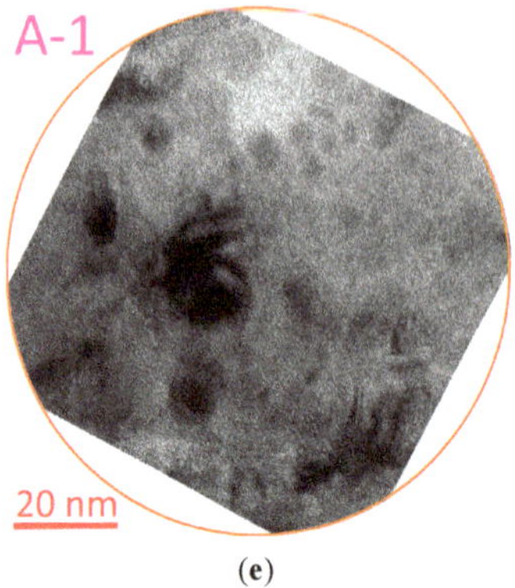

(**e**)

Figure 9. (**a**) General view of the lamella and localization of the areas for further studies; (**b**) Area B and HTEM image of the transition layer; (**c**) Area A, localization of line L2 and regions for SAED, results of the analysis of SAED 1 and SAED 2; (**d**) results of the study of change in the elemental composition along lines L1 and L2; (**e**) HTEM image of Area A-1. The localization of areas A, B and the line of study of the elemental composition L1 is shown in Figure 7a. The localization of area A-1 and the line of study of the elemental composition L2 are shown in Figure 7c.

To further study the composition of the transition layer in Coating M3, the change of the elemental composition in local Areas A-1 and B-1 was analyzed (Figure 10). The results of the analysis made it possible to determine with high reliability the presence of a sufficiently high oxygen content in the areas under consideration. An important feature of the graphs obtained is the practical congruence of the oxygen and chromium graphs, which assumes the presence of, namely, chromium oxide in the layer under consideration.

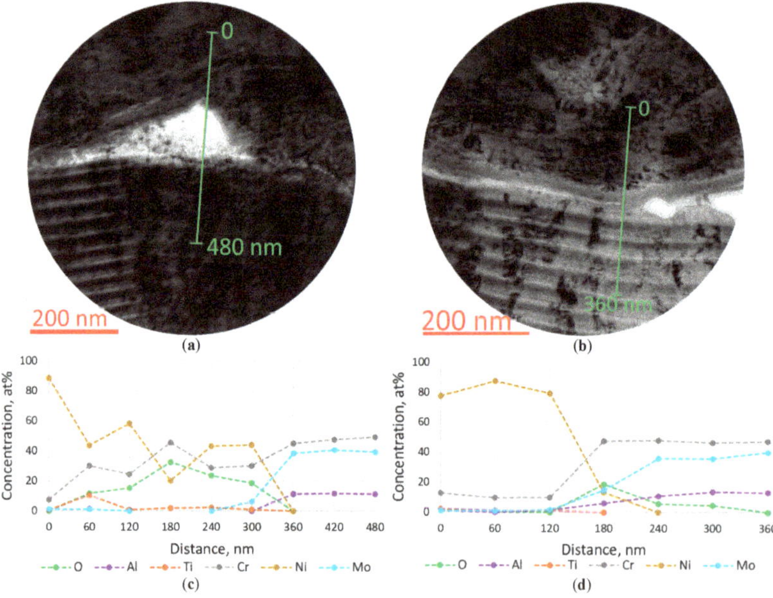

Figure 10. (**a**) Area A-1 and localization of the line for the study of elemental composition; (**b**) Area B-1 and localization of the line for the study of elemental composition, results of the study of change in elemental composition along lines in areas (**c**) A-1 and (**d**) B-1.

4. Conclusions

1. In the Zr-ZrN-(Zr,Mo,Al)N coating, two cubic nitride phases were formed—(Zr,Mo,Al)N (the ZrN-based solid solution) and (Mo,Zr,Al)N (the MoN-based solid solution)—while in the other coatings, there was only a single cubic nitride phase—(Ti,Mo,Al)N (the TiN-based solid solution) or (Cr,Mo,Al)N (the CrN-based solid solution).

2. At the cutting speed of 45 m/min, the tool with the Zr-ZrN-(Zr,Mo,Al)N coating had the highest wear intensity, while at the cutting speed of 60 m/min, the wear intensity was almost the same for all three samples. With a further increase in the cutting speed to 75 and 90 m/min, the tool with the Zr-ZrN-(Zr,Mo,Al)N coating demonstrated the lowest wear intensity. At the cutting speeds of 75 and 90 m/min, the average flank wear for the tool with the Zr-ZrN-(Zr,Mo,Al)N coating was approximately 30% lower compared to the tools with other coatings under consideration.

3. The investigation of the patterns of wear on the tools with the coatings under study after cutting at the speed of 90 m/min detected the presence of a reliably identified transition layer of zirconium oxide at the interface between the Zr-ZrN-(Zr,Mo,Al)N coating and the material being machined. No such transition layer was detected for the Ti-TiN-(Ti,Mo,Al)N coating, and for the Cr-CrN-(Cr,Mo,Al)N coating, a transition layer was detected, which, with a high degree of certainty, could be oxide of chromium.

4. It can be assumed that there is a certain relationship between the forming transition oxide layer and the depth of nickel diffusion into the coating. For the Zr-ZrN-(Zr,Mo,Al)N coating, no nickel diffusion was detected in the region with the transition layer. For the Cr-CrN-(Cr,Mo,Al)N coating, despite the identical thickness of the transition layer, the diffusion of nickel into the coating to a depth not exceeding 260 nm was detected. Thus, it can be assumed that the transition layer based on zirconium oxide (ZrO_2) had significantly better barrier properties with respect to the nickel diffusion flux compared to the transition layer based on chromium oxide.

Author Contributions: Conceptualization, S.G. and A.V.; methodology, A.V. and F.M.; validation, A.V., O.Y. and N.K.; investigation, F.M., N.A., A.S. (Alexander Shein), A.S. (Anton Seleznev) and O.K. and S.K.; resources, S.G.; data curation, N.A., A.S. (Alexander Shein), A.S. (Anton Seleznev), O.K., S.K. and P.P.; writing—original draft preparation, A.V. and F.M.; project administration, P.P.; funding acquisition, S.G. All authors have read and agreed to the published version of the manuscript.

Funding: This research was funded by the state assignment of the Ministry of Science and Higher Education of the Russian Federation, Project No. 0707-2020-0025.

Institutional Review Board Statement: Not applicable.

Informed Consent Statement: Not applicable.

Data Availability Statement: Not applicable.

Acknowledgments: The study used the equipment from the Centre for collective use of Moscow State Technological University STANKIN (agreement No. 075-15-2021-695, 26 July 2021). The coating structure was investigated using the equipment of the Centre for collective use of scientific equipment "Material Science and Metallurgy", purchased with the financial support of the Ministry of Science and Higher Education of the Russian Federation (GK 075-15-2021-696).

Conflicts of Interest: The authors declare no conflict of interest.

References

1. Bobzin, K. High-performance coatings for cutting tools. *CIRP J. Manuf. Sci. Technol.* **2017**, *18*, 1–9. [CrossRef]
2. Baptista, A.; Silva, F.; Porteiro, J.; Míguez, J.; Pinto, G. Sputtering physical vapour deposition (PVD) coatings: A critical review on process improvement andmarket trend demands. *Coatings* **2018**, *8*, 402. [CrossRef]
3. Qadir, M.; Li, Y.; Wen, C. Ion-substituted calcium phosphate coatings by physical vapor deposition magnetron sputtering for biomedical applications: A review. *Acta Biomater.* **2019**, *89*, 14–32. [CrossRef] [PubMed]
4. Selvakumar, N.; Barshilia, H.C. Review of physical vapor deposited (PVD) spectrally selective coatings for mid- and high-temperature solar thermal applications. *Sol. Energy Mater. Sol. Cells* **2012**, *98*, 1–23. [CrossRef]

5. Kamalisarvestani, M.; Saidur, R.; Mekhilef, S.; Javadi, F.S. Performance, materials and coating technologies of thermochromic thin films on smart windows. *Renew. Sustain. Energy Rev.* **2013**, *26*, 353–364. [CrossRef]

6. Mehran, Q.M.; Fazal, M.A.; Bushroa, A.R.; Rubaiee, S. A Critical Review on Physical Vapor Deposition Coatings Applied on Different Engine Components. *Crit. Rev. Solid State Mater. Sci.* **2018**, *43*, 158–175. [CrossRef]

7. Pogrebnjak, A.; Smyrnova, K.; Bondar, O. Nanocomposite multilayer binary nitride coatings based on transition and refractory metals: Structure and properties. *Coatings* **2019**, *9*, 155. [CrossRef]

8. Ezugwu, E.O.; Wang, Z.M.; Machado, A.R. The machinability of nickel-based alloys: A review. *J. Mater. Process. Technol.* **1998**, *86*, 1–16. [CrossRef]

9. Thakur, A.; Gangopadhyay, S. State-of-the-art in surface integrity in machining of nickel-based super alloys. *Int. J. Mach. Tools Manuf.* **2016**, *100*, 25–54. [CrossRef]

10. Ezugwu, E.O. Key improvements in the machining of difficult-to-cut aerospace superalloys. *Int. J. Mach. Tools Manuf.* **2005**, *45*, 1353–1367. [CrossRef]

11. Grigoriev, S.N.; Gurin, V.D.; Volosova, M.A.; Cherkasova, N.Y. Development of residual cutting tool life prediction algorithm by processing on CNC machine tool. *Mater. Werkst.* **2013**, *44*, 790–796. [CrossRef]

12. Grigoriev, S.N.; Sinopalnikov, V.A.; Tereshin, M.V.; Gurin, V.D. Control of parameters of the cutting process on the basis of diagnostics of the machine tool and workpiece. *Meas. Tech.* **2012**, *55*, 555–558. [CrossRef]

13. Devillez, A.; Schneider, F.; Dominiak, S.; Dudzinski, D.; Larrouquere, D. Cutting forces and wear in dry machining of Inconel 718 with coated carbide tools. *Wear* **2007**, *262*, 931–942. [CrossRef]

14. Hartung, P.D.; Kramer, B.M.; von Turkovich, B.F. Tool Wear in Titanium Machining. *CIRP Ann. Manuf. Technol.* **1982**, *31*, 75–80. [CrossRef]

15. Jawaid, A.; Sharif, S.; Koksal, S. Evaluation of wear mechanisms of coated carbide tools when face milling titanium alloy. *J. Mater. Process. Technol.* **2000**, *99*, 266–274. [CrossRef]

16. Grigoriev, S.; Vereschaka, A.; Milovich, F.; Andreev, N.; Bublikov, J.; Sitnikov, N.; Sotova, C.; Kutina, N. Investigation of wear mechanisms of multilayer nanostructured wear-resistant coatings during turning of steel. Part 2: Diffusion, oxidation processes and cracking in Ti-TiN-(Ti,Cr,Mo,Al)N coating. *Wear* **2021**, *486*, 204096. [CrossRef]

17. Aizawa, T.; Mitsuo, A.; Yamamoto, S.; Sumitomo, T.; Muraishi, S. Self-lubrication mechanism via the in situ formed lubricious oxide tribofilm. *Wear* **2005**, *259*, 708–718. [CrossRef]

18. Fox-Rabinovich, G.S.; Yamamoto, K.; Beake, B.D.; Kovalev, A.I.; Aguirre, M.H.; Veldhuis, S.C.; Dosbaeva, G.K.; Wainstein, D.L.; Biksa, A.; Rashkovskiy, A. Emergent behavior of nano-multilayered coatings during dry high-speed machining of hardened tool steels. *Surf. Coat. Technol.* **2010**, *204*, 3425–3435. [CrossRef]

19. Fox-Rabinovich, G.S.; Veldhuis, S.C.; Dosbaeva, G.K.; Yamamoto, K.; Kovalev, A.I.; Wainstein, D.L.; Gershman, I.S.; Shuster, L.S.; Beake, B.D. Nanocrystalline coating design for extreme applications based on the concept of complex adaptive behaviour. *J. Appl. Phys.* **2008**, *103*, 083510. [CrossRef]

20. Endrino, J.L.; Fox-Rabinovich, G.S.; Gey, C. Hard AlTiN, AlCrN PVD coatings for machining of austenitic stainless steel. *Surf. Coat. Technol.* **2006**, *200*, 6840–6845. [CrossRef]

21. Fox-Rabinovich, G.S.; Yamomoto, K.; Veldhuis, S.C.; Kovalev, A.I.; Dosbaeva, G.K. Tribological adaptability of TiAlCrN PVD coatings under high performance dry machining conditions. *Surf. Coat. Technol.* **2005**, *200*, 1804–1813. [CrossRef]

22. Beake, B.D.; Fox-Rabinovich, G.S. Progress in high temperature nanomechanical testing of coatings for optimising their performance in high speed machining. *Surf. Coat. Technol.* **2014**, *255*, 102–111. [CrossRef]

23. Salomon, C.J. Process for Machining Metals of Similar Acting Materials When Being Worked by Cutting Tools. German Patent No. 523594, 27 April 1931.

24. Boothroyd, G.; Knight, W.A. *Fundamentals of Machining and Machine Tools*; CRC Press: Boca Raton, FL, USA, 2006.

25. Kuzin, V.V.; Grigoriev, S.N.; Volosova, M.A. The role of the thermal factor in the wear mechanism of ceramic tools: Part 1. Macrolevel. *J. Frict. Wear* **2014**, *35*, 505–510. [CrossRef]

26. Loladze, T.N. Of the theory of diffusion wear. *CIRP Ann.-Manuf. Technol.* **1981**, *30*, 71–76. [CrossRef]

27. Loladze, T.N. Nature of brittle failure of cutting tool. *CIRP Ann.-Manuf. Technol.* **1975**, *24*, 13–16.

28. Singh, A.; Ghosh, S.; Aravindan, S. Flank wear and rake wear studies for arc enhanced HiPIMS coated AlTiN tools during high speed machining of nickel-based superalloy. *Surf. Coat. Technol.* **2020**, *381*, 125190. [CrossRef]

29. Bouzakis, K.D.; Michailidis, N.; Skordaris, G.; Bouzakis, E.; Biermann, D.; M'Saoubi, R. Cutting with coated tools: Coating technologies, characterization methods and performance optimization. *CIRP Ann.-Manuf. Technol.* **2012**, *61*, 703–723. [CrossRef]

30. Tkadletz, M.; Schalk, N.; Daniel, R.; Keckes, J.; Czettl, C.; Mitterer, C. Advanced characterization methods for wear resistant hard coatings: A review on recent progress. *Surf. Coat. Technol.* **2016**, *285*, 31–46. [CrossRef]

31. Grigoriev, S.; Peretyagin, P.; Smirnov, A.; Solis, W.; Diaz, L.A.; Fernandez, A.; Torrecillas, R. Effect of graphene addition on the mechanical and electrical properties of Al_2O_3-SiCw ceramics. *J. Eur. Ceram. Soc.* **2017**, *37*, 2473–2479. [CrossRef]

32. Suna, J.; Musil, J.; Dohnal, P. Control of macrostress s in reactively sputtered Mo−Al−N films by total gas pressure. *Vacuum* **2006**, *80*, 588–592. [CrossRef]

33. Xu, J.; Ju, H.; Yu, L. Microstructure, oxidation resistance, mechanical and tribological properties of Mo-Al-N films by reactive magnetron sputtering. *Vacuum* **2014**, *103*, 21–27. [CrossRef]

34. Klimashin, F.F.; Euchner, H.; Mayrhofer, P.H. Computational and experimental studies on structure and mechanical properties of Mo-Al-N. *Acta Mater.* **2016**, *107*, 273–278. [CrossRef]

35. Yang, J.F.; Yuan, Z.G.; Liu, Q.; Wang, X.P.; Fang, Q.F. Characterization of Mo–Al–N nanocrystalline films synthesized by reactive magnetron sputtering. *Mater. Res. Bull.* **2009**, *44*, 86–90. [CrossRef]

36. Tomaszewski, L.; Gulbinski, W.; Urbanowicz, A.; Suszko, T.; Lewandowski, A.; Gulbinski, W. TiAlN based wear resistant coatings modified by molybdenum addition. *Vacuum* **2015**, *121*, 223–229. [CrossRef]

37. Yousaf, M.I.; Pelenovicha, V.O.; Yangc, B.; Liua, C.S.; Fu, D.J. Effect of bilayer period on structural and mechanical properties of nanocomposite TiAlN/MoN multilayer films synthesized by cathodic arc ion-plating. *Surf. Coat. Technol.* **2015**, *282*, 94–102. [CrossRef]

38. Bobzin, K.; Brögelmann, T.; Kalscheuer, C.; Stahl, K.; Lohner, T.; Yilmaz, M. (Cr,Al)N and (Cr,Al,Mo)N hard coatings for tribological applications under minimum quantity lubrication. *Tribol. Int.* **2019**, *140*, 105817. [CrossRef]

39. Gilewicz, A.; Warcholinski, B. Deposition and characterisation of Mo2N/CrN multilayer coatings prepared by cathodic arc evaporation. *Surf. Coat. Technol.* **2015**, *279*, 126–133. [CrossRef]

40. Ju, H.; Yu, D.; Xu, J.; Yu, L.; Zuo, B.; Geng, Y.; Huang, T.; Shao, L.; Ren, L.; Du, C.; et al. Crystal structure and tribological properties of Zr-Al-Mo-N composite films deposited by magnetron sputtering. *Mater. Chem. Phys.* **2019**, *230*, 347–354. [CrossRef]

41. Sergevnin, V.S.; Blinkov, I.V.; Volkhonskii, A.O.; Belov, D.S.; Kuznetsov, D.V.; Gorshenkov, M.V.; Skryleva, E.A. Wear behaviour of wear-resistant adaptive nano-multilayered Ti-Al-Mo-N coatings. *Appl. Surf. Sci.* **2016**, *388*, 13–23. [CrossRef]

42. Glatz, S.A.; Koller, C.M.; Bolvardi, H.; Kolozsvári, S.; Riedl, H.; Mayrhofer, P.H. Influence of Mo on the structure and the tribomechanical properties of arc evaporated Ti-Al-N. *Surf. Coat. Technol.* **2017**, *311*, 330–336. [CrossRef]

43. Grigoriev, S.N.; Metel, A.S.; Fedorov, S.V. Modification of the structure and properties of high-speed steel by combined vacuum-plasma treatment. *Met. Sci. Heat Treat.* **2012**, *54*, 8–12. [CrossRef]

44. Yang, K.; Xian, G.; Zhaoc, H.; Fanb, H.; Wangb, J.; Wangc, H.; Du, H. Effect of Mo content on the structure and mechanical properties of TiAlMoN films deposited on WC–Co cemented carbide substrate by magnetron sputtering. *Int. J. Refract. Met. Hard Mater.* **2015**, *52*, 29–35. [CrossRef]

45. Glatz, S.A.; Moraes, V.; Koller, C.M.; Riedl, H.; Bolvardi, H.; Kolozsvári, S.; Mayrhofer, P.H. Effect of Mo on the thermal stability, oxidation resistance, and tribo-mechanical properties of arc evaporated Ti-Al-N coatings. *J. Vac. Sci. Technol. A Vac. Surf. Films* **2017**, *35*, 061515. [CrossRef]

46. Klimashin, F.F.; Mayrhofer, P.H. Ab initio-guided development of super-hard Mo–Al–Cr–N coatings. *Scr. Mater.* **2017**, *140*, 27–30. [CrossRef]

47. Bobzin, K.; Brögelmann, T.; Kalscheuer, C.; Stahl, K.; Lohner, T.; Yilmaz, M. Effects of (Cr,Al)N and (Cr,Al,Mo)N coatings on friction under minimum quantity lubrication. *Surf. Coat. Technol.* **2020**, *402*, 126154. [CrossRef]

48. Gu, B.; Tu, J.P.; Zheng, X.H.; Yang, Y.Z.; Peng, S.M. Comparison in mechanical and tribological properties of Cr–W–N and Cr–Mo–N multilayer films deposited by DC reactive magnetron sputtering. *Surf. Coat. Technol.* **2008**, *202*, 2189–2193. [CrossRef]

49. Bobzin, K.; Brögelmann, T.; Kalscheuer, C. Arc PVD (Cr,Al,Mo)N and (Cr,Al,Cu)N coatings for mobility applications. *Surf. Coat. Technol.* **2020**, *384*, 125046. [CrossRef]

50. Iram, S.; Wang, J.; Cai, F.; Zhang, J.; Ahmad, F.; Liang, J.; Zhang, S. Effect of bilayer number on mechanical and wear behaviours of the AlCrN/AlCrMoN coatings by AIP method. *Surf. Eng.* **2021**, *37*, 536–544. [CrossRef]

51. Metel, A.S.; Grigoriev, S.N.; Melnik, Y.A.; Panin, V.V. Filling the vacuum chamber of a technological system with homogeneous plasma using a stationary glow discharge. *Plasma Phys. Rep.* **2009**, *35*, 1058–1067. [CrossRef]

52. Grigoriev, S.; Vereschaka, A.; Milovich, F.; Tabakov, V.; Sitnikov, N.; Andreev, N.; Sviridova, T.; Bublikov, J. Investigation of multicomponent nanolayer coatings based on nitrides of Cr, Mo, Zr, Nb, and Al. *Surf. Coat. Technol.* **2020**, *401*, 126258. [CrossRef]

53. Vereschaka, A.; Grigoriev, S.; Milovich, F.; Sitnikov, N.; Migranov, M.; Andreev, N.; Bublikov, J.; Sotova, C. Investigation of tribological and functional properties of Cr,Mo-(Cr,Mo)N-(Cr,Mo,Al)N multilayer composite coating. *Tribol. Int.* **2021**, *155*, 106804. [CrossRef]

54. Grigoriev, S.; Vereschaka, A.; Milovich, F.; Sitnikov, N.; Andreev, N.; Bublikov, J.; Kutina, N. Investigation of the properties of the Cr,Mo-(Cr,Mo,Zr,Nb)N-(Cr,Mo,Zr,Nb,Al)N multilayer composite multicomponent coating with nanostructured wear-resistant layer. *Wear* **2021**, *468*, 203597. [CrossRef]

55. Vereschaka, A.A.; Grigoriev, S.N.; Volosova, M.A.; Batako, A.; Vereschaka, A.S.; Sitnikov, N.N.; Seleznev, A.E. Nano-scale multi-layered coatings for improved efficiency of ceramic cutting tools. *Int. J. Adv. Manuf. Technol.* **2017**, *90*, 27–43. [CrossRef]

56. Vereschaka, A.; Tabakov, V.; Grigoriev, S.; Sitnikov, N.; Oganyan, G.; Andreev, N.; Milovich, F. Investigation of wear dynamics for cutting tools with multilayer composite nanostructured coatings in turning constructional steel. *Wear* **2019**, *420*, 17–37. [CrossRef]

57. Vereschaka, A.; Milovich, F.; Migranov, M.; Andreev, N.; Alexandrov, I.; Muranov, A.; Mikhailov, M.; Tatarkanov, A. Investigation of the tribological and operational properties of $(Me_x,Mo_y,Al_{1-(x+y)})N$ (Me –Ti, Zr or Cr) coatings. *Tribol. Int.* **2022**, *165*, 107305. [CrossRef]

58. Grigoriev, S.; Vereschaka, A.; Milovich, F.; Migranov, M.; Andreev, N.; Bublikov, J.; Sitnikov, N.; Oganyan, G. Investigation of the tribological properties of Ti-TiN-(Ti,Al,Nb,Zr)N composite coating and its efficiency in increasing wear resistance of metal cutting tools. *Tribol. Int.* **2021**, *164*, 107236. [CrossRef]

59. Vereschaka, A.; Milovich, F.; Andreev, N.; Sitnikov, N.; Alexandrov, I.; Muranov, A.; Mikhailov, M.; Tatarkanov, A. Efficiency of application of (Mo, Al)N-based coatings with inclusion of ti, zr or cr during the turning of steel of nickel-based alloy. *Coatings* **2021**, *11*, 1271. [CrossRef]

60. Ju, H.; Wang, R.; Wang, W.; Wang, W.; Yu, L.; Luo, H. The microstructure and tribological properties of molybdenum and silicon nitride composite films. *Surf. Coat. Technol.* **2020**, *401*, 126238. [CrossRef]

61. Alexey, V. Improvement of working efficiency of cutting tools by modifying its surface properties by application of wear-resistant complexes. *Adv. Mater. Res.* **2013**, *712–715*, 347–351.

62. Ezugwu, E.O.; Wang, Z.M. Titanium alloys and their machinability—A review. *J. Mater. Process. Technol.* **1997**, *68*, 262–274. [CrossRef]

63. Ezugwu, E.O.; Bonney, J.; Yamane, Y. An overview of the machinability of aeroengine alloys. *J. Mater. Process. Technol.* **2003**, *134*, 233–253. [CrossRef]

64. Zhu, D.; Zhang, X.; Ding, H. Tool wear characteristics in machining of nickel-based superalloys. *Int. J. Mach. Tools Manuf.* **2013**, *64*, 60–77. [CrossRef]

65. Vereschaka, A.; Tabakov, V.; Grigoriev, S.; Aksenenko, A.; Sitnikov, N.; Oganyan, G.; Seleznev, A.; Shevchenko, S. Effect of adhesion and the wear-resistant layer thickness ratio on mechanical and performance properties of ZrN—(Zr,Al,Si)N coatings. *Surf. Coat. Technol.* **2019**, *357*, 218–234. [CrossRef]

66. Grigoriev, S.N.; Vereschaka, A.A.; Fyodorov, S.V.; Sitnikov, N.N.; Batako, A.D. Comparative analysis of cutting properties and nature of wear of carbide cutting tools with multi-layered nano-structured and gradient coatings produced by using of various deposition methods. *Int. J. Adv. Manuf. Technol.* **2017**, *90*, 3421–3435. [CrossRef]

67. Fu, C.-Q.; Huang, Y.-Z.; Li, S.-J.; Xiang, Y.-K. Study on Preparation and Corrosion Resistance of Ni-P/Ni-Zn-P Three-layer Composite Coating. *Surf. Technol.* **2021**, *50*, 400–407.

68. Movchan, B.A.; Malashenko, I.S.; Yakovchuk, K.Y.; Rybnikov, A.I.; Tchizhik, A.A. Two- and three-layer coatings produced by deposition in vacuum for gas turbine blade protection. *Surf. Coat. Technol.* **1994**, *67*, 55–63. [CrossRef]

69. Huang, J.-F.; Li, H.-J.; Zeng, X.-R.; Li, K.-Z. Preparation and oxidation kinetics mechanism of three-layer multi-layer-coatings-coated carbon/carbon composites. *Surf. Coat. Technol.* **2006**, *200*, 5379–5385. [CrossRef]

70. Kuzin, V.V.; Grigoriev, S.N. Method of Investigation of the Stress-Strain State of Surface Layer of Machine Elements from a Sintered Nonuniform Material. *Appl. Mech. Mater.* **2013**, *486*, 32–35. [CrossRef]

71. Fominski, V.Y.; Grigoriev, S.N.; Gnedovets, A.G.; Romanov, R.I. Pulsed laser deposition of composite Mo–Se–Ni–C coatings using standard and shadow mask configuration. *Surf. Coat. Technol.* **2012**, *206*, 5046–5054. [CrossRef]

72. Grigoriev, S.; Vereschaka, A.; Zelenkov, V.; Sitnikov, N.; Bublikov, J.; Milovich, F.; Andreev, N.; Mustafaev, E. Specific features of the structure and properties of arc-PVD coatings depending on the spatial arrangement of the sample in the chamber. *Vacuum* **2022**, *200*, 111047. [CrossRef]

73. Vereschaka, A.A.; Bublikov, J.I.; Sitnikov, N.N.; Oganyan, G.V.; Sotova, C.S. Influence of nanolayer thickness on the performance properties of multilayer composite nano-structured modified coatings for metal-cutting tools. *Int. J. Adv. Manuf. Technol.* **2018**, *95*, 2625–2640. [CrossRef]

74. Vereschaka, A.; Tabakov, V.; Grigoriev, S.; Sitnikov, N.; Milovich, F.; Andreev, N.; Sotova, C.; Kutina, N. Investigation of the influence of the thickness of nanolayers in wear-resistant layers of Ti-TiN-(Ti,Cr,Al)N coating on destruction in the cutting and wear of carbide cutting tools. *Surf. Coat. Technol.* **2020**, *385*, 125402. [CrossRef]

75. Oliver, W.C.; Pharr, G.M.J. An improved technique for determining hardness and elastic modulus using load and displacement sensing indentation. *J. Mater. Res.* **1992**, *7*, 1564–1583. [CrossRef]

76. *ASTM C1624-05(2015)*; Standard Test Method for Adhesion Strength and Mechanical Failure Modes of Ceramic Coatings by Quantitative Single Point Scratch Testing. ASTM International: West Conshohocken, PA, USA, 2010. [CrossRef]

77. *ISO 513:2012*; Classification and Application of Hard Cutting Materials for Metal Removal with Defined Cutting Edges—Designation of the Main Groups and Groups of Application. International Organization for Standardization: Geneva, Switzerland, 2012.

78. Abukhshim, N.A.; Mativenga, P.T.; Sheikh, M.A. Heat generation and temperature prediction in metal cutting: A review and implications for high speed machining. *Int. J. Mach. Tools Manuf.* **2006**, *46*, 782–800. [CrossRef]

79. Vereshchaka, A.S.; Vereshchaka, A.A.; Kirillov, A.K. Ecologically friendly dry machining by cutting tool from layered composition ceramic with nano-scale multilayered coatings. *Key Eng. Mater.* **2012**, *496*, 67–74. [CrossRef]

80. Grigoriev, S.N.; Kozochkin, M.P.; Sabirov, F.S.; Kutin, A.A. Diagnostic Systems as Basis for Technological Improvement. *Proc. CIRP* **2012**, *1*, 599–604. [CrossRef]

81. Shuai, J.; Zuo, X.; Wang, Z.; Guo, P.; Xu, B.; Zhou, J.; Wang, A.; Ke, P. Comparative study on crack resistance of TiAlN monolithic and Ti/TiAlN multilayer coatings. *Ceram. Int.* **2020**, *46*, 6672–6681. [CrossRef]

82. Bannister, M.; Shercliff, H.; Bao, G.; Zok, F.; Ashby, M.F. Toughening in brittle systems by ductile bridging ligaments. *Acta Metal. Mater.* **1992**, *40*, 1531–1537. [CrossRef]

83. Suresh, S. *Fatigue of Materials*, 2nd ed.; Cambridge University Press: Cambridge, UK, 2004; p. 701.

84. Faber, K.T.; Evans, A.G. Crack deflection processes—I. Theory. *Acta Metall.* **1983**, *31*, 565–576. [CrossRef]

85. Kumar, S.; Curtin, W.A. Crack interaction with microstructure. *Mater. Today* **2007**, *10*, 34–44. [CrossRef]

86. Dauskardt, R.H.; Lane, M.; Mab, Q.; Krishna, N. Adhesion and debonding of multi-layer thin film structures. *Eng. Fract. Mech.* **1998**, *61*, 141–162. [CrossRef]

Article

Frequency Effect on the Structure and Properties of Mo-Zr-Si-B Coatings Deposited by HIPIMS Using a Composite SHS Target

Philipp V. Kiryukhantsev-Korneev [1,*], **Alina D. Sytchenko** [1,*], **Pavel A. Loginov** [1], **Anton S. Orekhov** [1,2] and **Evgeny A. Levashov** [1]

[1] Scientific—Educational Center of SHS, National University of Science and Technology (MISiS), 119049 Moscow, Russia
[2] Federal Scientific Research Centre "Crystallography and Photonics", Shubnikov Institute of Crystallography, Russian Academy of Sciences, 119333 Moscow, Russia
* Correspondence: kiruhancev-korneev@yandex.ru (P.V.K.-K.); alina-sytchenko@yandex.ru (A.D.S.); Tel.: +7-(495)-638-46-59 (P.V.K.-K.)

Abstract: Mo-Zr-Si-B coatings were deposited by high-power impulse magnetron sputtering at a pulse frequency of 10, 50, and 200 Hz. The coating structure was studied by scanning electron microscopy, energy-dispersive spectroscopy, glow-discharge optical-emission spectroscopy, transmission electron microscopy, and X-ray diffraction. The mechanical characteristics, adhesive strength, coefficient of friction, wear resistance, resistance to cyclic-dynamic-impact loading, high-temperature oxidation resistance, and thermal stability of the coatings were determined. The coatings, obtained at 10 and 50 Hz, had an amorphous structure. Increasing the frequency to 200 Hz led to the formation of the h-MoSi$_2$ phase. As the pulse frequency increased from 10 to 50 and 200 Hz, the deposition rate rose by 2.3 and 9.0 times, while hardness increased by 1.9 and 2.9 times, respectively. The Mo-Zr-Si-B coating deposited at 50 Hz was characterized by better wear resistance, resistance to cyclic-dynamic-impact loading, and oxidation resistance at 1500 °C. Thermal stability tests of the coating samples heated in the transmission electron microscope column showed that the coating deposited at 50 Hz remained amorphous in the temperature range of 20–1000 °C. Long-term annealing in a vacuum furnace at 1000 °C caused partial recrystallization and the formation of a nanocomposite structure, as well as an increased hardness from 15 to 37 GPa and an increased Young's modulus from 250 to 380 GPa, compared to those of the as-deposited coatings.

Keywords: MoSi$_2$-based coatings; magnetron sputtering; microstructure; hardness; Young's modulus; tribological properties; impact wear resistance; oxidation resistance

Citation: Kiryukhantsev-Korneev, P.V.; Sytchenko, A.D.; Loginov, P.A.; Orekhov, A.S.; Levashov, E.A. Frequency Effect on the Structure and Properties of Mo-Zr-Si-B Coatings Deposited by HIPIMS Using a Composite SHS Target. *Coatings* **2022**, *12*, 1570. https://doi.org/10.3390/coatings12101570

Academic Editor: Michał Kulka

Received: 12 September 2022
Accepted: 10 October 2022
Published: 17 October 2022

Publisher's Note: MDPI stays neutral with regard to jurisdictional claims in published maps and institutional affiliations.

1. Introduction

One of the trends in surface engineering is designing high-temperature hard coatings based on transition metals for protecting parts exposed to high temperatures, including heaters, gas turbine blades, high-performance cutting tools, etc. [1,2]. Coatings based on molybdenum disilicide, which is characterized by a high melting point (2030 °C), high thermal conductivity (51.0 W·m^{-1}·K^{-1}), low coefficient of thermal expansion (7.0 × 10^{-6} K^{-1}) [3], and excellent oxidation and corrosion resistance [3–5], are promising from a practical standpoint. The high-temperature oxidation resistance of protective coatings can be enhanced by increasing the concentration of silicon [6–8], whose atoms facilitate the formation of a diffusion-barrier layer made of borosilicate glass when heated in air [9]. Boron doping of MoSi$_2$ increases the high-temperature oxidation resistance by reducing the viscosity of borosilicate glass and by healing surface defects. Zhestkov and Terent'eva [10] fabricated Mo-Si-B-based coatings that retained their protective properties at T = 1800–2100 °C for 100 s. In [11], the effect of oxide additives, namely Al$_2$O$_3$, SiO$_2$, and SiC, on the properties of coatings was studied. It was shown that the Mo-Si-B coating prepared with SiO$_2$ showed the smallest weight gain of 0.5 mg·cm^{-2} at 1250 °C for 100 h, which is 5–7 times less than the other samples. Doping the Mo-Si-B coating with Ni-Cr to improve its erosion and corrosion resistance was tested in [12]. Significant high-temperature oxidation resistance at temperatures up to 1700 °C was observed for the Mo-Si-B coating with high silicon content in [13]. Doping the Mo-Si-B coating with Zr and Hf increased fracture toughness during high-temperature heating due to a reduction in the grain size of the h-MoSi$_2$ phase [14].

Such technologies as diffusion saturation, the slurry method, and thermal spraying are commonly used in the industry to produce MoSi$_2$-based coatings. Increasing attention is being paid to physical vapor deposition (PVD) methods, such as cathodic arc vapor deposition [15] and magnetron sputtering [16]. The magnetron sputtering technique has a number of advantages, including the simplicity of controlling the coating composition, structure, and properties due to flexibility of sputtering parameter tuning, as well as the low concentration of defects and impurities in the coating, low surface roughness, lack of restrictions with respect to the substrate material, lack of effect on the geometry of end products, and high output of the deposition process. High-power impulse magnetron sputtering (HIPIMS) offers additional opportunities for ceramic coating deposition [17]. Due to the higher power, the HIPIMS method significantly increases plasma density from ~10^{10} ions/cm^3 for direct current magnetron sputtering (DCMS) to 10^{13}–10^{14} ions/cm^3 for HIPIMS [18]. In the case of HIPIMS, sputtered atoms are intensely ionized as they pass through plasma, and the flux predominantly consists of ions, rather than atoms as it is for the conventional DCMS. The elevated ion/atom ratio in the flux in the case of HIPIMS significantly increases the adhesion strength of deposited coatings due to the formation of pseudo-diffusion layers and ion implantation effects at the stage of substrate-surface pre-etching [19]. The mechanical properties, wear resistance, and high-temperature oxidation resistance of coatings are improved due to the increased structure density and adhesion strength [20]. The structure and properties of HIPIMS coatings can be controlled by varying the deposition parameters, such as the current and pulse frequency [21,22]. Nedfors [22] studied the effect of pulse frequency on the mechanical properties of Ti-B coatings. As pulse frequency was reduced from 1000 to 200 Hz, hardness increased from 35 to 49 GPa, while residual stress changed from +0.5 to −3.8 GPa due to the increased intensity of ion flow onto the surface at low-pulse frequencies. Meanwhile, the high-pulse frequency is known to enhance target etching and inhibit the reaction between the chemically active gas and the target, thus efficiently preventing target poisoning and giving rise to a smooth coating surface [23,24]. Importantly, coating deposition by HIPIMS during sputtering multiphase-ceramic targets has been studied insufficiently. Several targets are usually employed when producing coatings with a complex composition; the HIPIMS mode is implemented for a metal target, while ceramic targets are sputtered in the DC or pulsed DC modes [25]. Coatings based on Mo-Si-B have not previously been obtained by the HIPIMS method. Therefore, the study on the effect of HIPIMS energy parameters on the properties of ceramic coatings based on Mo-Si-B is an unexplored and urgent task.

This work aims to study the structure and properties of the Mo-Zr-Si-B coatings deposited by high-power impulse magnetron sputtering at varied pulse frequencies, including the in situ assessment of thermal stability of coating lamellae heated to 1000 °C in the column of a transmission electron microscope.

2. Materials and Methods

Mo-Zr-Si-B coatings were deposited by high-power impulse magnetron sputtering (HIPIMS) using a TruPlasma Highpulse 4002 power supply (Hüttinger, Freiburg, Germany) in a setup based on an UVN-2M pumping system [14]. The (90%MoSi$_2$ + 10%MoB) + 5%ZrB$_2$ sputtering target was fabricated by self-propagating, high-temperature synthesis. The target diameter and thickness were 120 and 10 mm, respectively. The deposition was carried out in an Ar atmosphere (99.9995%): the total pressure was ~0.1 Pa. Plates of aluminum oxide of VOK-100-1 grade (structural studies, mechanical characteristics, high-temperature oxidation resistance, and thermal stability) and molybdenum of MCh-1 grade (adhesion strength, tribological characteristics, and resistance to dynamic impact loading) were used as substrates. The distance between the target and the substrate was 80 mm. No bias voltage was supplied to the substrate during deposition, and the substrate temperature was no higher than 250 °C. Prior to sputtering, the substrates were cleaned with Ar^+ ions (energy, 2 keV) for 10 min using an ion source to remove contamination. The duration of coating deposition was 40 min.

The elemental composition and structure of the coatings were studied by S3400 (Hitachi, Tokyo, Japan) scanning electron microscope (SEM) and energy-dispersive analysis using Noran-7 (Thermo Fisher Scientific, Waltham, MA, USA). The profiles of element distribution across the thickness were recorded by glow-discharge optical emission spectroscopy (GDOES) using a Profiler-2 spectrometer (Horiba Jobin Yvon, Longjumeau, France) [26]. X-ray diffraction (XRD) of the coatings was performed on a Phaser D2 (Bruker, Karlsruhe, Germany) diffractometer using $CuK\alpha$ radiation (wavelength, 0.154 nm). The mechanical properties of the coatings were determined using a nanohardness tester (CSM Instruments, Peseux, Switzerland) with a Berkovich indenter: the indentation load was 4 mN. For each of the samples, at least 9 measurements were carried out. The indentation depth was 70–100 nm, which did not exceed 10% of the coating thickness [27]. Elastic recovery W was calculated using the following formula: $W = (h_m - h_p)/h_m$, where h_m was indenter penetration depth and h_p was the indenter residual depth. The adhesion strength of the coatings was studied using a Revertest scratch tester (CSM Instruments). The maximum operating load was 70 N for all the samples. The tests for determining the coefficient of friction were carried out using an HT-Tribometer (CSM Instruments) according to the "pin-on-disk" scheme with an Al_2O_3 counter body (load, 1 N; linear speed, 10 cm/s; radius of the wear, 5 mm). Resistance to cyclic-dynamic-impact loading was measured using an impact tester (CemeCon, Würselen, Germany) (load, 100 N; number of impacts, 10^5; a 5 mm ball made of hard alloy was used as a counter body). After the tribological and dynamic impact tests, the coating surface was examined using a WYKO-NT1100 optical profilometer (Veeco, Plainview, NY, USA). Annealing in air in a SNOL-7.2/1200 (AB "UMEGA", Utena, Lithuania) muffle furnace was carried out to record the oxidation kinetics at 1000 °C and for an exposure duration of 10, 40, 100, and 280 min. Single-step annealing of the coatings in air was performed at 1300 and 1500 °C in a TK 15.1800.DM.1F ("Termokeramika", Moscow, Russia) muffle furnace. The thermal stability and structural phase transformations in the coatings were studied during heating in the column of the JEM-2100 (Jeol, Tokyo, Japan) transmission electron microscope (TEM) at 200–1000 °C. Lamellae were prepared by the mechanical polishing of the samples, followed by ion etching (FIB and PIPS techniques). Vacuum annealing in a Thermionic T1 furnace was performed at temperatures identical to those used for heating in the TEM column. The coating samples were exposed to each temperature for 30 min. After annealing, the coating samples were studied by XRD.

3. Results and Discussion

3.1. Deposition Parameters

As the pulse frequency (f_p) rose from 10 to 50 and 200 Hz, the pulse voltage increased linearly from 940 to 950 and 963 V (Table 1). The maximum pulsed current (42 A) was observed at a frequency of 10 Hz, while the minimum pulse current (22 A) was recorded at f_p = 50 Hz. The peak current decreased linearly from 33 to 30 A with the increasing pulse frequency. Similar findings were obtained in [22], in which the characteristics of the HIPIMS discharge pulse depending on the pulse frequency were studied. The average power was found to increase with pulse frequency, resulting in a higher coating growth rate. Meanwhile, the degree of ionization of target components declined. Thus, it can be predicted that the coating growth rate increases as the average power and, therefore, the pulse frequency, rise.

Table 1. Coating deposition parameters.

Deposition Parameters	Coating 1	Coating 2	Coating 3
Frequency, Hz	10	50	200
Pulse duration, µs	200	200	200
Pulse voltage, V	940	950	963
Pulse current, A	42	22	27
Average power, kW	0.1	0.2	1
Maximum peak power, kW	35	30	30
Maximum peak current, A	33	25–30	30

3.2. Composition and Structure

Table 2 summarizes the average thickness concentrations of the main elements of the coatings determined by GDOES.

Table 2. The composition (according to the GDOES data), thickness, and growth rate of coatings.

Frequency, Hz	Elemental Composition, at.%					Thickness, µm	Growth Rate, nm/min
	Mo	Si	B	Zr	C + O		
10	30.1 ± 1.3	63.2 ± 1.2	5.3 ± 0.2	1.2 ± 0.1	0.2 ± 0.1	0.3	7.5
50	25.8 ± 1.3	60.2 ± 1.4	9.1 ± 0.4	3.8 ± 0.2	1.1 ± 0.4	0.7	17.5
200	28.0 ± 2.0	56.8 ± 1.5	9.6 ± 0.4	4.5 ± 0.4	1.1 ± 0.4	2.7	67.5

The sum of the concentrations of carbon and oxygen impurities in the coatings was no higher than 1.1 at.%. Similar concentrations of Mo (25.8–30.1 at.%) and Si (56.8–63.2 at.%) were observed for all the coatings; the Mo/Si ratio was near-stoichiometric. As the pulse frequency increased from 10 to 50 and 200 Hz, the boron concentration rose 1.7–1.8-fold and the zirconium concentration increased 3.2- and 3.8-fold, respectively. This phenomenon can be attributed to the reduction in the peak target current at high-pulse frequencies [21]. A similar change in concentration can take place due to the difference in atom-scattering energy during sputtering [28].

SEM images (Figure 1a) show that the coatings had a dense homogeneous structure, and their thickness increased by 2.3 and 9.0 times as frequency rose from 10 to 50 and 200 Hz, respectively (Table 2).

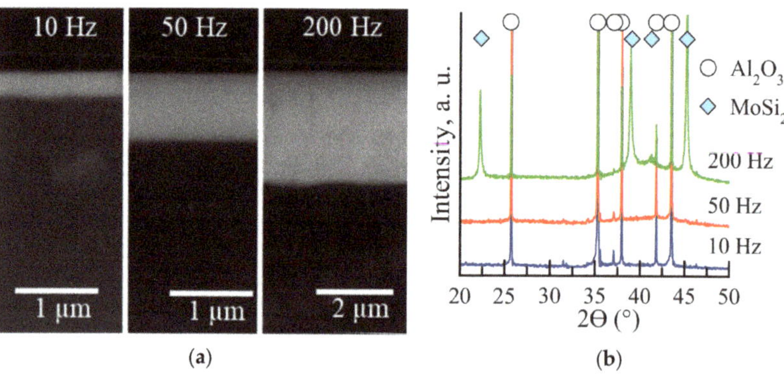

Figure 1. Cross-section SEM images of coatings (**a**) and the respective XRD patterns (**b**).

The minimal thickness (0.3 µm) and growth rate (7.5 nm/min) were observed for the sample deposited at 10 Hz and the maximum peak current (Table 1). It was reported earlier that the high-peak target current increased the ionization rate and plasma density, and the absolute number of ions failing to reach the substrate increased because of the magnetic confinement and return effects [29,30]. As the frequency rose to 50 and 200 Hz, the growth rate increased to 17.5 and 67.5 nm/min, respectively. This may be attributed to the increased sputtering rate of the target and the total flux of sputtered species to the substrate at a high frequency [22]. The low ionization rate at the low-peak current reduced the counter-attraction to the target surface, so a greater number of metal ions and metals were transferred to the substrate, and the deposition rate increased [29]. Furthermore, the reflection or desorption of atoms and ions on the substrate surface decreased because of a weakened ion bombardment at high-pulse frequencies, which also increased the deposition rate [31].

Figure 1b shows the XRD patterns of the coatings. Only peaks belonging to the Al_2O_3 substrate (ICDD 46-1212) were detected for the samples deposited at pulse frequencies of 10 and 50 Hz because of the low coating thickness < 1 μm. Peaks belonging to h-$MoSi_2$ (ICDD 80-4771) at 2θ = 22.3, 38.9, 41.2, and 45.1° were identified for the sample deposited at 200 Hz. The size of the h-$MoSi_2$ grains determined using the Scherrer formula lay within the range of 40–60 nm. The lattice parameters were a = 0.460 and c = 0.655 nm.

3.3. Mechanical Properties

The nanoindentation tests revealed a linear dependence between the mechanical characteristics and frequency. The coating deposited at 10 Hz was characterized by the minimal hardness H = 8 GPa, Young's modulus E = 196 GPa, and elastic recovery W = 35% (Table 3).

Table 3. Mechanical and tribological properties of the coatings.

Frequency, Hz	H, GPa	E, GPa	W, %	H/E	H^3/E^2, GPa	f	Vw, $\times 10^{-4}$ $mm^3 \cdot N^{-1} \cdot m^{-1}$	V, $\times 10^3$ μm^3
10	8 ± 1	196 ± 29	35 ± 4	0.041	0.013	1.01	3.0	309
50	15 ± 3	251 ± 30	44 ± 6	0.060	0.054	0.90	1.9	281
200	23 ± 3	299 ± 18	65 ± 4	0.077	0.136	0.81	2.3	561

H—hardness; E—Young's modulus; W—elastic recovery; H/E—elastic strain to failure; H^3/E^2—resistance to plastic deformation; f—coefficient of friction; Vw—wear rate; V—crater volume.

As the frequency increased, the mechanical properties were improved (H = 23 GPa) for the coating deposited at 200 Hz. The elastic strain to failure H/E and resistance to plastic deformation H^3/E^2 were also determined using the nanoindentation data, which are important from the perspective of wear resistance and deformation mechanisms [32–34]. By analyzing the H/E and H^3/E^2 ratios, one can assume that the coating deposited at 200 Hz will be characterized by the maximum wear resistance due to the high parameters of H/E = 0.077 and H^3/E^2 = 0.136 GPa. In this study, the increase in hardness can be attributed to the rising zirconium and boron concentrations in the coatings. The hardness of ZrB_2 is known to be higher than that of $MoSi_2$ [35,36]. It can be assumed that the high zirconium and boron contents increase the number of Zr-B bonds and, therefore, increase hardness. Furthermore, the effect of coating thickness on the results of measuring mechanical properties should not be ruled out [37].

3.4. Adhesion Strength

Figure 2 shows the scratch-test parameters as a function of indentation load and the images of scratches recorded using an optical profilometer. An abrupt jump in acoustic emission (AE) at a load of Lc_1 = 2 N was observed in the scratch-test curve for the Mo-Zr-Si-B coating deposited on molybdenum at a frequency of 10 Hz, which can be associated with the cracking of the coating. The load Lc_3 = 3.2 N can be attributed to the point where the molybdenum substrate started to be deformed by the indenter. In the rest of the region, the AE signal remained constant and was near-zero. No jump was observed in the diagram, showing the dependence between the penetration depth (*h*) and indentation load; the h value increased linearly from 0 to ~40 μm. For the sample deposited at 50 Hz, the jump in AE associated with cracking was revealed at a load of Lc_1 = 2.2 N. The critical load at which the indenter touched the substrate was Lc_3 = 4.5 N. The penetration depth increased smoothly from 0 to ~21 μm. For the coating deposited at 200 Hz, the acoustic emission changed jump-wise as the indentation load increased from 1 to 25 N. In this case, Lc_1 = 3.1 N. After that, AE declined to zero and remained constant in the region of 25–70 N. The reduction in AE at 25 N was accompanied by a jump in the diagram, showing the dependence between penetration depth and indentation load. This load corresponds to Lc_3. By the end of the test, the h value was 34 μm. The difference in penetration depths at the maximum indentation load can be attributed to the different thicknesses and structural features of the coatings.

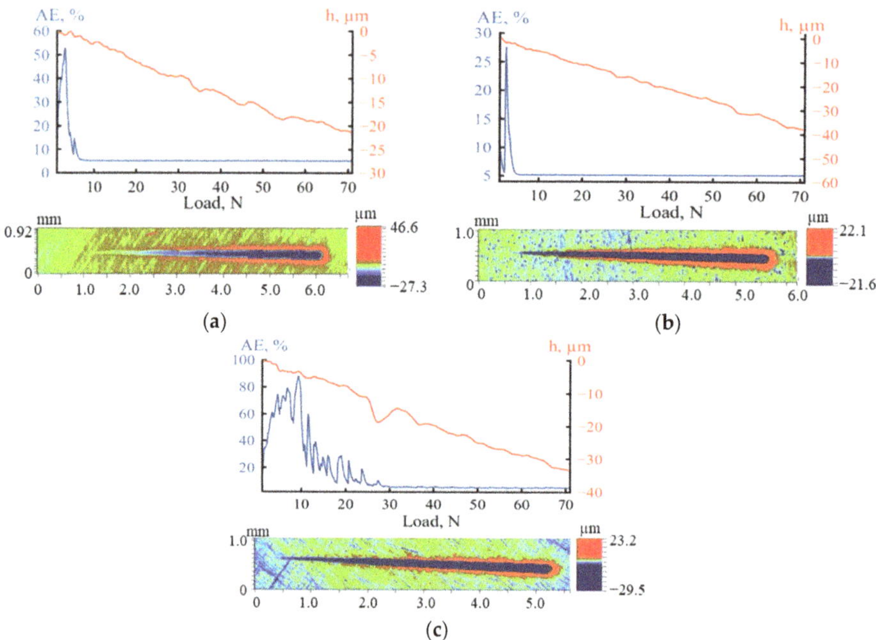

Figure 2. Scratch-test parameters as a function of indenter load and the scratch profiles after testing of Mo-Zr-Si-B coatings deposited by HIPIMS at frequencies of 10 (**a**), 50 (**b**), and 200 Hz (**c**).

The minimal penetration depth was observed for the coating deposited at 50 Hz. Radial cracks usually result from the plastic deformation of the substrate and the low strength of the substrate/coating interface [38]. According to the profiles of scratches detected after the scratch tests, in which there was p to a load of 70 N for all the Mo-Zr-Si-B coatings deposited by the HIPIMS method, no delamination was observed, which indicates that these coatings were characterized by high adhesion strength. It should be noted that an adhesion strength of 70 N for Mo-Zr-Si-B HIPIMS coatings is higher than that of the typical coatings obtained on metal substrates by conventional direct current magnetron sputtering [35–37,39].

3.5. Tribological Characteristics

The tribology tests revealed that, for all coatings, there was a breaking-in region (at distances up to 15 m), characterized by a high coefficient of friction (f~1.0) (Figure 3). The shortest running-in period was observed for the coating deposited at a frequency of 200 Hz. The coating deposited at 10 Hz was characterized by a high coefficient of friction f~1

(Table 3). As the frequency was increased to 50 Hz, the f value declined to 0.90. The coating deposited at 200 Hz had the minimal f value. The friction coefficients agreed with the previously published data for Mo-Si-B-based coatings [14,40]. The wear rate (Vw) (Table 3) was determined according to the wear track profiles (Figure 3), which was 3.0, 1.9, and 2.3×10^{-4} mm$^3 \cdot$N$^{-1} \cdot$m^{-1} for the coatings deposited at 10, 50, and 200 Hz, respectively.

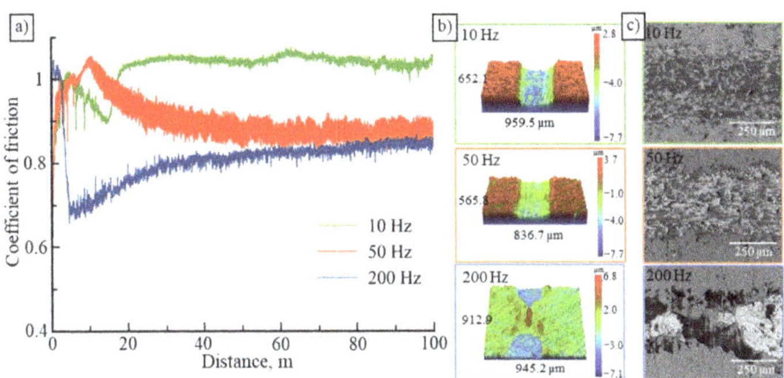

Figure 3. The coefficient of friction (**a**) as a function of distance, the 3D profiles (**b**) and SEM images of tribocontact zone (**c**) of the Mo-Zr-Si-B coatings deposited at 10, 50, and 200 Hz.

The coatings deposited at 10 and 50 Hz had a similar shape of wear tracks. The coating deposited at 200 Hz was characterized by non-uniform wear. During tribological contact, a counter body was pressed into the ductile molybdenum substrate material until the maximum stress was reached. As a result, different depths were observed along the entire length of the wear track, similarly to the situation reported in [41].

3.6. Resistance to Dynamic Impact Loading

Figure 4 shows the 3D profiles of the craters after the dynamic impact tests of the coatings deposited onto Mo substrates. The crater volumes (V) calculated according to the 2D profiles are summarized in Table 3. The craters of the coatings deposited at frequencies of 10 and 50 Hz had a regular shape, thus indicating that plastic deformation typical of molybdenum prevailed. Signs of brittle failure were observed for the coating deposited at 200 Hz, which is typical of coatings with a high H^3/E^2 ratio [42]. The crater volumes for the coatings deposited at 10, 50, and 200 Hz were 309, 281, and 561 $\times 10^3$ μm^3, respectively.

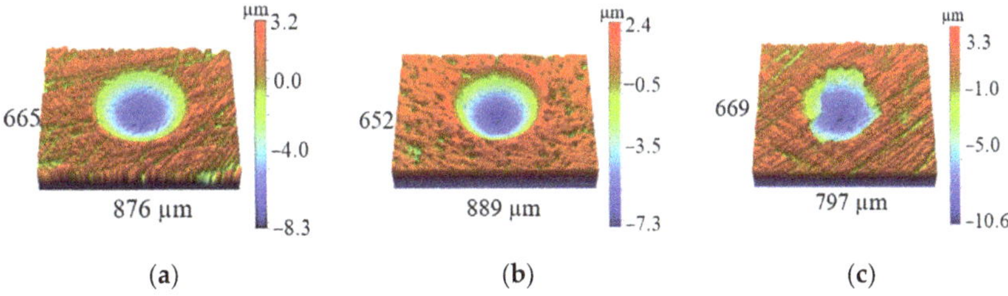

Figure 4. The 3D profiles of craters after dynamic impact tests of the Mo-Zr-Si-B samples deposited by HIPIMS at frequencies of 10 (**a**), 50 (**b**), and 200 Hz (**c**).

The sample deposited at a frequency of 50 Hz exhibited the best wear resistance and resistance to cyclic-dynamic-impact loading, which can be attributed to the low coating thickness, as well as the influence of the substrate. In [42], it is shown that, during impact testing of Zr-Si-B coatings deposited on a molybdenum alloy with H/E = 0.0035, plastic deformation prevailed. At the same time, the substrate was characterized by a smaller crater depth compared to the Zr-Si-B coating, which is associated with the high ductility of Mo-alloys.

3.7. High-Temperature Oxidation Resistance

Figure 5 shows the results of studies focusing on the oxidation kinetics at 1000 °C.

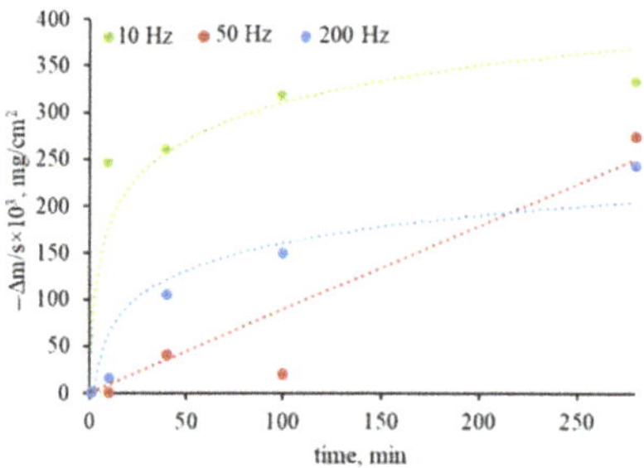

Figure 5. Changes in specific weight of the Mo-Zr-Si-B coatings deposited by HIPIMS onto an Al$_2$O$_3$ substrate at frequencies of 10, 50, and 200 Hz as a function of time of annealing at 1000 °C.

The mass of the coating deposited at 10 Hz decreased abruptly at an exposure time of 0–60 min due to the rapid oxidation of molybdenum and the evaporation of MoO$_x$ [43]. The mass loss ($\Delta m/s$) for coating one was the maximum (-0.33 mg/cm^2) for an exposure time of 280 min. The mass of the coating deposited at 50 Hz increased ($\Delta m/s = 0.042$ mg/cm^2) at an exposure time of 0–40 min because of the formation of a SiO$_2$-based oxide layer. The decline in $\Delta m/s$ to -0.27 mg/cm^2 observed in the region of 40–280 min was also caused by the evaporation of molybdenum oxide. Similar results ($\Delta m/s = -0.24$ mg/cm^2) were obtained for the coating deposited at 200 Hz, in which the evaporation of MoO$_x$ was observed at an exposure time of 10–280 min. Note that the coatings obtained at 10 and 200 Hz were oxidized according to parabolic law, which corresponds to the kinetic curves for the bulky samples and Mo-Si-B coatings [44,45]; meanwhile, the coating at an average frequency of 50 Hz oxidized according to linear law. When evaluating the appearance of the annealed samples of the coating deposited at the minimal frequency of 10 Hz, we clearly observed coating degradation resulting from complete oxidation. The oxidation of other samples was accompanied by the formation of a dense protective oxide layer, rather than delamination or critical damage. The best resistance to oxidation at T = 1000 °C, in terms of appearance and kinetic curves, was shown by the coating obtained at 200 Hz.

Annealing at 1300 °C showed that the coating which was deposited at 10 Hz and characterized by minimal thickness was completely oxidized and delaminated. Therefore, SEM studies were performed only for the samples deposited at 50 and 200 Hz (Figure 6). Blistering caused by the evaporation of MoO$_x$ was observed on the surface of the coating deposited at 50 Hz. Coating-delamination zones resulted from the collapse of blistered regions. The EDS maps recorded for the cross-section fracture showed that a ~500 nm thick oxide layer was formed on the surface of the coating deposited at 50 Hz (Figure 6). The layer consisted of amorphous silicon oxide with inclusions of ZrO$_2$ crystallites (light-colored regions in the microimages) that preferentially resided in the upper section of the layer (Figure 6). No blistering was observed on the surface of the coating deposited at the maximum frequency (200 Hz), although some local delamination was detected (Figure 6). The structure and thickness of the oxide layer observed at the transverse failure were similar to those of the coating deposited at 50 Hz. Small ZrO$_2$ grains were observed on the surface. It is known that the introduction of zirconium often increases oxidation resistance due to the formation of zirconium oxide particles in a dense oxide layer [46].

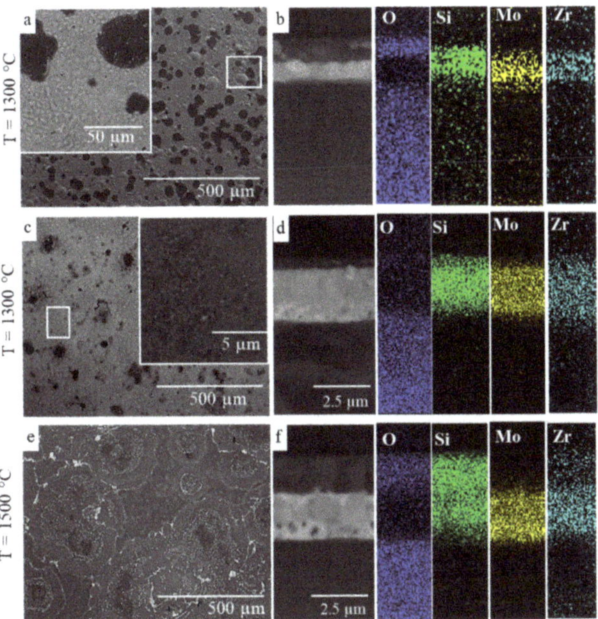

Figure 6. SEM images of the surface (**a**,**c**,**e**) and cross-section fracture (**b**,**d**,**f**) of the coatings deposited at frequencies of 50 (**a**,**b**,**e**,**f**) and 200 Hz (**c**,**d**) after annealing in air at 1300 and 1500 °C.

Figure 7a shows the XRD data for the coatings annealed at 1300 °C. These findings prove that the coating deposited at the minimal frequency of 10 Hz was completely oxidized: the XRD patterns contained only the peaks corresponding to the Al_2O_3 substrate and MoO_x oxide (ICDD 73-1536). Along with the peaks listed above, lines belonging to the tetragonal $t\text{-}Mo_5Si_3$ phase (ICDD 34-0371) were identified for the remaining samples. The following phase transformations are known to take place in the Mo-Si-B-based coatings during heating in air: the $h\text{-}MoSi_2$ phase is transformed into the $t\text{-}MoSi_2$ phase (ICDD 41-0612) at temperatures of ~1000–1200 °C; and the $t\text{-}Mo_5Si_3$ phase is formed at higher temperatures [13].

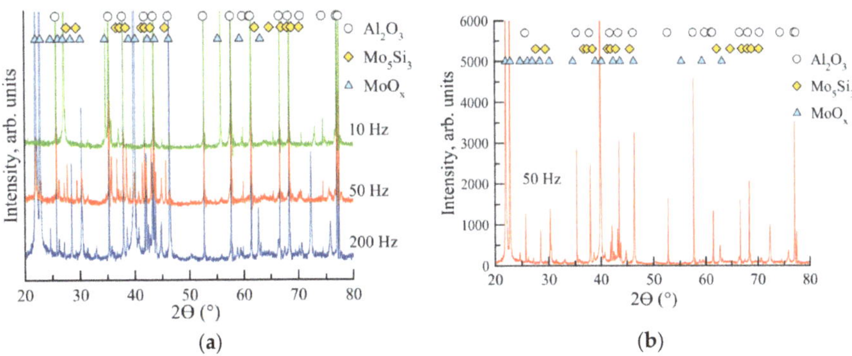

Figure 7. X-ray diffraction patterns of the Mo-Zr-Si-B coatings annealed in air at 1300 °C (**a**) and the coating deposited at 50 Hz and annealed at 1500 °C (**b**).

Annealing at 1500 °C demonstrated that the coating deposited at a medium-pulse frequency of 50 Hz exhibited good oxidation resistance.

Meanwhile, thermal treatment resulted in the cracking, partial delamination, and oxidation of the coating deposited at 200 Hz. Figure 6 shows the SEM images for the coating deposited at 50 Hz. The agglomeration of irregularly shaped light-colored grains corresponding to zirconium oxide was observed on the coating surface. One can see in the SEM image of the cross-section fracture of the coating that ZrO_2 grains are predominantly located in the upper portion of the SiO_2-based oxide layer. The thickness of the oxide layer was ~5 μm. The XRD pattern of the coating recorded at temperatures above 1500 °C (Figure 7b) was identical to those observed after annealing at 1300 °C.

Hence, the best high-temperature oxidation resistance was observed for the Mo-Zr-S-B coating deposited by HIPIMS at a medium frequency of 50 Hz, which can be attributed to the relatively small thickness of the coating (~1 μm), which is sufficient to ensure the formation of a protective a-SiO_2/nc-ZrO_2 film surface during heating, while not causing an accumulation of internal stresses that results in sample degradation during rapid heating or cooling.

3.8. Thermal Stability

An in situ examination of the structural phase transformations occurring in the Mo-Zr-Si-B coating deposited at 50 Hz during stepwise heating in the TEM column was performed. Figure 8 shows the bright-field TEM images of the structure and electron-diffraction patterns of the coating heated within the temperature range of 20–1000 °C.

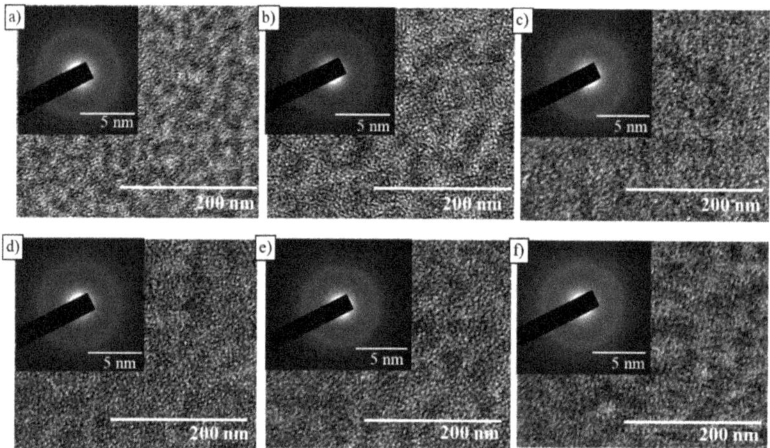

Figure 8. TEM images and electron-diffraction patterns of the Mo-Zr-Si-B coating deposited by HIPIMS (50 Hz) at temperatures 20 (**a**), 200 (**b**), 400 (**c**), 600 (**d**), 800 (**e**) and 1000 °C (**f**).

The as-deposited coating had an amorphous structure, as indicated by the presence of a broad ring in the electron-diffraction pattern. The interplanar distance estimated using the ImageJ software was ~0.270 nm. This value was close to that of the $MoSi_2$ crystalline phase. No noticeable structural phase transformations took place as the temperature was increased to 1000 °C, and the coating remained amorphous.

The next stage of studying the structural phase transformations in Mo-Zr-Si-B coatings involved vacuum annealing at 600, 800, and 1000 °C at an isothermal exposure time of 30 min. No changes in the XRD patterns were observed for the coating deposited at 10 Hz after heating to 1000 °C. For the X-ray amorphous coating deposited at 50 Hz, heating to 600 °C resulted in the emergence of broad peaks at 2Θ = 22.4, 26.3, 35.4, 39.2, 41.6, and 45.5 °, which corresponded to the hexagonal h-$MoSi_2$ phase (Figure 9a).

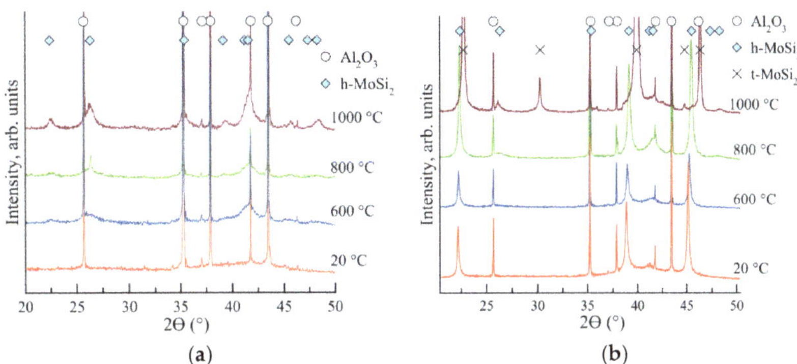

Figure 9. X-ray diffraction patterns of the Mo-Zr-Si-B coatings deposited by HIPIMS at 50 (**a**) and 200 Hz (**b**), in the as-deposited state and after vacuum annealing at T = 600, 800, and 1000 °C.

The intensity of the peaks corresponding to h-MoSi$_2$ increased as the temperature rose to 800 and 1000 °C. The grain size of this phase, which was determined using the Scherrer formula, increased from 7 to 9 and 13 nm as the temperature rose from 600 to 800 and 1000 °C, respectively. The lattice parameters of the hexagonal h-MoSi$_2$ phase remained unchanged as the temperature increased from 600 to 1000 °C and were a = 0.460 and c = 0.655 nm, which corresponded to the tabular data for the h-MoSi$_2$ phase. Heating of the coating deposited at 200 Hz to 600 °C did not cause any qualitative changes in the XRD pattern (Figure 9b). A broad peak (101) at 2θ = 26.1°, corresponding to the hexagonal h-MoSi$_2$ phase, emerged at 800 °C, in addition to the peaks detected earlier. These changes may be associated with the onset of recrystallization. As the temperature rose to 800 °C, the lattice parameters a and c declined to 0.459 and 0.652 nm, respectively. Peaks at 2θ = 22.7, 30.3, 39.9, 44.8, and 46.3°, corresponding to the t-MoSi$_2$ phase, were detected at T = 1000 °C. The grain size of the tetragonal t-MoSi$_2$ phase was ~70 nm. The low-intensity peaks belonging to the hexagonal h-MoSi$_2$ phase were observed at 2θ = 26.1 and 41.2°.

The differences in the TEM and XRD data for the coating deposited at 50 Hz can be explained by the different exposure times and cooling rates during thermal treatment in the TEM column and in the vacuum furnace. In the case of stepwise heating in the TEM column to temperatures up to 1000 °C, the exposure time was 30–40 min, while it was 10 min for the temperature of 1000 °C; the cooling rate of the sample was ~100 °C/min. In these temperature modes, the coating had an amorphous structure in the entire temperature range of 20–1000 °C. Vacuum annealing was carried out for 30 min for all the temperatures. The cooling rate was ~10 °C/min, i.e., the sample was exposed to higher temperatures for a longer time. This is a plausible reason for the recrystallization detected in the annealed coatings. A similar effect of the cooling rate on the structure of Zn-Mg-Al coatings was reported in [47]. Additionally, the reason for the difference between the TEM and XRD data could be related to the local recrystallization of the coating as a result of vacuum annealing, which is not easy to observe in TEM.

The mechanical characteristics of the coatings subjected to vacuum annealing at 1000 °C were studied (Figure 10).

The hardness and elastic recovery of the coating deposited at 10 Hz were H = 10 GPa and W = 46%, being higher than those of the initial coating by 25% and 30%, respectively. Young's modulus decreased by 15% as the temperature was increased from 20 to 1000 °C. For the coating deposited at 50 Hz, vacuum annealing at 1000 °C made H and E increase to 37 GPa and 380 GPa, respectively. The self-hardening effect observed for the Mo-Zr-Si-B coating deposited at 50 Hz could be related to the partial recrystallization of the amorphous coating and the formation of a nanocomposite structure at 1000 °C [30]. For the coating deposited at 200 Hz and characterized by the maximum mechanical characteristics in its initial state, the H and E parameters decreased by 56 and 25%, respectively, as the temperature was increased to 1000 °C, while W increased by 35%. This effect was related to phase transformations and the fact that the predominating phase in the structure after annealing was the tetragonal t-MoSi$_2$ phase, which was characterized by a lower hardness compared to that of the hexagonal h-MoSi$_2$ phase [48]. The samples before and after annealing were studied using a profilometer to determine the radius of the curvature of the coated lamellae and to subsequently calculate the internal stresses using the Stoney formula. The levels of internal stresses affecting the mechanical properties of the coatings were similar at all the frequencies used in the temperature range of 20–1000 °C.

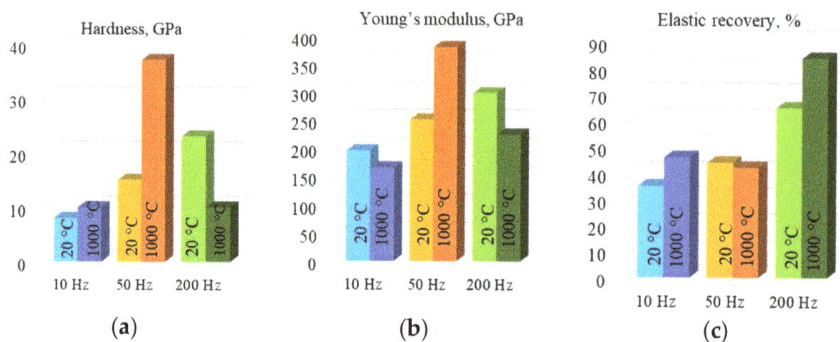

Figure 10. Hardness (**a**), Young's modulus (**b**), and elastic recovery (**c**) of Mo-Zr-Si-B coatings in the as-deposited state and after vacuum annealing at 1000 °C.

4. Conclusions

1. The Mo-Zr-Si-B coatings deposited by HIPIMS at frequencies of 10, 50, and 200 Hz had a dense and homogeneous structure. The minimal growth rate (7.5 nm/min) was observed at 10 Hz for the highest maximum peak current. As frequency was increased to 50 and 200 Hz, the growth rate rose to 17.5 and 67.5 nm/min, respectively.

2. The hardness of the Mo-Zr-Si-B coatings deposited at 10, 50, and 200 Hz was 8, 15, and 23 GPa, respectively. The coating deposited at 50 Hz was characterized by a higher resistance to wear and cyclic-dynamic-impact loading, as well as high-temperature oxidation resistance at 1300 and 1500 °C.

3. Heating in the transmission electron microscope column at temperatures of 20–1000 °C did not result in structural changes, and the coating deposited at 50 Hz remained amorphous. The segregation and growth of MoSi$_2$ phase grains were observed in the temperature range of 600–1000 °C during vacuum annealing because of the longer exposure time and low heating/cooling rates. The hardness and Young's modulus of the coatings deposited at frequencies of 10 and 200 Hz decreased after vacuum annealing. The self-hardening effect was detected for the coating deposited at 50 Hz. As the temperature rose from 20 to 1000 °C, the hardness and Young's modulus increased from 15 and 250 GPa to 37 and 380 GPa, respectively, due to the crystallization process and the transformation of the amorphous structure into a nanocomposite one.

Author Contributions: Conceptualization, P.V.K.-K.; Formal analysis, A.D.S.; Investigation, P.V.K.-K., A.D.S., P.A.L. and A.S.O.; Project administration, E.A.L.; Resources, E.A.L.; Writing—review & editing, P.V.K.-K. and A.S.O. All authors have read and agreed to the published version of the manuscript.

Funding: This work was performed with financial support from the Russian Science Foundation (project No. 19-19-00117-П).

Institutional Review Board Statement: Not applicable.

Informed Consent Statement: Not applicable.

Data Availability Statement: Not applicable.

Acknowledgments: The authors are grateful to M.I. Petrzhik for the nanoindentation tests and N.V. Shvyndina for the SEM-EDS study.

Conflicts of Interest: The authors declare no conflict of interest.

References

1. Li, H.; Yao, D.; Fu, Q.; Liu, L.; Zhang, Y.; Yao, X.; Wang, Y.; Li, H. Anti-Oxidation and Ablation Properties of Carbon/Carbon Composites Infiltrated by Hafnium Boride. *Carbon* **2013**, *52*, 418–426. [CrossRef]
2. Grigoriev, S.N.; Vereschaka, A.A.; Volosova, M.A.; Sitnikov, N.N.; Sotova, E.S.; Seleznev, A.E.; Bublikov, J.I.; Batako, A.D. Improved Efficiency of Ceramic Cutting Tools in Machining Hardened Steel—An Application with Nanostructured Multilayered Coatings. *Handb. Mod. Coat. Technol.* **2021**, 381–433. [CrossRef]
3. Kim, Y.H.; Shin, T.H.; Myung, J. ha MoSi2-Based Cylindrical Susceptor for Rapid High-Temperature Induction Heating in Air. *Ceram. Int.* **2020**, *46*, 23636–23642. [CrossRef]
4. Feng, P.; Wu, J.; Islam, S.H.; Liu, W.; Niu, J.; Wang, X.; Qiang, Y. Effects of Boron Addition on the Formation of MoSi2 by Combustion Synthesis Mode. *J. Alloys Compd.* **2010**, *494*, 161–165. [CrossRef]
5. Zhang, L.; Tong, Z.; He, R.; Xie, C.; Bai, X.; Yang, Y.; Fang, D. Key Issues of MoSi2-UHTC Ceramics for Ultra High Temperature Heating Element Applications: Mechanical, Electrical, Oxidation and Thermal Shock Behaviors. *J. Alloys Compd.* **2019**, *780*, 156–163. [CrossRef]
6. Xin, L.; Chen, Q.; Teng, Y.; Wang, W.; Sun, A.; Zhu, S.; Wang, F. Effects of Silicon and Multilayer Structure of TiAl(Si)N Coatings on the Oxidation Resistance of Ti6Al4V. *Surf. Coat. Technol.* **2013**, *228*, 48–58. [CrossRef]
7. Bae, K.E.; Chae, K.W.; Park, J.K.; Lee, W.S.; Baik, Y.J. Oxidation Behavior of Amorphous Boron Carbide–Silicon Carbide Nano-Multilayer Thin Films. *Surf. Coat. Technol.* **2015**, *276*, 55–58. [CrossRef]
8. Niu, Y.; Wang, H.; Li, H.; Zheng, X.; Ding, C. Dense ZrB2-MoSi2 Composite Coating Fabricated by Low Pressure Plasma Spray (LPPS). *Ceram. Int.* **2013**, *39*, 9773–9777. [CrossRef]
9. Ren, X.; Li, H.; Chu, Y.; Li, K.; Fu, Q. ZrB2–SiC Gradient Oxidation Protective Coating for Carbon/Carbon Composites. *Ceram. Int.* **2014**, *40*, 7171–7176. [CrossRef]
10. Zhestkov, B.E.; Terent'eva, V.S. Multifunctional Coating MAI D5 Intended for the Protection of Refractory Materials. *Russ. Metall.* **2010**, *2010*, 33–40. [CrossRef]
11. Yang, J.; Liu, H.; Gao, W.; Su, L.; Jiang, K. Effect of Different Fillers on the Microstructural Evolution and High Temperature Oxidation Resistance of Mo-Si-B Coatings Prepared by Pack Cementation. *Int. J. Refract. Met. Hard Mater.* **2021**, *100*, 105625. [CrossRef]
12. Shrestha, S.; Hodgkiess, T.; Neville, A. Erosion–Corrosion Behaviour of High-Velocity Oxy-Fuel Ni–Cr–Mo–Si–B Coatings under High-Velocity Seawater Jet Impingement. *Wear* **2005**, *259*, 208–218. [CrossRef]
13. Kiryukhantsev-Korneev, P.V.; Iatsyuk, I.V.; Shvindina, N.V.; Levashov, E.A.; Shtansky, D.V. Comparative Investigation of Structure, Mechanical Properties, and Oxidation Resistance of Mo-Si-B and Mo-Al-Si-B Coatings. *Corros. Sci.* **2017**, *123*, 319–327. [CrossRef]
14. Kiryukhantsev-Korneev, P.V.; Sytchenko, A.D.; Sviridova, T.A.; Sidorenko, D.A.; Andreev, N.V.; Klechkovskaya, V.V.; Polčak, J.; Levashov, E.A. Effects of Doping with Zr and Hf on the Structure and Properties of Mo-Si-B Coatings Obtained by Magnetron Sputtering of Composite Targets. *Surf. Coat. Technol.* **2022**, *442*, 128141. [CrossRef]
15. Veprek, S.; Jilek, M. Super- and Ultrahard Nanacomposite Coatings: Generic Concept for Their Preparation, Properties and Industrial Applications. *Vacuum* **2002**, *67*, 443–449. [CrossRef]
16. Kelly, P.J.; Beevers, C.F.; Henderson, P.S.; Arnell, R.D.; Bradley, J.W.; Bäcker, H. A Comparison of the Properties of Titanium-Based Films Produced by Pulsed and Continuous DC Magnetron Sputtering. *Surf. Coat. Technol.* **2003**, *174–175*, 795–800. [CrossRef]
17. Elmkhah, H.; Attarzadeh, F.; Fattah-alhosseini, A.; Kim, K.H. Microstructural and Electrochemical Comparison between TiN Coatings Deposited through HIPIMS and DCMS Techniques. *J. Alloys Compd.* **2018**, *735*, 422–429. [CrossRef]
18. Helmersson, U.; Lattemann, M.; Bohlmark, J.; Ehiasarian, A.P.; Gudmundsson, J.T. Ionized Physical Vapor Deposition (IPVD): A Review of Technology and Applications. *Thin Solid Films* **2006**, *513*, 1–24. [CrossRef]
19. Lattemann, M.; Ehiasarian, A.P.; Bohlmark, J.; Persson, P.Å.O.; Helmersson, U. Investigation of High Power Impulse Magnetron Sputtering Pretreated Interfaces for Adhesion Enhancement of Hard Coatings on Steel. *Surf. Coat. Technol.* **2006**, *200*, 6495–6499. [CrossRef]
20. Lu, C.Y.; Diyatmika, W.; Lou, B.S.; Lu, Y.C.; Duh, J.G.; Lee, J.W. Influences of Target Poisoning on the Mechanical Properties of TiCrBN Thin Films Grown by a Superimposed High Power Impulse and Medium-Frequency Magnetron Sputtering. *Surf. Coat. Technol.* **2017**, *332*, 86–95. [CrossRef]
21. Mei, H.; Ding, J.C.; Xiao, X.; Luo, Q.; Wang, R.; Zhang, Q.; Gong, W.; Wang, Q. Influence of Pulse Frequency on Microstructure and Mechanical Properties of Al-Ti-V-Cu-N Coatings Deposited by HIPIMS. *Surf. Coat. Technol.* **2021**, *405*, 126514. [CrossRef]
22. Nedfors, N.; Mockute, A.; Palisaitis, J.; Persson, P.O.Å.; Näslund, L.Å.; Rosen, J. Influence of Pulse Frequency and Bias on Microstructure and Mechanical Properties of TiB2 Coatings Deposited by High Power Impulse Magnetron Sputtering. *Surf. Coat. Technol.* **2016**, *304*, 203–210. [CrossRef]

23. Dai, W.; Kwon, S.H.; Wang, Q.; Liu, J. Influence of Frequency and C2H2 Flow on Growth Properties of Diamond-like Carbon Coatings with AlCrSi Co-Doping Deposited Using a Reactive High Power Impulse Magnetron Sputtering. *Thin Solid Films* **2018**, *647*, 26–32. [CrossRef]

24. Samuelsson, M.; Sarakinos, K.; Högberg, H.; Lewin, E.; Jansson, U.; Wälivaara, B.; Ljungcrantz, H.; Helmersson, U. Growth of Ti-C Nanocomposite Films by Reactive High Power Impulse Magnetron Sputtering under Industrial Conditions. *Surf. Coat. Technol.* **2012**, *206*, 2396–2402. [CrossRef]

25. Polaček, M.; Souček, P.; Alishahi, M.; Koutná, N.; Klein, P.; Zábranský, L.; Czigány, Z.; Balázsi, K.; Vašina, P. Synthesis and Characterization of Ta–B–C Coatings Prepared by DCMS and HiPIMS Co-Sputtering. *Vacuum* **2022**, *199*, 110937. [CrossRef]

26. Kiryukhantsev-Korneev, F.V. Possibilities of Glow Discharge Optical Emission Spectroscopy in the Investigation of Coatings. *Russ. J. Non-Ferrous Met.* **2014**, *55*, 494–504. [CrossRef]

27. Daghbouj, N.; Sen, H.S.; Čížek, J.; Lorinčík, J.; Karlík, M.; Callisti, M.; Čech, J.; Havránek, V.; Li, B.; Krsjak, V.; et al. Characterizing Heavy Ions-Irradiated Zr/Nb: Structure and Mechanical Properties. *Mater. Des.* **2022**, *219*, 110732. [CrossRef]

28. Shimizu, T.; Teranishi, Y.; Morikawa, K.; Komiya, H.; Watanabe, T.; Nagasaka, H.; Yang, M. Impact of Pulse Duration in High Power Impulse Magnetron Sputtering on the Low-Temperature Growth of Wurtzite Phase (Ti,Al)N Films with High Hardness. *Thin Solid Films* **2015**, *581*, 39–47. [CrossRef]

29. Papa, F.; Gerdes, H.; Bandorf, R.; Ehiasarian, A.P.; Kolev, I.; Braeuer, G.; Tietema, R.; Krug, T. Deposition Rate Characteristics for Steady State High Power Impulse Magnetron Sputtering (HIPIMS) Discharges Generated with a Modulated Pulsed Power (MPP) Generator. *Thin Solid Films* **2011**, *520*, 1559–1563. [CrossRef]

30. Sarakinos, K.; Alami, J.; Konstantinidis, S. High Power Pulsed Magnetron Sputtering: A Review on Scientific and Engineering State of the Art. *Surf. Coat. Technol.* **2010**, *204*, 1661–1684. [CrossRef]

31. Bobzin, K.; Brögelmann, T.; Brugnara, R.H. Aluminum-Rich HPPMS (Cr1−xAlx)N Coatings Deposited with Different Target Compositions and at Various Pulse Lengths. *Vacuum* **2015**, *122*, 201–207. [CrossRef]

32. Musil, J.; Novák, P.; Čerstvý, R.; Soukup, Z. Tribological and Mechanical Properties of Nanocrystalline-TiC/a-C Nanocomposite Thin Films. *J. Vac. Sci. Technol. Vacuum Surfaces Film.* **2010**, *28*, 244. [CrossRef]

33. Leyland, A.; Matthews, A. Optimization of Nanostructured Tribological Coatings. In *Nanostructured Coatings. Nanostructure Science and Technology*; Cavaleiro, A., de Hosson, J.T.M., Eds.; Springer: New York, NY, USA, 2006; pp. 511–538.

34. Ge, F.F.; Sen, H.S.; Daghbouj, N.; Callisti, M.; Feng, Y.J.; Li, B.S.; Zhu, P.; Li, P.; Meng, F.P.; Polcar, T.; et al. Toughening Mechanisms in V-Si-N Coatings. *Mater. Des.* **2021**, *209*, 109961. [CrossRef]

35. Berg, G.; Friedrich, C.; Broszeit, E.; Berger, C. Data Collection of Properties of Hard Material. *Handb. Ceram. Hard Mater.* **2008**, 965–995. [CrossRef]

36. Gutman, M.B. *Materials for Electrothermal Plants: Handbook*; Energoatomizdat: Moscow, Russia, 1987. (in Russian)

37. Veprek, S.; Mukherjee, S.; Karvankova, P.; Männling, H.D.; He, J.L.; Moto, K.; Prochazka, J.; Argon, A.S. Hertzian Analysis of the Self-Consistency and Reliability of the Indentation Hardness Measurements on Superhard Nanocomposite Coatings. *Thin Solid Films* **2003**, *436*, 220–231. [CrossRef]

38. Rezakhanlou, R.; von Stebut, J. Paper VII (Iii) Damage Mechanisms of Hard Coatings on Hard Substrates: A Critical Analysis of Failure in Scratch and Wear Testing. *Tribol. Ser.* **1990**, *17*, 183–192. [CrossRef]

39. Kiryukhantsev-Korneev, P.V.; Sheveyko, A.N.; Vorotilo, S.A.; Levashov, E.A. Wear-Resistant Ti–Al–Ni–C–N Coatings Produced by Magnetron Sputtering of SHS-Targets in the DC and HIPIMS Modes. *Ceram. Int.* **2020**, *46*, 1775–1783. [CrossRef]

40. Kudryashov, A.E.; Lebedev, D.N.; Potanin, A.Y.; Levashov, E.A. Structure and Properties of Coatings Produced by Pulsed Electrospark Deposition on Nickel Alloy Using Mo-Si-B Electrodes. *Surf. Coat. Technol.* **2018**, *335*, 104–117. [CrossRef]

41. Xi, H.H.; He, P.F.; Wang, H.D.; Liu, M.; Chen, S.Y.; Xing, Z.G.; Ma, G.Z.; Lv, Z.L. Microstructure and Mechanical Properties of Mo Coating Deposited by Supersonic Plasma Spraying. *Int. J. Refract. Met. Hard Mater.* **2020**, *86*, 105095. [CrossRef]

42. Kiryukhantsev-Korneev, P.V.; Sytchenko, A.D. Influence of Physical and Mechanical Properties of the Substrate on the Behavior of Zr–Si–B Coatings under Sliding Friction and Cyclic Impact-Dynamic Loading. *Prot. Met. Phys. Chem. Surfaces* **2020**, *56*, 981–989. [CrossRef]

43. Perepezko, J.H. High Temperature Environmental Resistant Mo-Si-B Based Coatings. *Int. J. Refract. Met. Hard Mater.* **2018**, *71*, 246–254. [CrossRef]

44. Deng, X.; Zhang, G.; Wang, T.; Ren, S.; Shi, Y.; Bai, Z.; Cao, Q. Microstructure and Oxidation Resistance of a Multiphase Mo-Si-B Ceramic Coating on Mo Substrates Deposited by a Plasma Transferred Arc Process. *Ceram. Int.* **2019**, *45*, 415–423. [CrossRef]

45. Wang, F.; Shan, A.; Dong, X.; Wu, J. Oxidation Behavior of Mo–12.5Si–25B Alloy at High Temperature. *J. Alloys Compd.* **2008**, *459*, 362–368. [CrossRef]

46. Yang, Y.; Bei, H.; Chen, S.; George, E.P.; Tiley, J.; Chang, Y.A. Effects of Ti, Zr, and Hf on the Phase Stability of Mo_ss + Mo3Si + Mo5SiB2 Alloys at 1600 °C. *Acta Mater.* **2010**, *58*, 541–548. [CrossRef]

47. Ha, Y.; Kim, T.C.; Baeg, J.H.; Kim, J.S.; Shon, M.; Cho, Y.R. Effect of Cooling Rate on Surface Properties of ZnMgAl Coating and Adhesion to Epoxy Adhesive. *Int. J. Adhes. Adhes.* **2022**, *117*, 103182. [CrossRef]

48. Guo, Z.; Zhang, L.; Qiao, Y.; Gao, Q.; Xiao, Z. A New C11b-Type High Entropy Refractory Metal Silicide to Improve MoSi2 Mechanical Properties More Easily. *Scr. Mater.* **2022**, *218*, 114798. [CrossRef]

Article

Influence of Interlayer Materials on the Mechanical Properties and Thermal Stability of a CrAlN Coating on a Tungsten Carbide Substrate

Hoe-Kun Kim [1], Sung-Min Kim [2,*] and Sang-Yul Lee [1,*]

[1] Center for Surface Technology and Applications, Department of Materials Engineering, Korea Aerospace University, Goyang 10540, Korea
[2] Heat & Surface Technology R&D Department, Korea Institute of Industrial Technology (KITECH), Incheon 21999, Korea
* Correspondence: sungminkim@kitech.re.kr (S.-M.K.); sylee@kau.ac.kr (S.-Y.L.)

Abstract: CrAlN coatings have earned significant attention for use in cutting tool coating applications due to their excellent properties such as high hardness and superb oxidation resistance. It is well known that the interlayer between the coating and the substrate can influence the mechanical properties of the coatings. In this work, three interlayers—CrN, CrZrN, and CrN/CrZrSiN—were synthesized between a CrAlN coating and a tungsten carbide substrate to improve the mechanical properties and thermal stability of the CrAlN coating. All the CrAlN coatings with their respective interlayers showed high hardness values in the range of 34.5 to 35.1 GPa, and they were not significantly affected by the interlayer type. However, wear and scratch tests showed that the CrAlN coatings with CrN and CrN/CrZrSiN interlayers exhibited an improved friction coefficient of 0.33 and adhesion strength (Lc2) of 69 N compared to the CrAlN coating with the CrZrN interlayer. These improved wear properties were attributed to the H/E ratio of the interlayer between the coating and the substrate, in that the CrN and CrZrSiN interlayers effectively induced a smooth transition of the coating stress under a loading condition. During the thermal stability tests, the hardness of the CrAlN coating with the CrN/CrZrSiN interlayer was maintained up to 1000 °C due to the excellent oxidation resistance of the CrZrSiN layer, which contained an amorphous Si_xN_y phase.

Keywords: CrAlN coating; interlayer; friction coefficient; adhesion strength; thermal stability

Citation: Kim, H.-K.; Kim, S.-M.; Lee, S.-Y. Influence of Interlayer Materials on the Mechanical Properties and Thermal Stability of a CrAlN Coating on a Tungsten Carbide Substrate. *Coatings* 2022, 12, 1134. https://doi.org/10.3390/coatings12081134

Academic Editor: Sergey N. Grigoriev

Received: 10 June 2022
Accepted: 3 August 2022
Published: 6 August 2022

Publisher's Note: MDPI stays neutral with regard to jurisdictional claims in published maps and institutional affiliations.

1. Introduction

For the past decade, Cr-based ternary hard coatings such as CrAlN have been applied in industrial fields due to their superior wear and oxidation resistance at high temperatures compared to binary CrN coatings [1–3]. Their excellent thermal stability has been theoretically and experimentally reported in many publications, suggesting that both chromium and aluminum in CrAlN coatings are able to form protective oxide layers to inhibit oxygen diffusion [2,3]. Typically, the microstructure and phase of CrAlN coatings are correlated with their hardness and oxidation resistance, which can be controlled by precisely tuning the elemental composition [4–6]. It is well established that increasing the Al content up to the solid solubility limitation for a cubic B1 structure leads to enhanced mechanical and tribological properties via a strengthening effect. However, in the case of excess Al content over the cubic solubility of $Cr_{1-x}Al_xN$ (x_{max}, ~0.7), a mixed structure of cubic B1 and hexagonal B4 (wurzite AlN) is formed, resulting in reduced hardness and wear performance [4]. Importantly, improvement in CrAlN coatings is still required in order to meet the strict criteria of a fast-changing industry.

In general, it is well known that interlayers are very effective in enhancing the adhesive strength between substrates and materials to be deposited. Recent research has demonstrated that the application of interlayers to hard coatings enables the modification of the structure and internal stress to enhance the mechanical and tribological properties [7–12]. In particular, the modulation ratio, i.e., H/E or H^3/E^2, is a key factor in the enhanced mechanical and adhesive strength in the presence of an interlayer, since it plays a simultaneous role of hindering dislocation glide to some extent and relieving a high stress gradient between the substrates and coatings. Therefore, it is believed that the mechanical and tribological properties of CrAlN coatings could be enhanced by the introduction of appropriate interlayer materials. Nevertheless, the effects of various interlayers for CrAlN coatings have never been reported in depth.

In our previous study, CrAlN monolithic layer coatings showed great mechanical properties such as high hardness (35 GPa) and high adhesion (46 N for SKD61 steel), but the coatings also showed relatively poor wear resistance and thermal stability [4]. In addition, other monolithic layers such as CrZrN and CrZrSiN coatings with a low friction coefficient of approximately 0.25 and excellent thermal stability up to 800 °C have been reported [13–17]. To deposit a CrAlN coating on a very hard WC ($\approx$20 GPa) substrate without delamination, the use of an interlayer of these Cr-based nitride compounds could be advantageous for adhesive strength by reducing the gap of hardness and Young's modulus between the CrAlN coating and the WC substrate. Although the formation of Si_xN_y compounds in the coating shows a strong oxidation resistance, nitride interlayers containing Si have not been researched yet. Accordingly, in this work, CrAlN coatings with three interlayer materials of CrN, CrZrN, and CrZrSiN were designed and synthesized in order to obtain coatings with excellent hardness values, friction coefficients, adhesion properties, and high-temperature properties. This work clarifies the importance of structuring an interlayer between the CrAlN top coating and the substrate, which will be a significant benefit for practical applications in the cutting tool industry.

2. Experimental Details

CrAlN coatings with the selected interlayers were synthesized on two types of substrates—Si (100) wafers and WC-6 wt.% Co substrates—using a closed-field unbalanced magnetron sputtering system. The CrN, CrZrN, and CrN/CrZrSiN interlayers were deposited between the CrAlN coating and the substrate. The thickness of each interlayer was fixed at 300 nm, which was approximately 10% of the total thickness, for avoiding unexpected fracture generation due to shear stress [8]. Pure Cr single and CrAl, CrZr, and CrZrSi segment targets were used as target materials. Before the deposition process, the Si (100) wafers and WC substrates were ultrasonicated in acetone for 30 min and then dried well. After evacuating a base pressure below 2.8×10^{-3} Pa, the substrates were further cleaned by Ar plasma etching for 20 min at an Ar pressure of 0.4 Pa with a DC power of 0.5 kW. The deposition of the coatings was performed in an Ar–N_2 mixed atmosphere, and the working pressure was maintained at 0.52 Pa. All the layers were deposited using pulsed DC power (freq.: 25 kHz, duty ratio: 70%). The distance of applied DC bias and the deposition temperature during deposition were fixed at -100 V and 400 °C, respectively.

The microstructures of the CrAlN coatings with various interlayers were examined via a field-emission scanning electron microscope (FE-SEM, JEOL, JSM-7100F, Tokyo, Japan) operating at an accelerated voltage of 15 kV, and the chemical compositions of each layer were determined using energy dispersive spectroscopy (EDS, JEOL, JED-2300). The hardness and elastic modulus of the individual layers—e.g., the CrN, CrZrN, and CrN/CrZrSiN interlayers—and the CrAlN coating were measured using nanoindentation (Helmut Fischer, HM2000, Sindelfingen, Germany) with a load of 25 mN and a dwell time of 30 s. These measurements were carried out 12 times. In order to alleviate the thickness effect, the indentation depth did not exceed approximately 0.24 μm and was kept to less than 10% of the total coating thickness [18]. To obtain reliable results, nanoindentation tests were carried out 10 times on each sample. The friction coefficients of the coatings were measured using a ball-on-disk-type wear tester with an alumina counter ball (Al_2O_3, $\varnothing = 9.25$ mm) at room temperature after annealing at 500 °C. The sliding velocity was 0.25 m/s, and the total sliding distance was 1000 m with an applied load of 5 N. Tribological tests on each sample were carried out 4 times to obtain precise and reproducible results, and the representative average results were used. A scratch tester (CSM, Revetest, Corcelles, Switzerland), with a Rockwell C diamond stylus with a radius 200 μm, was used to determine the quantitative value of the adhesion strength by measuring the place to be initially delaminated, denoted as critical load (Lc2), during a progressive loading up to 100 N. The load rate and sliding speed were maintained at 100 N/min and 10 mm/min, respectively. The track of the coatings was investigated using optical microscopy (OM, Olympus, BX51M, Tokyo, Japan), and the failure mode was determined. The surface morphologies were examined using atomic force microscopy (AFM, Park Systems, XE-100, Suwon, Korea), and the scan area was fixed at 10 μm × 10 μm. All the coatings were annealed at temperatures ranging from 500 to 1000 °C in air for 30 min, and the hardness values were investigated using nanoindentation (Helmut Fischer, HM2000, Hünenberg, Switzerland) to evaluate the thermal stability.

3. Results and Discussion

3.1. Characteristics of the CrAlN Coatings

The CrAlN coatings with the three interlayers were synthesized to enhance their hardness values, friction coefficients, adhesion properties, and thermal stability. Cross-sectional images of the CrAlN coatings with different interlayers are shown in Figure 1. The total thickness of all the CrAlN coatings was fixed at 3 μm and was controlled by the deposition duration based on the sputtering yield. Most of the layers, e.g., the CrN, CrZrN, and CrAlN layers, exhibited a columnar structure, while the CrZrSiN layer showed a featureless structure due to high Si content (Table 1). This phenomenon could be attributed to the fact that a nanocrystalline or amorphous phase with high Si content (Si > 7%) was formed, which is in good agreement with previous work [19]. The chemical composition of the CrAlN top-layer coating was controlled to be 1:1 to obtain the optimal solid solution effect [4]. As can be seen in Figure 2, no significant difference was observed in the hardness and elastic modulus dependent on the interlayer material, and they were measured to have a hardness and elastic modulus ranging from 34.5 to 35.8 GPa and from 423.6 to 425.0 GPa, respectively.

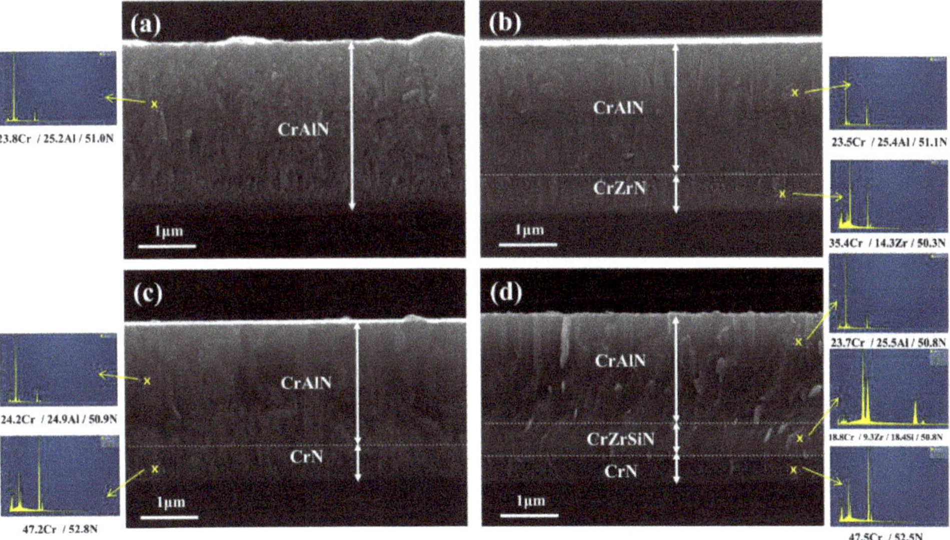

Figure 1. Cross-sectional images of the CrAlN coatings with (**a**) no interlayer, (**b**) CrZrN interlayer, (**c**) CrN interlayer, and (**d**) CrN/CrZrN interlayer.

Table 1. Chemical composition of CrZrN, CrN, and CrZrSiN interlayers and the CrAlN coating.

Layer	Chemical Composition (at%)
CrZrN	35.4Cr-14.3Zr-50.3N
CrN	49.2Cr-50.8N
CrZrSiN	18.8Cr-9.3Zr-18.4Si-53.5N
CrAlN	23.8Cr-25.2Al-51.0N

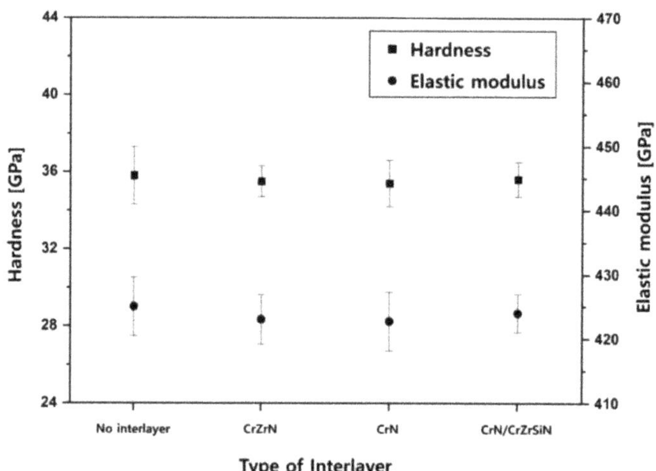

Figure 2. Hardness and elastic modulus of the CrAlN coatings with three interlayers.

3.2. Tribological Properties of the CrAlN Coatings

The tribological properties of the CrAlN coatings were evaluated by carrying out ball-on-disk-type wear tests, the results of which are shown in Figure 3a. The friction coefficient curves were relatively stable with little scattering. The CrAlN coating with the CrZrN interlayer exhibited a high friction coefficient of 0.41; this value was similar to that of the CrAlN coating without an interlayer. However, the friction coefficient values of the CrAlN coatings with CrN and CrN/CrZrSiN interlayers drastically decreased, and the coating with the CrN/CrZrSiN interlayer produced the lowest friction coefficient of 0.33. To further investigate the tribological properties, the wear rate (K) was calculated [20].

$$K = V/F{\cdot}d \tag{1}$$

where K is the wear rate, V is the wear volume, F the normal load, and d the sliding distance. As shown in Figure 3b, it was revealed that the wear rate was affected by the interlayer material, and the results showed an identical trend to that of the friction coefficient. These wear resistance differences in the CrAlN coatings dependent on the interlayer could mainly be associated with the gradient of the modulation ratio (H/E) of the substrate/interlayer/coating. As reported, an interlayer with an intermediate H/E value between individual values of the coating and the substrate can increase the coating toughness improvements involving the wear resistance of the coating [21,22]. In this work, the H/E ratio of the CrZrN interlayer exhibited a much higher value (0.119) than that of the CrAlN (0.089), as listed in Table 2. This could be explained by the fact that a large difference in the H/E ratio between the CrAlN and CrZrN interlayers exerted a non-uniform distribution of loading stress, which resulted in the poor wear properties of the coating [7]. In view of the CrAlN coating with the CrN/CrZrSiN interlayer structure, however, there was a gradual increase in the H/E ratio from the WC substrate (H/E, 0.040) to the CrN interlayer (H/E, 0.076), CrZrSiN interlayer (H/E, 0.083), and the CrAlN coating

(H/E, 0.089). Therefore, the enhanced wear properties could be attributed to the fact that the CrN and CrZrSiN interlayers effectively induced the stress relief between the interfaces by the sequential H/E ratio (WC < CrN or CrZrSiN < CrAlN) during the wear test.

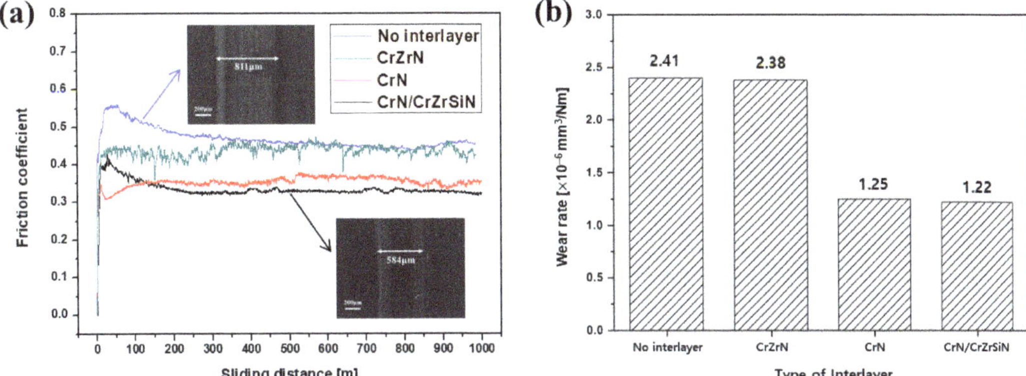

Figure 3. (**a**) Friction curve and (**b**) wear rate of the CrAlN coatings with three interlayers.

Table 2. Hardness, elastic modulus, and H/E ratio of the WC substrate; the CrZrN, CrN, and CrZrSiN interlayers; and the CrAlN coating.

Layer	Hardness (GPa)	Elastic Modulus (GPa)	H/E Ratio
WC (Substrate)	19.6	510.6	0.040
CrZrN	35.1	293.1	0.119
CrN	23.3	305.6	0.076
CrZrSiN	24.5	288.6	0.083
CrAlN	35.8	425.0	0.089

3.3. Adhesion Properties of the CrAlN Coatings

The scratch test is an informative method for evaluating the cohesion and adhesion properties of coatings. During the scratch test with progressive loading on a coating, three failure modes were identified, which could be as critical loads of Lc0, Lc1, and Lc2. We defined a semi-circular crack inside the scratch track as Lc0, a partial fragment at a track edge as Lc1, and an initial spallation or delamination of the entire coating surface as Lc2 [23,24]. To quantify the adhesion strength, the Lc2 was used, and the average values were calculated from the five measurements made for each coating. Figure 4 shows the entire scratch tracks of all the CrAlN coatings. The Lc2 of the CrAlN coating with the CrZrN interlayer exhibited a slightly increased value of 23.6 N compared to 17.8 N for the CrAlN coating without an interlayer. In the scratch tracks of the CrAlN coatings, chipping cracks were observed, and the coatings were totally delaminated over Lc2. It has been reported that chipping will often occur during a scratch test of a high-hardness coating on a hard and brittle substrate [23,25]. However, the Lc2 of the CrAlN coatings with CrN and CrN/CrZrSiN interlayers showed much improved adhesion strength (approximately 69 N), and there were no obvious chipping cracks or signs of delamination in any of the scratch tracks.

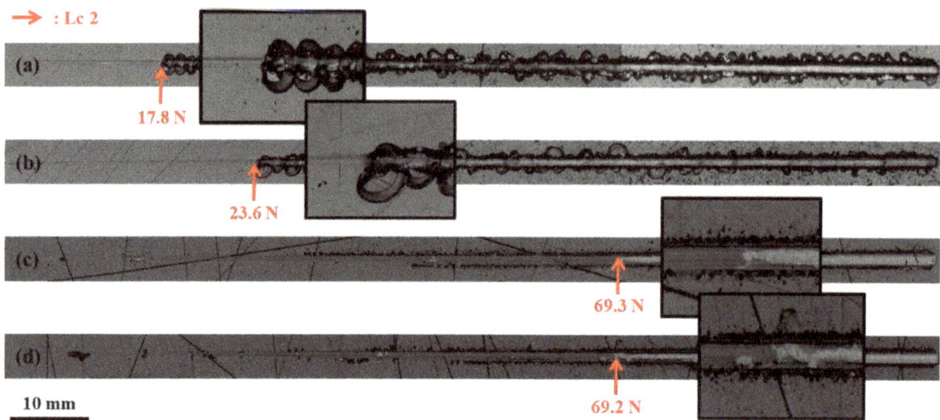

Figure 4. Optical microscopy images of the scratch tracks of the CrAlN coatings with (**a**) no interlayer, (**b**) CrZrN interlayer, (**c**) CrN interlayer, and (**d**) CrN/CrZrSiN interlayer.

The CrAlN coatings with the CrZrN interlayers showed poor crack resistance, as evidenced by a lower Lc2 relative to the CrN and CrN/CrZrSiN interlayers. This phenomenon is correlated with the resistance to plastic deformation of these coatings [25]. In general, once a crack is initiated during a scratch test, it propagates promptly and failure subsequently occurs due to the low toughness of the coating. Conversely, the CrAlN coating with the CrN/CrZrSiN interlayer exhibited greater toughness because the CrN and CrZrSiN interlayers effectively induced a smooth transition of the coating stress. Thus, crack propagation was inhibited during the wear test, and this led to an improvement in wear resistance for the CrAlN coating.

3.4. High Temperature Properties of the CrAlN Coatings

Figure 5 presents the hardness variation in the CrAlN coatings dependent on the interlayer after annealing in air at temperatures ranging from 500 to 1000 °C for 30 min. In the case of the CrAlN coating with the CrN/CrZrSiN interlayer, the hardness value was maintained above approximately 28 GPa up to 1000 °C, whereas the hardness values for the other coatings decreased drastically to 20 GPa over 800 °C. This revealed that the CrAlN coating with the CrN/CrZrSiN interlayer had an excellent thermal stability compared to the coatings with other interlayers. In previous studies, CrZrSiN monolithic coatings with a high Si content led to an improved thermal stability of the coatings compared to other hard coatings such as CrN and CrZrN [19]. To enhance the thermal stability of the CrAlN coating, a CrZrSiN monolithic layer with a high Si content (18 at%) was deposited between the coating and the substrate. The enhanced thermal stability of the CrZrSiN interlayer could be attributed to the outward diffusion of Si since the interlayer structurally existed below the CrAlN coatings. It is well known that the standard Gibbs free energy for oxide stability increased in the sequence of Cr_2O_3 ($\Delta G = -552.296$ kJ·mol^{-1} at 900 °C), SiO_2 ($\Delta G = -693.833$ kJ·mol^{-1} at 900 °C), Al_2O_3 ($\Delta G = -866.907$ kJ·mol^{-1} at 900 °C), and ZrO_2 ($\Delta G = -880.833$ kJ·mol^{-1} at 900 °C). Thus, the outmost surface could be mainly composed of a dense layer of Al_2O_3 and a coarse layer of ZrO_2 [26]. It might be postulated that the formation of nano-scaled Al_2O_3 dense film could block the oxygen diffusion, whereas ZrO_2 layers enable the coating to be coarse, providing a pathway for oxygen diffusion into the coating. However, more importantly, it is likely that the strong protective SiO_2 layer was competitively formed, which inhibited the diffusion of oxygen into the coatings [27]. As evidenced by EDS (see Figure 5), it was found that the oxygen content of the CrAlN coating with an interlayer containing the Si element is much lower than that of the coatings with CrN and CrZrN interlayers.

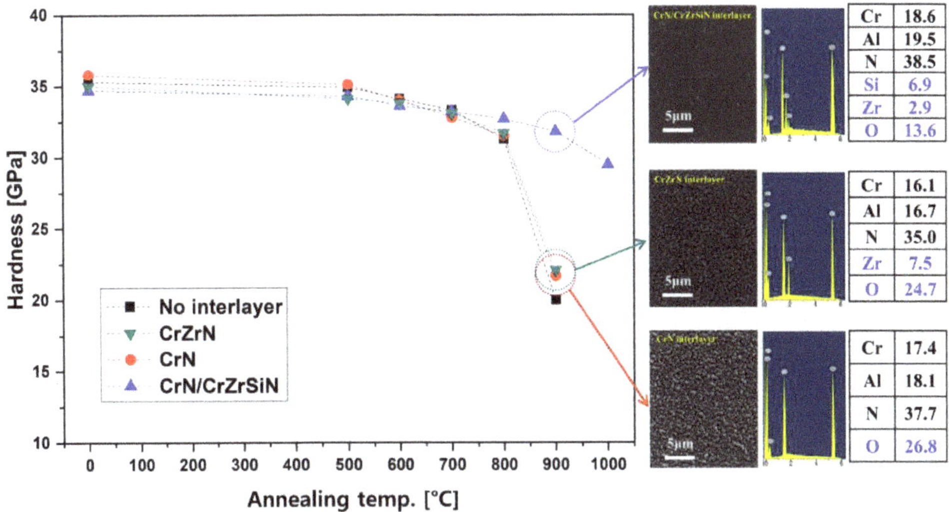

Figure 5. Hardness variation in the CrAlN coatings with three interlayers after annealing test in air for 30 min, and the surface EDS data of each coating after annealing.

The friction coefficient values of all the CrAlN coatings with various interlayers at room temperature and at 500 °C in air are shown in Figure 6a. The friction coefficients of all the coatings at room temperature increased at 500 °C. It was noted that, despite the similar hardness after annealing at 500 °C (Table 3), the friction coefficient values of every CrAlN coating increased, and the CrAlN coating with the CrN/CrZrSiN interlayer showed the lowest increment from 0.33 to 0.42. Theoretically, it was revealed that the wear properties are closely associated with the hardness to the wear resistance of a surface. Hard coatings must have high resistance to plastic deformation as well as a low Young's modulus (E) during a wear test. According to Johnson's analysis, the load required to initiate plastic deformation is proportional to H^3/E^2 when the contact event occurs [28]. Therefore, the H^3/E^2 parameter controlling the resistance of materials to plastic deformation is closely related to the wear resistance of the coating. Moreover, the H/E ratio must be a strong indicator of a coating's resistance to plastic deformation [25]. The wear resistance is related to the elastic strain to coating failure, i.e., the ability to recover from elastic deformation without plastic deformation. For the present study, the hardness, H^3/E^2 parameters, and H/E ratios of the coatings with various interlayers after annealing at 500 °C for 30 min are shown in Table 3. However, there was not much difference between the values, and these results are not sufficient to explain the different friction coefficients at 500 °C. This suggested that another factor was affecting the friction behaviors of the coatings at 500 °C.

The surface morphologies and roughness (Rms, root mean square) values of all the CrAlN coatings with three interlayers after annealing at 500 °C were examined and compared to those at room temperature. The results are summarized in Figure 6b. The surface of all the coatings changed to a cone-like pebble structure at 500 °C, and the grains of the coatings were found to grow and coarsen. This phenomenon, known as a recovery process, corresponds to the deposition-induced lattice point defects caused by increased diffusivity at the annealing temperature, and this leads to grain growth. For these reasons, the surface roughness values increased compared to those at room temperature, and a morphology change was observed.

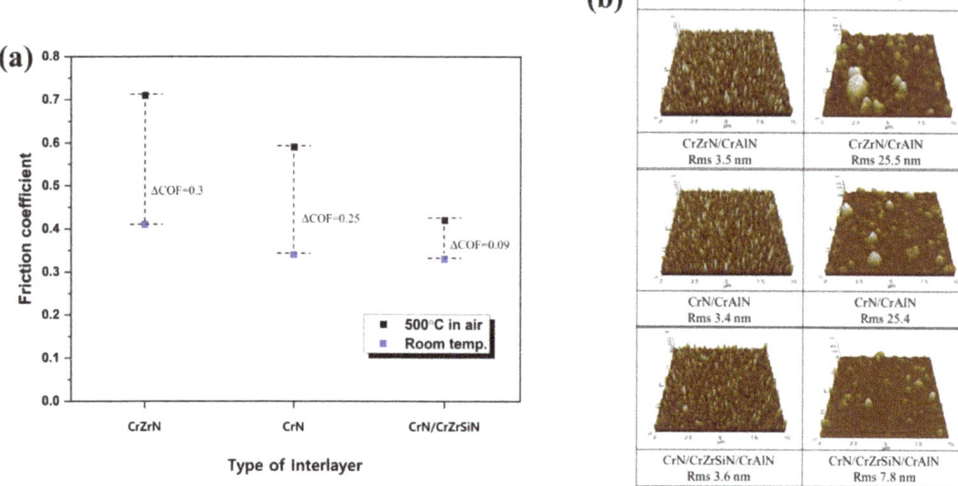

Figure 6. (**a**) Friction coefficient and (**b**) surface roughness values of the CrAlN coatings with three interlayers at room temperature and 500 °C.

Table 3. Hardness, H^3/E^2 parameter, and H/E ratio of the CrAlN coatings with three interlayer types after an annealing test at 500 °C in air for 30 min.

Parameters	Interlayer		
	CrZrN	**CrN**	**CrN/CrZrSiN**
Hardness [GPa]	35.5	35.4	35.6
H^3/E^2 [GPa]	0.255	0.250	0.248
H/E ratio	0.084	0.084	0.084

The surface roughness values of all the CrAlN coatings with various interlayers at 500 °C were examined and compared to those at room temperature. The results are summarized in Figure 6b. The surface roughness values of the CrAlN coatings with the CrZrN and CrN interlayers varied from 3.4 nm at room temperature to 25.4 nm after annealing at 500 °C. However, the surface roughness value of the annealed CrAlN coating with the CrN/CrZrSiN interlayer (7.8 nm) became only twice the value at room temperature (3.6 nm).

The reason there existed a large increase in the surface roughness of the coatings with the CrZrN and CrN interlayers could be attributed to the residual oxygen deposited into the films during the coating process. This oxygen could easily diffuse and react to form oxides with the CrZrN and CrN phases, resulting in an increased surface roughness in the coatings with the CrZrN and CrN interlayers. Conversely, the CrAlN coating with the CrN/CrZrSiN interlayer showed only a small increase in the surface roughness because the CrZrSiN interlayer, which consists of a Si_xN_y amorphous phase, is expected to work as an oxygen diffusion barrier to inhibit oxidation by the residual oxygen during the annealing process at 500 °C. These results indicated that the variation in the surface roughness of the coating due to the oxidation by the residual oxygen during the annealing process could affect the friction behaviors of the coating at 500 °C. The lowest increment in the friction coefficient variation in the CrAlN coating with the CrN/CrZrSiN interlayer, Δ0.09, could be attributed to the lowest Rms variation before and after the annealing process at 500 °C.

4. Conclusions

In this work, CrAlN coatings with three different interlayers were synthesized using a closed-field unbalanced magnetron sputtering system on WC-6 wt.% Co substrates to improve their mechanical properties. The hardness and elastic modulus of each coating showed similar values. However, the friction coefficient and adhesion strength of the CrAlN coating with the CrN/CrZrSiN interlayer showed greater values than those of the other coatings due to the smooth transition of the coating stress between the coating and the substrate via the median H/E ratio of the CrN and CrZrSiN interlayers. After the thermal stability test, the hardness value of the CrAlN coating with the CrN/CrZrSiN interlayer was approximately 28 GPa up to 1000 °C without drastic changes in hardness, and there were no significant increases in the surface roughness and friction coefficient at 500 °C. The variation in the surface roughness of the coating due to the oxidation by the residual oxygen could affect the friction behaviors of the coating, and the lowest increment in the friction coefficient variation in the CrAlN coating with the CrN/CrZrSiN interlayer could be attributed to the lowest Rms variation. These results indicated that the mechanical properties and thermal stability of the CrAlN hard coatings could be improved via optimal design and structuring interlayers. This work suggests that a CrAlN coating designed with a multi-interlayer will be a promising candidate as a protective hard coating for the cutting, milling, and machining tool industry.

Author Contributions: Conceptualization, H.-K.K.; methodology, H.-K.K.; software, H.-K.K.; validation, H.-K.K. and S.-M.K.; formal analysis, H.-K.K.; investigation, H.-K.K.; resources, H.-K.K.; data curation, H.-K.K.; writing—original draft preparation, H.-K.K.; writing—review and editing, S.-M.K. and S.-Y.L.; visualization, H.-K.K.; supervision, S.-M.K. and S.-Y.L.; project administration, S.-M.K. and S.-Y.L.; funding acquisition, S.-M.K. and S.-Y.L. All authors have read and agreed to the published version of the manuscript.

Funding: This research was funded by the support of the Korea Institute of Industrial Technology as "Development of root technology for multi-product flexible production [KITECH EO-22-0006]" and was supported by 2022 Korea Aerospace University Faculty Research Grant.

Institutional Review Board Statement: Not applicable.

Informed Consent Statement: Not applicable.

Data Availability Statement: Not applicable.

Conflicts of Interest: The authors declare no conflict of interest.

References

1. Sousa, V.F.C.; Silva, F.J.G. Recent Advances in Turning Processes Using Coated Tools—A Comprehensive Review. *Metals* **2020**, *10*, 170. [CrossRef]
2. Chim, Y.C.; Ding, X.Z.; Zeng, X.T.; Zhang, S. Oxidation Resistance of TiN, CrN, TiAlN and CrAlN Coatings Deposited by Lateral Rotating Cathode Arc. *Thin Solid Film.* **2009**, *517*, 4845–4849. [CrossRef]
3. Barshilia, H.C.; Selvakumar, N.; Deepthi, B.; Rajam, K.S. A Comparative Study of Reactive Direct Current Magnetron Sputtered CrAlN and CrN Coatings. *Surf. Coat. Technol.* **2006**, *201*, 2193–2201. [CrossRef]
4. Kim, G.S.; Lee, S.Y. Microstructure and Mechanical Properties of AlCrN Films Deposited by CFUBMS. *Surf. Coat. Technol.* **2006**, *201*, 4361–4366. [CrossRef]
5. Ding, X.Z.; Zeng, X.T. Structural, Mechanical and Tribological Properties of CrAlN Coatings Deposited by Reactive Unbalanced Magnetron Sputtering. *Surf. Coat. Technol.* **2005**, *200*, 1372–1376. [CrossRef]
6. Tang, J.F.; Lin, C.Y.; Yang, F.C.; Chang, C.L. Influence of Nitrogen Content and Bias Voltage on Residual Stress and the Tribological and Mechanical Properties of CrAlN Films. *Coatings* **2020**, *10*, 546. [CrossRef]
7. Kim, H.K.; La, J.H.; Kim, K.S.; Lee, S.Y. The Effects of the H/E Ratio of Various Cr-N Interlayers on the Adhesion Strength of CrZrN Coatings on Tungsten Carbide Substrates. *Surf. Coat. Technol.* **2015**, *284*, 230–234. [CrossRef]
8. Kim, K.S.; Kim, H.K.; La, J.H.; Lee, S.Y. Effects of Interlayer Thickness and the Substrate Material on the Adhesion Properties of CrZrN Coatings. *Jpn. J. Appl. Phys.* **2016**, *55*, 1–6. [CrossRef]
9. Ni, W.; Cheng, Y.T.; Lukitsch, M.; Weiner, A.M.; Lev, L.C.; Grummon, D.S. Novel Layered Tribological Coatings Using a Superelastic NiTi Interlayer. *Wear* **2005**, *259*, 842–848. [CrossRef]

10. Özkan, D.; Alper Yılmaz, M.; Szala, M.; Türküz, C.; Chocyk, D.; Tunç, C.; Göz, O.; Walczak, M.; Pasierbiewicz, K.; Barış Yağcı, M. Effects of Ceramic-Based CrN, TiN, and AlCrN Interlayers on Wear and Friction Behaviors of AlTiSiN+TiSiN PVD Coatings. *Ceram. Int.* **2021**, *47*, 20077–20089. [CrossRef]
11. Borawski, B.; Todd, J.A.; Singh, J.; Wolfe, D.E. The influence of ductile interlayer material on the particle erosion resistance of multilayered TiN based coatings. *Wear* **2011**, *271*, 2890–2898. [CrossRef]
12. Tool, H.; Chowdhury, S.; Bose, B.; Yamamoto, K.; Veldhuis, S.C. Effect of Interlayer Thickness on Nano-Multilayer Coating Performance during High Speed Dry Milling of H13 Tool Steel. *Coatings* **2019**, *9*, 737. [CrossRef]
13. Kim, S.M.; Kim, B.S.; Kim, G.S.; Lee, S.Y.; Lee, B.Y. Evaluation of the High Temperature Characteristics of the CrZrN Coatings. *Surf. Coat. Technol.* **2008**, *202*, 5521–5525. [CrossRef]
14. Kim, G.S.; Kim, B.S.; Lee, S.Y.; Hahn, J.H. Structure and Mechanical Properties of Cr-Zr-N Films Synthesized by Closed Field Unbalanced Magnetron Sputtering with Vertical Magnetron Sources. *Surf. Coat. Technol.* **2005**, *200*, 1669–1675. [CrossRef]
15. Kim, D.J.; La, J.H.; Kim, K.S.; Kim, S.M.; Lee, S.Y. Tribological Properties of CrZr-Si-N Films Synthesized Using Cr-Zr-Si Segment Targets. *Surf. Coat. Technol.* **2014**, *259*, 71–76. [CrossRef]
16. Kim, H.; Kim, S.; Lee, S. Mechanical Properties and Thermal Stability of CrZrN / CrZrSiN Multilayer Coatings with Different Bilayer Periods. *Coatings* **2022**, *12*, 1025. [CrossRef]
17. Kim, D.; La, J.; Kim, S.; Lee, S.Y. The Tribological Performance of Laser Surface Treated CrZrSiN Thin Films. *Mater. Res. Bull.* **2014**, *58*, 39–43. [CrossRef]
18. Pharr, G.M.; Oliver, W.C. Measurement of Thin Film Mechanical Properties Using Nanoindentation. *MRS Bull.* **1992**, *17*, 28–33. [CrossRef]
19. Kim, Y.S.; Kim, G.S.; Lee, S.Y. Thermal Stability and Electrochemical Properties of CrZr-Si-N Films Synthesized by Closed Field Unbalanced Magnetron Sputtering. *Surf. Coat. Technol.* **2009**, *204*, 978–982. [CrossRef]
20. Ludema, K.C. Classes of Wear. *Ind. Eng. Chem. Prod. Res. Dev.* **1980**, *19*, 335–337. [CrossRef]
21. Bemporad, E.; Sebastiani, M.; Staia, M.H.; Puchi Cabrera, E. Tribological Studies on PVD/HVOF Duplex Coatings on Ti6Al4V Substrate. *Surf. Coat. Technol.* **2008**, *203*, 566–571. [CrossRef]
22. Lin, C.K.; Hsu, C.H.; Kung, S.C. Effect of Electroless Nickel Interlayer on Wear Behavior of CrN/ZrN Multilayer Films on Cu-Alloyed Ductile Iron. *Appl. Surf. Sci.* **2013**, *284*, 59–65. [CrossRef]
23. Bull, S.J. Failure Modes in Scratch Adhesion Testing. *Surf. Coat. Technol.* **1991**, *50*, 25–32. [CrossRef]
24. Bull, S.J.; Rickerby, D.S.; Matthews, A.; Leyland, A.; Pace, A.R.; Valli, J. The Use of Scratch Adhesion Testing for the Determination of Interfacial Adhesion: The Importance of Frictional Drag. *Surf. Coat. Technol.* **1988**, *36*, 503–517. [CrossRef]
25. Leyland, A.; Matthews, A. On the Significance of the H/E Ratio in Wear Control: A Nanocomposite Coating Approach to Optimised Tribological Behaviour. *Wear* **2000**, *246*, 1–11. [CrossRef]
26. Kim, D.J.; Kim, S.M.; La, J.H.; Lee, S.Y.; Hong, Y.S.; Lee, M.H. Synthesis and Characterization of CrZrAlN Films Using Unbalanced Magnetron Sputtering with Segment Targets. *Met. Mater. Int.* **2013**, *19*, 1295–1299. [CrossRef]
27. Barshilia, H.C.; Deepthi, B.; Rajam, K.S. Deposition and Characterization of CrN/Si_3N_4 and $CrAlN/Si_3N_4$ Nanocomposite Coatings Prepared Using Reactive DC Unbalanced Magnetron Sputtering. *Surf. Coat. Technol.* **2007**, *201*, 9468–9475. [CrossRef]
28. Mayrhofer, P.H.; Mitterer, C.; Musil, J. Structure-Property Relationships in Single- and Dual-Phase Nanocrystalline Hard Coatings. *Surf. Coat. Technol.* **2003**, *174–175*, 725–731. [CrossRef]

Article

Mechanical Properties and Thermal Stability of CrZrN/CrZrSiN Multilayer Coatings with Different Bilayer Periods

Hoe-Kun Kim [1], Sung-Min Kim [2],* and Sang-Yul Lee [1],*

[1] Center for Surface Technology and Applications, Department of Materials Engineering, Korea Aerospace University, Goyang 10540, Korea; ndkim2@kau.kr
[2] Heat & Surface Technology R&D Department, Korea Institute of Industrial Technology (KITECH), Incheon 21999, Korea
* Correspondence: sungminkim@kitech.re.kr (S.-M.K.); sylee@kau.ac.kr (S.-Y.L.)

Abstract: The CrZrN/CrZrSiN multilayer coatings at a bilayer period range decreasing from 1.35 μm to 0.45 μm were synthesized on a Si (100) wafer and WC-6 wt.% Co substrate using a closed-field unbalanced magnetron sputter, and the thickness effects on the mechanical properties and thermal stability were investigated. The CrZrN/CrZrSiN multilayer coatings showed high hardness and elastic modulus in the ranges of 28 to 33 GPa and 255 to 265 GPa, respectively, and the friction coefficient showed the lowest value of 0.24 on the multilayer coating with a bilayer period of 0.54 μm. The bilayer periods affected the adhesion strength of the multilayer coatings. From the scratch test, the critical load (Lc2) steadily increased with the decreasing of the bilayer period, and the CrZrN/CrZrSiN multilayer coating with a bilayer period of 0.45 μm showed the highest critical load (Lc2) of 79 N. In the case of the annealing test, the bilayer periods affected the thermal stability of the multilayer coatings, and the CrZrN/CrZrSiN multilayer coatings with 0.54 μm showed a maximum hardness value of approximately 30 GPa up to 800 °C.

Keywords: CrZrSiN coating; bilayer period; hardness; friction coefficient; adhesion strength

Citation: Kim, H.-K.; Kim, S.-M.; Lee, S.-Y. Mechanical Properties and Thermal Stability of CrZrN/CrZrSiN Multilayer Coatings with Different Bilayer Periods. *Coatings* **2022**, *12*, 1025. https://doi.org/10.3390/coatings12071025

Academic Editor: Sergey N. Grigoriev

Received: 10 June 2022
Accepted: 17 July 2022
Published: 19 July 2022

Publisher's Note: MDPI stays neutral with regard to jurisdictional claims in published maps and institutional affiliations.

1. Introduction

During the past decade, ternary nitride materials such as TiAlN [1,2], TiZrN [3,4], AlCrN [5,6], and CrZrN [7,8] have been studied to improve tool life due to their excellent mechanical and tribological properties such as high hardness and a low friction coefficient. Recently, transition metal nitride coatings with nanocomposite structure that are defined as a coexistence of the nanocrystalline and amorphous phases with complete immiscibility have emerged [9]. In particular, Si-containing nanocomposite coatings have received much attention due to the incorporation of Si, which led to the grain refinement of the nanocrystalline phase and the formation of amorphous Si_xN_y phase [9–14]. Furthermore, the use of nanocomposites containing Si has been expanded since amorphous Si_xN_y phase suppresses the diffusion of the oxygen ion [15].

Most recently, multilayered hard coatings have gained increasing attention from the industrial and scientific fields because of the enhanced mechanical properties compared to those usually revealed in the corresponding monolithic coatings. As reported in many publications, the multilayer coatings exhibited superior mechanical and tribological properties relative to single-layer hard coatings [16–22]. These improved properties could be attributed to a multilayered structure, consisting of alternating layers with different interface number, composition, and thickness. Generally, multilayer coatings usually show enhanced hardness as the bilayer period decreases. This improved property was attributed to the resistance against dislocation glide across interfaces in multilayer structures [23,24].

From our previous works, the CrZrN coatings showed great mechanical properties such as high hardness (33 GPa), low surface roughness (Rms 0.82 nm), and low friction coefficient (0.25) but relatively poor thermal stability over 500 °C [7,8]. The CrZrSiN coatings, however, showed moderate hardness (30 GPa) and friction coefficient (0.30), but also excellent thermal stability up to 800 °C [25,26]. Thus, in this work, CrZrN/CrZrSiN multilayer coatings with various bilayer periods were designed and synthesized in order to obtain the coatings with excellent mechanical properties such as hardness, friction coefficient, adhesion properties, and thermal stability. This study provides an insight into the underlying bilayer periods for enhanced mechanical properties and thermal stability of CrZrN/CrZrSiN multilayer coatings.

2. Experimental Details

The CrZrN/CrZrSiN multilayer coatings with different bilayer periods were synthesized on two types of substrate, Si (100) wafer and WC-6 wt.% Co substrate, using a closed-field unbalanced magnetron sputtering process. The CrN adhesion interlayer of 300 nm was deposited between the CrZrN/CrZrSiN upper layer and the WC substrate to improve the adhesion, and the bilayer periods were controlled in a range decreasing from 1.35 to 0.45 μm. A pure Cr, Zr, and Si single target (99.99%) was used as target material. Before sputtering, the surface of the Si (100) wafer and WC substrate were cleaned by ultrasonication in acetone for 10 min. After a base pressure was evacuated below 2.8×10^{-3} Pa, the substrates were etched in the presence of Ar plasma for 20 min at a pressure of 0.4 Pa. All the layers were deposited by a pulsed DC power with applied power of 0.5 kW, frequency of 25 kHz, and duty ratio of 70%. The deposition temperature was maintained at 400 °C. The detailed deposition conditions of the CrN interlayer and CrZrN, CrZrSiN upper layers are listed in Table 1.

Table 1. Deposition conditions of the CrN interlayer and CrZrN, CrZrSiN upper layers.

Parameters	CrN	CrZrN	CrZrSiN
Base pressure (Pa)	2.8×10^{-3}	2.8×10^{-3}	2.8×10^{-3}
Working pressure (Pa)	4.2×10^{-1}	6.0×10^{-1}	6.0×10^{-1}
Ar gas flow (sccm)	6	15	15
N_2 gas flow (sccm)	10	8	10
Target current (A)	Cr 1.8	Cr 1.2/Zr 1.8	Cr 1.2/Zr 1.6/Si 0.6

The microstructures of the as-deposited coatings were examined by field-emission scanning electron microscopy (FE-SEM, JEOL, JSM-7100F, Tokyo, Japan) operating at an accelerated voltage of 15 kV, and the elemental compositions of individual layers were detected using Oxford Instruments X-Max^N energy dispersive spectroscopy (EDS). The hardness and elastic modulus of the individual interlayers and coatings were measured using nanoindentation (Helmut Fischer, HM2000, Sindelfingen, Germany) with a load of 25 mN and dwell time of 30 s. Taking into consideration the thickness effect, the indentation depth was maintained at less than approximately 0.16 μm, which was controlled at less than 10% of the total coating thickness [27]. For accurate and reproducible results, both the hardness and elastic modulus tests were carried out 9 times on each sample. The friction coefficient of the coatings was measured using a customized ball-on-disk-type wear tester (Figure 1) with an alumina counter ball (Al_2O_3, Ø = 9.25 mm). The sliding velocity was 0.25 m/s and the total sliding distance was 1000 m with a normal load of 5 N. The adhesion strength values were evaluated by a scratch tester (CSM, Revetest, Corcelles, Switzerland) with Rockwell C diamond stylus with radius 200 um, which was drawn across the surface of coatings with a progressively increasing load up to 100 N and a constant sliding speed of 10 mm/min. The critical load (Lc2) and the failure mode of the coatings were determined by observing optical microscopy (Olympus, BX51M, Tokyo, Japan). To evaluate the thermal stability of all the multilayer coatings, annealing was carried out at temperatures ranging from 500 to 800 °C in air for 30 min, and the hardness values of annealed samples were measured using nanoindentation.

Figure 1. Ball-on-disk-type wear tester used for the wear test.

3. Results and Discussion

The CrZrN/CrZrSiN multilayer coatings were prepared by varying bilayer periods to enhance their mechanical, tribological, and adhesion properties, and thermal stabilities. As can be seen in the cross-sectional FE-SEM images of the CrZrN/CrZrSiN multilayer coatings (Figure 2), the thickness of all the coatings was adjusted to 3 μm by sputtering duration. In previous work, by means of restructuring the coating with an optimized gradient of the H/E ratio of the WC substrate/CrN interlayer/CrZrN coating, i.e., the minimization of stress gradient between the coating and WC substrate, the adhesion strength was enhanced [28]. Therefore, the CrN interlayer was chosen as the adhesion layer and deposited between the CrZrN/CrZrSiN upper layer and the WC substrate to improve their adhesion strength in this work. Regardless of bilayer periods, the CrZrN in a multilayered structure shows a columnar growth, while the CrZrSiN tends to have a smooth and featureless morphology. However, importantly, with decreasing the bilayer period from 1.35 μm to 0.45 μm, the columnar structure of CrZrN layers transformed from a coarse columnar structure to a dense columnar structure.

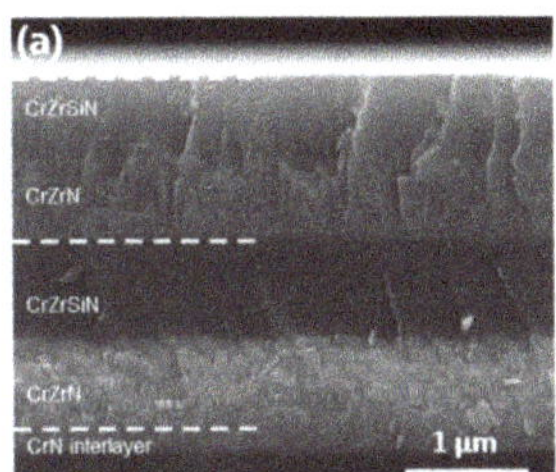
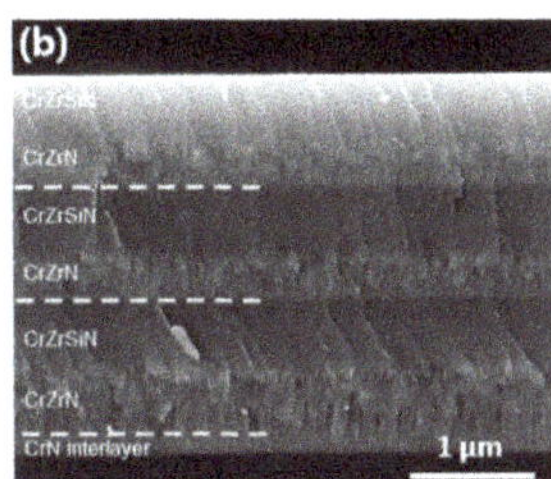
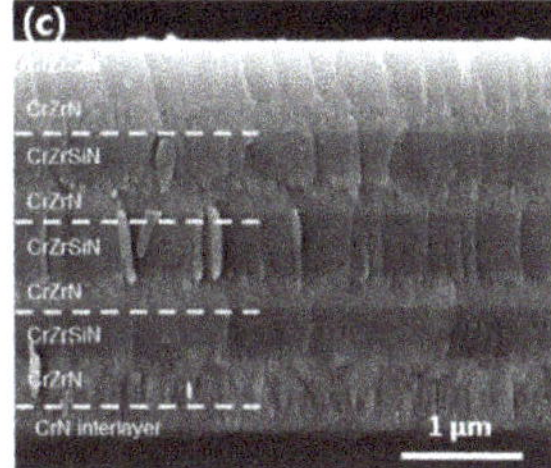
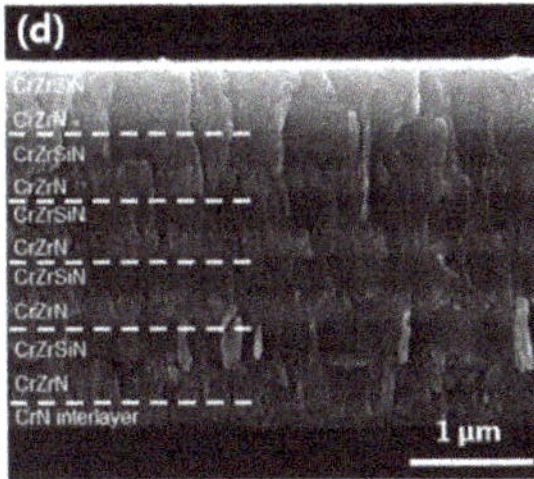
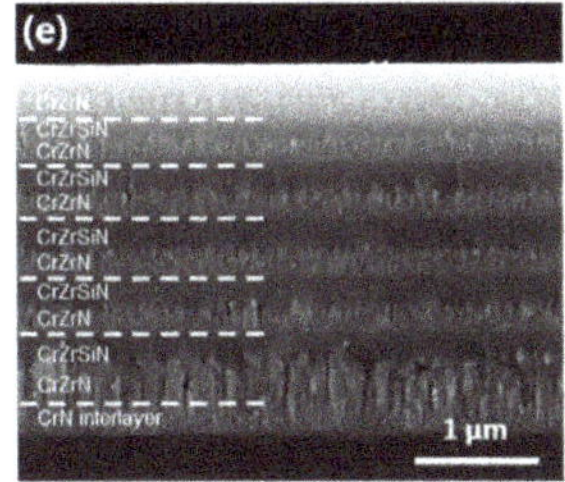

Figure 2. Cross-sectional images of the CrZrN/CrZrSiN multilayer coatings with various bilayer periods: (**a**) 1.35 μm, (**b**) 0.90 μm, (**c**) 0.67 μm, (**d**) 0.54 μm, and (**e**) 0.45 μm.

The average elemental composition of the WC substrate, CrN interlayer, and CrZrN, CrZrSiN upper layers was measured using SEM-equipped EDS, as summarized in Table 2.

The ratios of non-metallic to metallic elements for all coatings were calculated to be approximately 1.08, which suggests that the ratio of metals to N is almost stoichiometric, 1:1. The hardness (H), elastic modulus (E), and H/E ratio of individual WC substrate, CrN interlayer, and CrZrN, CrZrSiN upper layer are summarized in Table 2, and the hardness and elastic modulus of the CrZrN/CrZrSiN multilayer coatings with various bilayer periods are shown in Figure 3. Both the hardness and elastic modulus increased with the bilayer period decrement. The highest hardness and elastic modulus were approximately 33 and 265 GPa, respectively.

Table 2. Elemental compositions, hardness, elastic modulus, and H/E ratio of the WC substrate, CrN interlayer, and CrZrN, CrZrSiN upper layers.

Layer	Composition (at.%)	Hardness (GPa)	Elastic Modulus (GPa)	H/E Ratio
WC substrate	-	19.7 ± 1.6	510.6 ± 8.4	0.045
CrN	48Cr-52N	23.3 ± 1.8	305.6 ± 10.2	0.071
CrZrN	32Cr-16Zr-52N	32.1 ± 1.5	269.3 ± 8.7	0.113
CrZrSiN	21Cr-9Zr-17Si-53N	25.4 ± 1.9	191.5 ± 9.5	0.138

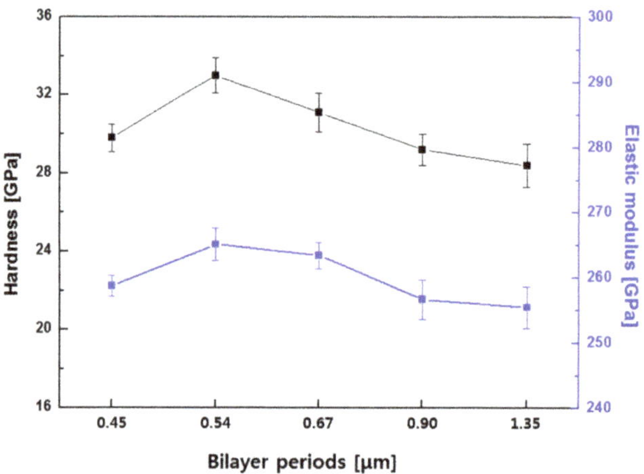

Figure 3. The hardness and elastic modulus of the CrZrN/CrZrSiN multilayer coatings with various bilayer periods.

Generally, the multilayered structure of coatings could enhance the hardness on the basis of the dislocation blocking by layer interfaces and lead to the strengthening effect [20,23,24]. The dislocation blocking effect occurred at the interface between layers due to differences in the shear modulus of the individual materials of layers, and the stress required for dislocation glide across interfaces increases. As a result, the high interface density of multilayer structure could contribute to interrupting dislocation glide across the interface between layers. When the bilayer period was decreased to 0.45 µm, however, there was a decrease in the hardness and elastic modulus. These decreased properties could be explained by the loss of the characteristics of a multilayer structure, and the lack of interface effect in the coatings [23,29]. These results show that there is a critical bilayer period of the CrZrN/CrZrSiN multilayer coating required to obtain optimized hardness and elastic modulus.

The friction coefficients of all the CrZrN/CrZrSiN multilayer coatings were evaluated by using the ball-on-disk-type wear tester, and the results are shown in Figure 4. All the friction curves were stable with little variation. It was revealed that the friction coefficient decreased gradually with the decrease in the bilayer period from 1.35 to 0.54 μm, and the value increased when the bilayer period was 0.45 μm. The coatings with bilayer period of 0.54 μm exhibited the lowest friction coefficient of 0.24, and this follows the same trend as the hardness and elastic modulus of the multilayer coatings.

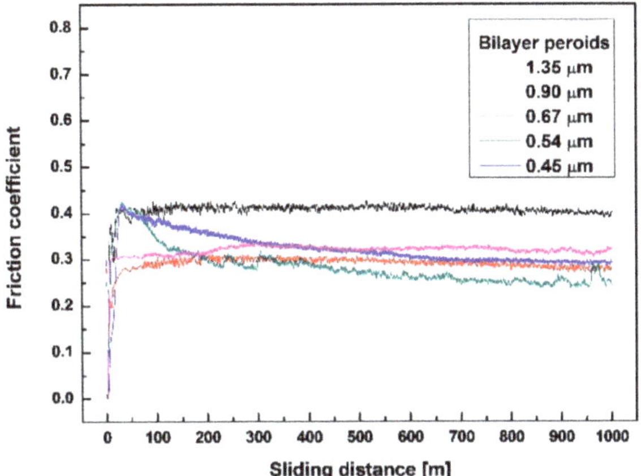

Figure 4. Friction curve of the CrZrN/CrZrSiN multilayer coatings with various bilayer periods.

From the point of view of classical wear theories, the hardness is closely related to the wear resistance of the coating surface. It has been widely accepted that a material with high hardness has high wear resistance. Many reports have also suggested the importance of the elastic modulus correlation with wear behavior [30,31], and wear resistance could be related to the elastic strain to failure of the coating, representing the ability of a material to deform elastically and subsequently recover without plastic deformation. This approach has been applied to the surfaces of coatings, and the elastic strain to failure has been explained by the correlation with the H/E ratio. Therefore, the H/E ratio is a more suitable parameter for predicting the wear resistance of a coating. As can be seen in Figure 5, the highest H/E ratio of the CrZrN/CrZrSiN multilayer coating with a bilayer period of 0.54 μm indicates its best wear resistance, and the friction coefficient of the multilayer coating tends to be proportional to the H/E ratio.

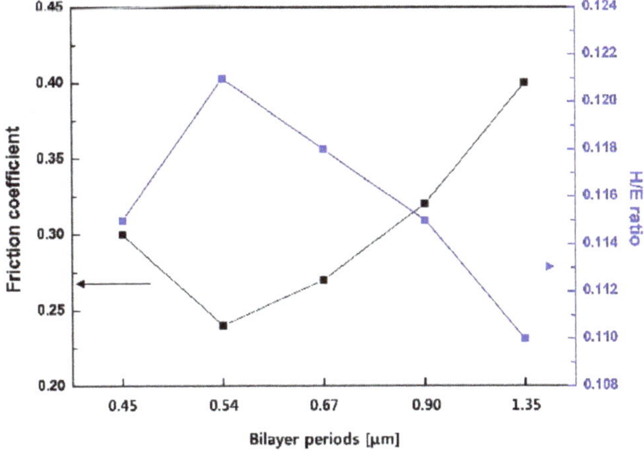

Figure 5. Friction coefficient and H/E ratio as a function of various bilayer periods.

The scratch test provides an informative clue for the adhesion and cohesion properties between a substrate and coating. In the process of a scratch test, three main failures classified as critical load of Lc0, Lc1, and Lc2 are typically derived from a progressive loading on a coating. More specifically, the classification of critical loads is phenomenologically explained as a half-circular crack inside the scratched track induced by plastic deformation (Lc0), an initial fragment or spallation of coating at a track edge (Lc1), and an initial delamination of coating (Lc2) [32,33]. The Lc2 was chosen to comparatively investigate the adhesion strength of the CrZrN/CrZrSiN coating with the different bilayer periods, and the average values calculated from the five measurements made for each coating in this work. The scratch tracks of each coating are shown in Figure 6. It can be observed that the Lc2 increased significantly as the bilayer period decreased from 1.35 to 0.45 μm, and the highest Lc2 was 79 N when the bilayer period was 0.45 μm, as shown in Table 3. In the scratch tracks of the multilayer coating with the bilayer period of 1.35 and 0.90 μm, the chipping cracks were observed and the multilayer coatings were totally delaminated over the Lc2.

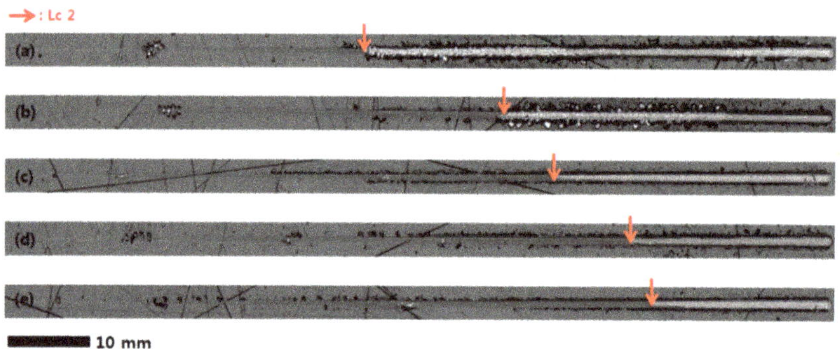

Figure 6. OM observation from the scratch tracks of the CrZrN/CrZrSiN multilayer coatings with various bilayer periods: (**a**) 1.35 μm, (**b**) 0.90 μm, (**c**) 0.67 μm, (**d**) 0.54 μm, and (**e**) 0.45 μm.

Table 3. The hardness, elastic modulus, H/E ratio, and critical load (Lc2) of the CrZrN/CrZrSiN multilayer coatings with various bilayer periods.

Bilayer Periods (μm)	Hardness (GPa)	Elastic Modulus (GPa)	H/E Ratio	Critical Load (Lc2) (N)
1.35	28.4 ± 2.1	255.5 ± 7.2	0.110	44
0.90	29.2 ± 1.7	256.7 ± 6.7	0.115	60
0.67	31.1 ± 2.2	263.4 ± 8.2	0.118	67
0.54	33.0 ± 2.0	265.1 ± 7.8	0.121	75
0.45	29.8 ± 1.9	258.8 ± 7.3	0.115	79

As reported in the literature [34], the chipping would be often observed in a high hardness coating deposited on the brittle substrate with high hardness. However, there were no obvious chipping cracks or delamination in whole scratch tracks below a bilayer period of 0.67 μm. These improved adhesion properties could be ascribed to the large number of the interfaces in the CrZrN/CrZrSiN multilayer structure [35,36]. During the scratch test, the interfaces in the multilayer coatings played a crucial role as the sites of stress dissipation, and this led to residual stress relaxation and crack propagation prevention in the coating [37]. Therefore, the CrZrN/CrZrSiN multilayer coatings with high density of the interfaces effectively prevented crack propagation during the progressive scratch test, and showed distinctly improved adhesion to the WC substrate with decreasing of the bilayer period.

Figure 7 presents the hardness variation of the CrZrN/CrZrSiN multilayer coatings depending on the bilayer period after annealing at a temperature ranging from 500 to 800 °C in air for 30 min. In all conditions, the hardness values of the multilayer coatings were almost constant above approximately 27 GPa up to 800 °C without drastic change in the hardness, unlike the monolayered CrZrN coating. In addition, the CrZrN/CrZrSiN multilayer coatings exhibited higher hardness compared to the monolithic CrZrSiN, coatings even after an annealing process at 800 °C. This means that all the CrZrN/CrZrSiN multilayer coatings showed excellent thermal stability.

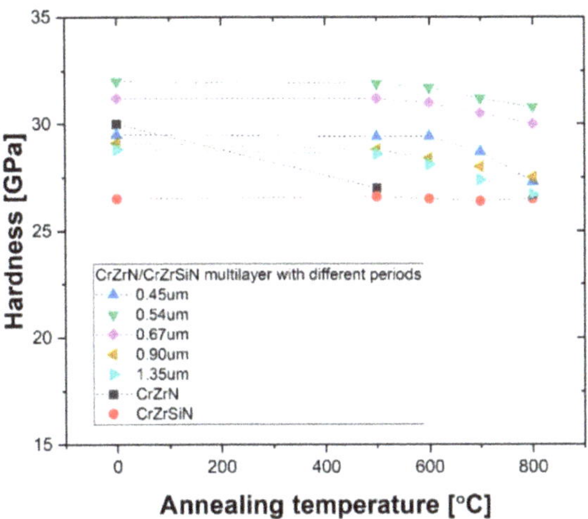

Figure 7. Hardness variation of the monolithic CrZrN, CrZrSiN, and multilayered CrZrN/CrZrSiN coatings with various bilayer periods after annealing test in air for 30 min.

Nevertheless, the thermal stability of the multilayer coatings seemed to be influenced by the bilayer periods. It has been well established that the CrZrSiN layer, which consists of Si_xN_y amorphous phase, plays an important role as an oxygen diffusion barrier to inhibit oxidation by residual oxygen during the annealing process, while the CrZrN component is easily oxidized. Interestingly, the hardness change dependent on the bilayer periods was observed after the annealing process, although it was expected that the hardness reduction should be constant due to an almost identical Si content regardless of various bilayer periods. In the case of a bilayer period of 0.45 and 1.35 μm, the hardness decreased approximately 7.4% with increasing annealing temperature from room temperature to 800 °C, relative to the bilayer periods of 0.54, 0.67, and 0.90 μm (3.7~5.5%). This deterioration of thermal stability could be mainly attributed to the structure induced by the bilayer period. Typically, a coarse microstructure of coatings tends to provide the oxygen pathway along the grain boundaries [38]. In addition, the inter-diffusion could occur with the decreased bilayer period, which led to the unexpected monolayered structure [20,21]. Thus, it is believed that the dense microstructure and moderate bilayer period in the range of 0.45–0.90 μm seems to be beneficial to the inhibition of oxygen diffusion and inter-diffusion between layers, leading to improved oxidation resistance.

4. Conclusions

In this work, CrZrN/CrZrSiN multilayer coatings with different bilayer periods were successfully prepared using a closed-field unbalanced magnetron sputtering process on WC-6 wt.% Co substrate to improve their mechanical properties and thermal stability. The hardness and elastic modulus of the CrZrN/CrZrSiN multilayer coatings gradually increased with the decreasing of the bilayer period from 1.35 to 0.54 μm; however, the values decreased on the multilayer coating with a bilayer period of 0.45 μm. The friction coefficient followed the same trend as the hardness, and the CrZrN/CrZrSiN multilayer coating with a bilayer period of 0.54 μm showed the lowest friction coefficient of 0.24. The result of the scratch test showed that the adhesion strength increased with the decreasing of the bilayer period, and obvious chipping or delamination were not found when the bilayer period decreased from 0.67 to 0.45 μm. After annealing, the hardness values of all the CrZrN/CrZrSiN multilayer coatings were almost constant above approximately 27 GPa up to 800 °C without drastic change in the hardness, and all the multilayer coatings also showed excellent thermal stability. Therefore, the CrZrN/CrZrSiN multilayer coatings with a moderate bilayer period showed much-improved mechanical properties (hardness, friction coefficient, adhesion) and thermal stability compared to those of each monolithic coating. This work experimentally revealed an optimal period of multilayered coatings for high-performance cutting-tool application.

Author Contributions: Conceptualization, H.-K.K.; methodology, H.-K.K.; software, H.-K.K.; validation, H.-K.K. and S.-M.K.; formal analysis, H.-K.K.; investigation, H.-K.K.; resources, H.-K.K.; data curation, H.-K.K.; writing—original draft preparation, H.-K.K.; writing—review and editing, S.-M.K. and S.-Y.L.; visualization, H.-K.K.; supervision, S.-M.K. and S.-Y.L.; project administration, S.-M.K. and S.-Y.L.; funding acquisition, S.-M.K. and S.-Y.L. All authors have read and agreed to the published version of the manuscript.

Funding: This study was conducted with the support of the Korea Institute of Industrial Technology as "Development of root technology for multi-product flexible production (KITECH EO-22-0006)", and this work was supported by a 2022 Korea Aerospace University Faculty Research Grant.

Institutional Review Board Statement: Not applicable.

Informed Consent Statement: Not applicable.

Data Availability Statement: Not applicable.

Conflicts of Interest: The authors declare no conflict of interest.

References

1. Chen, L.; Paulitsch, J.; Du, Y.; Mayrhofer, P.H. Thermal Stability and Oxidation Resistance of Ti-Al-N Coatings. *Surf. Coat. Technol.* **2012**, *206*, 2954–2960. [CrossRef]
2. Wang, S.; Ji, W.; Wang, Y.; Wei, J.; Qiu, L.; Chen, C.; Jiang, X.; Ran, Q.; Han, R. Comparative Study of Corrosion Behavior of LPCVD-Ti0.17Al0.83N and PVD-Ti1−xAlxN Coatings. *Coatings* **2022**, *12*, 835. [CrossRef]
3. Tung, H.M.; Wu, P.H.; Yu, G.P.; Huang, J.H. Microstructures, Mechanical Properties and Oxidation Behavior of Vacuum Annealed TiZrN Thin Films. *Vacuum* **2015**, *115*, 12–18. [CrossRef]
4. Volosova, M.; Grigoriev, S.; Metel, A.; Shein, A. The Role of Thin-Film Vacuum-Plasma Coatings and Their Influence on the Efficiency of Ceramic Cutting Inserts. *Coatings* **2018**, *8*, 287. [CrossRef]
5. Lomello, F.; Sanchette, F.; Schuster, F.; Tabarant, M.; Billard, A. Influence of Bias Voltage on Properties of AlCrN Coatings Prepared by Cathodic Arc Deposition. *Surf. Coat. Technol.* **2013**, *224*, 77–81. [CrossRef]
6. Warcholinski, B.; Gilewicz, A.; Myslinski, P.; Dobruchowska, E.; Murzynski, D. Structure and Properties of AlCrN Coatings Deposited Using Cathodic Arc Evaporation. *Coatings* **2020**, *10*, 793. [CrossRef]
7. Kim, S.M.; Kim, B.S.; Kim, G.S.; Lee, S.Y.; Lee, B.Y. Evaluation of the High Temperature Characteristics of the CrZrN Coatings. *Surf. Coat. Technol.* **2008**, *202*, 5521–5525. [CrossRef]
8. Kim, G.S.; Kim, B.S.; Lee, S.Y.; Hahn, J.H. Structure and Mechanical Properties of Cr-Zr-N Films Synthesized by Closed Field Unbalanced Magnetron Sputtering with Vertical Magnetron Sources. *Surf. Coat. Technol.* **2005**, *200*, 1669–1675. [CrossRef]
9. Lin, J.; Wang, B.; Ou, Y.; Sproul, W.D.; Dahan, I.; Moore, J.J. Structure and Properties of CrSiN Nanocomposite Coatings Deposited by Hybrid Modulated Pulsed Power and Pulsed Dc Magnetron Sputtering. *Surf. Coat. Technol.* **2013**, *216*, 251–258. [CrossRef]
10. Diserens, M.; Patscheider, J.; Lévy, F. Mechanical Properties and Oxidation Resistance of Nanocomposite TiN-SiN$_x$ Physical-Vapor-Deposited Thin Films. *Surf. Coat. Technol.* **1999**, *120–121*, 158–165. [CrossRef]
11. Mae, T.; Nose, M.; Zhou, M.; Nagae, T.; Shimamura, K. The Effects of Si Addition on the Structure and Mechanical Properties of ZrN Thin Films Deposited by an r.f. Reactive Sputtering Method. *Surf. Coat. Technol.* **2001**, *142–144*, 954–958. [CrossRef]
12. Xiang, Y.; Zou, C. Effect of Arc Currents on the Mechanical, High Temperature Oxidation and Corrosion Properties of CrSiN Nanocomposite Coatings. *Coatings* **2022**, *12*, 40. [CrossRef]
13. Liu, Y.; Wang, T.G.; Lin, W.; Zhu, Q.; Yan, B.; Hou, X. Microstructure and Properties of the AlCrSi(O)N Tool Coatings by Arc Ion Plating. *Coatings* **2020**, *10*, 841. [CrossRef]

14. Chang, L.C.; Zheng, Y.Z.; Chen, Y.I. Mechanical Properties of Zr-Si-N Films Fabricated through HiPIMS/RFMS Co-Sputtering. *Coatings* **2018**, *8*, 263. [CrossRef]
15. Barshilia, H.C.; Deepthi, B.; Rajam, K.S. Deposition and Characterization of CrN/Si_3N_4 and $CrAlN/Si_3N_4$ Nanocomposite Coatings Prepared Using Reactive DC Unbalanced Magnetron Sputtering. *Surf. Coat. Technol.* **2007**, *201*, 9468–9475. [CrossRef]
16. Araujo, J.A.; Araujo, G.M.; Souza, R.M.; Tschiptschin, A.P. Effect of Periodicity on Hardness and Scratch Resistance of CrN/NbN Nanoscale Multilayer Coating Deposited by Cathodic Arc Technique. *Wear* **2015**, *330–331*, 469–477. [CrossRef]
17. Al-Bukhaiti, M.A.; Al-Hatab, K.A.; Tillmann, W.; Hoffmann, F.; Sprute, T. Tribological and Mechanical Properties of Ti/TiAlN/TiAlCN Nanoscale Multilayer PVD Coatings Deposited on AISI H11 Hot Work Tool Steel. *Appl. Surf. Sci.* **2014**, *318*, 180–190. [CrossRef]
18. Bao, M.; Xu, X.; Zhang, H.; Liu, X.; Tian, L.; Zeng, Z.; Song, Y. Tribological Behavior at Elevated Temperature of Multilayer TiCN/TiC/TiN Hard Coatings Produced by Chemical Vapor Deposition. *Thin Solid Film.* **2011**, *520*, 833–836. [CrossRef]
19. Ding, X.Z.; Zeng, X.T.; Liu, Y.C. Structure and Properties of CrAlSiN Nanocomposite Coatings Deposited by Lateral Rotating Cathod Arc. *Thin Solid Film.* **2011**, *519*, 1894–1900. [CrossRef]
20. Kim, S.-M.; Kim, G.-S.; Lee, S.-Y.; Lee, J.-W.; Lee, J.-Y.; Lee, B.-Y. Mechanical Properties and Thermal Stability of CrSiN/AlN Multilayer Coatings Using a Closed-Field Unbalanced Magnetron Sputtering System. *J. Korean Phys. Soc.* **2009**, *54*, 1109–1114. [CrossRef]
21. Kim, S.; Kim, E.; Kim, D.; La, J.; Lee, S. Effects of Bilayer Period on the Microhardness and Its Strengthening Mechanism of CrN/AlN Superlattice Coatings. *J. Korean Inst. Surf. Eng.* **2012**, *45*, 257–263. [CrossRef]
22. Bartsch, H.; Grieseler, R.; Mánuel, J.; Pezoldt, J.; Müller, J. Magnetron Sputtered AlN Layers on LTCC Multilayer and Silicon Substrates. *Coatings* **2018**, *8*, 289. [CrossRef]
23. Zhou, S.Y.; Yan, S.J.; Han, B.; Yang, B.; Lin, B.Z.; Zhang, Z.D.; Ai, Z.W.; Pelenovich, V.O.; Fu, D.J. Influence of Modulation Period and Modulation Ratio on Structure and Mechanical Properties of TiBN/CrN Coatings Deposited by Multi-Arc Ion Plating. *Appl. Surf. Sci.* **2015**, *351*, 1116–1121. [CrossRef]
24. Chang, Y.Y.; Wu, C.J. Mechanical Properties and Impact Resistance of Multilayered TiAlN/ZrN Coatings. *Surf. Coat. Technol.* **2013**, *231*, 62–66. [CrossRef]
25. Kim, D.J.; La, J.H.; Kim, K.S.; Kim, S.M.; Lee, S.Y. Tribological Properties of CrZr-Si-N Films Synthesized Using Cr-Zr-Si Segment Targets. *Surf. Coat. Technol.* **2014**, *259*, 71–76. [CrossRef]
26. Kim, Y.S.; Kim, G.S.; Lee, S.Y. Thermal Stability and Electrochemical Properties of CrZr-Si-N Films Synthesized by Closed Field Unbalanced Magnetron Sputtering. *Surf. Coat. Technol.* **2009**, *204*, 978–982. [CrossRef]
27. Klemm, S.O.; Schauer, J.C.; Schuhmacher, B.; Hassel, A.W. High Throughput Electrochemical Screening and Dissolution Monitoring of Mg-Zn Material Libraries. *Electrochim. Acta* **2011**, *56*, 9627–9636. [CrossRef]
28. Kim, H.K.; La, J.H.; Kim, K.S.; Lee, S.Y. The Effects of the H/E Ratio of Various Cr-N Interlayers on the Adhesion Strength of CrZrN Coatings on Tungsten Carbide Substrates. *Surf. Coat. Technol.* **2015**, *284*, 230–234. [CrossRef]
29. Kot, M.; Rakowski, W.A.; Major, R.; Morgiel, J. Effect of Bilayer Period on Properties of Cr/CrN Multilayer Coatings Produced by Laser Ablation. *Surf. Coat. Technol.* **2008**, *202*, 3501–3506. [CrossRef]
30. Guo, J.; Wang, H.; Meng, F.; Liu, X.; Huang, F. Tuning the H/E* Ratio and E* of AlN Coatings by Copper Addition. *Surf. Coat. Technol.* **2013**, *228*, 68–75. [CrossRef]
31. Leyland, A.; Matthews, A. On the Significance of the H/E Ratio in Wear Control: A Nanocomposite Coating Approach to Optimised Tribological Behaviour. *Wear* **2000**, *246*, 1–11. [CrossRef]
32. Bull, S.J.; Rickerby, D.S.; Matthews, A.; Leyland, A.; Pace, A.R.; Valli, J. The Use of Scratch Adhesion Testing for the Determination of Interfacial Adhesion: The Importance of Frictional Drag. *Surf. Coat. Technol.* **1988**, *36*, 503–517. [CrossRef]
33. Valli, J.; Mäkelä, U.; Matthews, A.; Murawa, V. TiN Coating Adhesion Studies Using the Scratch Test Method. *J. Vac. Sci. Technol. A Vac. Surf. Film.* **1985**, *3*, 2411–2414. [CrossRef]
34. Bull, S.J. Failure Modes in Scratch Adhesion Testing. *Surf. Coat. Technol.* **1991**, *50*, 25–32. [CrossRef]
35. Bai, W.Q.; Cai, J.B.; Wang, X.L.; Wang, D.H.; Gu, C.D.; Tu, J.P. Mechanical and Tribological Properties of A-C/a-C:Ti Multilayer Films with Various Bilayer Periods. *Thin Solid Film.* **2014**, *558*, 176–183. [CrossRef]
36. Voevodin, A.A.; Capano, M.A.; Laube, S.J.P.; Donley, M.S.; Zabinski, J.S. Design of a Ti/TiC/DLC Functionally Gradient Coating Based on Studies of Structural Transitions in Ti-C Thin Films. *Thin Solid Film.* **1997**, *298*, 107–115. [CrossRef]
37. Holleck, H.; Schier, V. Multilayer PVD Coatings for Wear Protection. *Surf. Coat. Technol.* **1995**, *76–77*, 328–336. [CrossRef]
38. Kuo, Y.L.; Kencana, S.D. Mechanism of Oxygen Ion Diffusion in Gd-Doped Ceria Electrolyte Films Deposited via Reactive and Direct Sputtering. *Surf. Coat. Technol.* **2017**, *320*, 47–52. [CrossRef]

Article

Wear Resistance, Patterns of Wear and Plastic Properties of Cr,Mo-(Cr,Mo,)N-(Cr,Mo,Al)N Composite Coating with a Nanolayer Structure

Alexey Vereschaka [1],*, Anton Seleznev [2] and Vladislav Gaponov [2]

[1] Institute of Design and Technological Informatics of the Russian Academy of Sciences (IDTI RAS), 127055 Moscow, Russia

[2] VTO Department, Moscow State Technological University STANKIN, 127994 Moscow, Russia; a.seleznev@stankin.ru (A.S.); gaponov-vlad@inbox.ru (V.G.)

* Correspondence: dr.a.veres@yandex.ru

Abstract: This paper discusses the results of studies focused on the wear resistance, patterns of wear and plastic properties of Cr,Mo-(Cr,Mo,)N-(Cr,Mo,Al)N coating, containing 20 at.% Mo. The coating had a nanolayer structure with a modulation period $\lambda = 50$ nm. The studies revealed the hardness, fracture resistance in scratch testing, as well as elemental and phase composition of the coating. The studies of the tool life of carbide cutting tools with the Cr,Mo-(Cr,Mo,)N-(Cr,Mo,Al)N coating proved their longer tool life compared to that of uncoated tools and tools with the reference Cr-(Cr,Al)N coating of equal thickness and equal content of aluminum (Al). The studies included the comparison of the tools coated with Cr,Mo-(Cr,Mo,)N-(Cr,Mo,Al)N and Cr-(Cr,Al)N. The experiments focused on the specific features of the coating nanostructure and were conducted using a transmission electron microscope (TEM), revealing the different mechanisms of fracture. The penetration of particles of the material being machined between nanolayers of the coating results in interlayer delamination. When exposed to a moving flow of the material being machined, plastic deformation (bending) of the coating nanolayers occurs. The diffusion of iron into the coating (up to 200 nm) and diffusion of Cr and Mo into the cut material to a depth of up to 250 nm are observed. The presented information can help in the design of metal cutting tools and the choice of coatings for them.

Keywords: multilayer composite multicomponent coating; diffusion; wear; metal-cutting tool; plastic deformation

Citation: Vereschaka, A.; Seleznev, A.; Gaponov, V. Wear Resistance, Patterns of Wear and Plastic Properties of Cr,Mo-(Cr,Mo,)N-(Cr,Mo,Al)N Composite Coating with a Nanolayer Structure. *Coatings* **2022**, *12*, 758. https://doi.org/10.3390/coatings12060758

Academic Editor: Alina Vladescu

Received: 5 May 2022
Accepted: 31 May 2022
Published: 31 May 2022

Publisher's Note: MDPI stays neutral with regard to jurisdictional claims in published maps and institutional affiliations.

1. Introduction

One of the most efficient ways to enhance the performance properties of materials is to modify their surface parameters by the deposition of special coatings [1–4]. Similar materials, composite in nature, including substrates and coatings, are increasingly used in products for various purposes (cutting tools, friction pairs, medical implants, etc.).

The production of metal-cutting tools is one of the spheres where modifying coatings is widely used. Modern machine equipment allows machining with high efficiency. In particular, the cutting speed can be increased considerably without loss of system rigidity. The cutting tool is a weak link in cutting system technology, which limits the further growth of cutting speeds and machining [5–7]. New tool materials, including composite materials consisting of a substrate and a coating, are introduced in order to improve the characteristics of cutting tools. At the same time, the coatings themselves are improved due to the development of a more complex structure and composition [8–10]. In particular, the introduction of additional elements into the coating composition can increase its hardness, heat resistance, and, finally, its general ability to resist wear [11–18].

Coatings based on the $(Cr_x,Al_{1-x})N$ system are widely used, and they provide good wear properties for tools during cutting [19–22]. The additional increase in the characteristics of the considered coating can be obtained due to the introduction into its composition of such elements as molybdenum (Mo), niobium (Nb), boron (B), vanadium (V), tantalum (Ta), yttrium (Y), as well as silicon (Si) [23–28]. Several studies demonstrate that the most favorable properties can be ensured with the introduction of molybdenum, simultaneously providing an increase in hardness, heat resistance, and an improvement of its tribological properties due to the formation of MoO_3 oxides at high temperatures [28–31].

On the other hand, coatings based on the $(Cr_x,Mo_{1-x})N$ system are characterized not only by sufficient hardness and heat resistance but also by excellent tribological properties at high temperatures due to the possibility of the formation of MoO_3 oxide films [32–34]. In the considered system, molybdenum performs the functions described above by providing thermal stability and improving tribological properties. In particular, with the introduction of 17 at.% Mo into the considered system, the temperature threshold of the start of active oxidation increases considerably in comparison with the CrN coating [32]. The presence of molybdenum also contributes to the preservation of a dense columnar structure upon heating to at least 700 °C, while at the same temperatures, the columnar structure of the CrN coating begins to fracture with the growth of internal stresses, which results in the destruction of the coating on separate fragments [32,34]. On the coating surface, a film of MoO_3 begins to form at a temperature of 600 °C, performing at the same time the protective (prevention of further oxidation) and lubrication (decrease in the coefficient of friction) functions [32,34]. With the introduction of molybdenum into the composition of the $(Cr_x,Mo_{1-x})N$ system, a substitution solid solution based on the cubic structure of CrN is formed, but with an increase in the content of Mo, a phase of γ-Mo_2N is also detected. When the content of Mo is prevailing (69.3 at.%), a certain amount of the bcc-Mo phase also arises [35]. Changes in the phase composition of the coating also affect its mechanical properties. In particular, the introduction of molybdenum into the composition of the system increases the hardness of the coating, but with an increase in the Mo content to 45 at.% and above, a decrease in hardness is detected [35,36].

When comparing the available data on the hardness of CrN and Mo_2N, it can be concluded that despite a wide variety of data on the hardness of these materials, the hardness of Mo_2N is still slightly higher (19–51 GPa [37–42]) than the hardness of CrN (16–27 GPa [43–45]).

The introduction of aluminum into the composition of the coatings provides a considerable improvement in their properties due to the formation of a substitution solid solution with the distortion of the crystal lattice, as the atomic radius of aluminum is noticeably smaller than those of titanium (Ti), chromium (Cr), or molybdenum (Mo). In particular, the experiments reveal that the (Cr,Al,Mo)N coating is characterized by higher heat resistance and hardness than the (Cr,Mo)N coating [28–31,46]. In addition to the described advantages, the (Cr,Al,Mo)N coating also demonstrates good tribological properties at high temperatures due to the formation of oxide films based on Cr_2O_3, Al_2O_3, and MoO_3 [28–31,46]. The coating hardness slightly decreases with an increase in the content of Mo [47–49] (37 GPa with the Mo content of 7 at.% [47], 35 GPa with the Mo content of 24 at.% [48], and 34 GPa with the Mo content of 41 at.% [48]). Meanwhile, the hardness is still much higher than the hardness of the (Cr,Al)N coating (29 GPa [47,49]). It is reported that with the introduction of a small amount of molybdenum (4 at.%), there is a slight decrease (by 1 GPa) in the coating hardness compared to the coating containing no molybdenum [49].

A decrease in the coefficient of friction is even more noticeable (from 0.90 for the (Cr,Al)N coating to 0.48 for the (Cr,Al,Mo)N coating with 33.2 at.% Mo [48]), which can be associated with the active formation of the above-mentioned phase of MoO_3. It is also important that the formation of tribologically active oxides in the (Cr,Al,Mo)N coating starts at lower temperatures than in the (Cr,Al)N coating. This fact can be explained by the lower temperature of molybdenum nitride oxidation and, accordingly, the formation of MoO_3, compared to the temperatures of oxidation of (Cr,Al)N and formation of Cr_2O_3 and Al_2O_3.

In addition to the elemental composition, the properties of the coatings are significantly affected by the parameters of the coating structure. The transition to the multilayer structure additionally increases the key properties of the coatings [50–52]. The application of the multilayer structure with a nanometric thickness of layers improves the properties of the coating, such as resistance to cracking and brittle fracture [51,52], as well as tribological properties [52]. Earlier studies of the coatings based on the (Cr,Mo,Al)N system (in particular, Ti-TiN-(Ti,Cr,Mo,Al)N [53], Cr,Mo-(Cr,Mo,Zr,Nb)N-(Cr,Mo,Zr,Nb,Al)N [11], and Cr,Mo-(Cr,Mo)N-(Cr,Mo,Al)N [10]) proved their high efficiency as coatings for cutting tools.

The choice of molybdenum content in the coating was associated with the need to provide a balanced combination of properties. On the one hand, if the content of molybdenum is too low, there is no noticeable change in the tribological properties since a sufficient amount of molybdenum oxide is not formed [48]. On the other hand, an excessive increase in the content of molybdenum (more than 45 at.%) leads to a decrease in the hardness of the coating [35,36,48]. The temperature threshold for the onset of active oxidation noticeably increases at a content of 17 at.% Mo [32]. Based on the above, the molybdenum content was chosen to be about 20 at.%. Thus, the Cr,Mo-(Cr,Mo)N-(Cr,Mo,Al)N coating with three-layer architecture was chosen for the present study [54,55]. The adhesive layer of Cr,Mo provides high adhesion to the substrate and smooths out the surfaces of the substrate by filling places of microroughness. The transition layer of (Cr,Mo)N solves the problem of a smooth transition of the properties from the adhesive layer to the wear-resistance layer, and the wear-resistant layer of (Cr,Mo,Al)N, which, in turn, has a nanolayer structure, provides resistance to wear and optimization of the contact conditions in the cutting zone [16].

2. Materials and Methods

The coating of Cr,Mo-(Cr,Mo)N-(Cr,Mo,Al)N was deposited with the specialized VIT-2 unit (IDTI RAS—MSTU STANKIN, Moscow, Russia) using evaporators operating under filtered cathodic vacuum arc deposition (FCVAD) technology [54,56–61] (for an aluminum cathode) and Controlled Accelerated Arc (CAA-PVD) [62] (for cathodes of chromium and molybdenum). Three cathode systems were used, including cathodes of Al (99.8%, installed in the FCVAD system), cathodes of Cr,Mo (50 + 50%), and cathodes of Cr (99.9%, installed in the CAA-PVD systems). The cathode of Al (99.8%) and two cathodes of Cr (99.9%) were used to deposit the reference coating of Cr-(Cr,Al)N.

The Cr,Mo-(Cr,Mo)N-(Cr,Mo,Al)N coating had the following functional layer parameters: the Cr,Mo adhesive layer, 20–50 nm thick, the (Cr,Mo)N transition layer, about 700 nm thick, and the (Cr,Mo,Al)N wear-resistant layer, about 2000 nm thick. Thus, the total thickness of the Cr,Mo-(Cr,Mo)N-(Cr,Mo,Al)N coating was about 2750 nm. For the reference Cr-(Cr,Al)N coating, the thickness of the Cr adhesive layer was 20–50 nm, and the thickness of the (Cr,Al)N layer was 2600–2800 nm. Thus, the coatings under comparison had equal total thickness. The layers of (Cr,Mo,Al)N and (Cr,Al)N had a nanolayer structure with a modulation period λ = 45–50 nm.

The conditions of the coating deposition are presented in Table 1.

Table 1. The conditions of the coating deposition.

	Gas Pressure p_N (Pa)	Voltage on Substrate U (V)	Cathode Arc Current (A)		
			Al	Cr-Mo	Cr
Heating and subsequent cleaning of samples with gas plasma	2.0	100 DC	75	110	75
Coating deposition	0.42	−150 DC	160	110	75
Product cooling	0.06	-	-	-	-

The characteristics of the nanolayer structure of the coatings (the thickness of the nanolayers and the nanolayer period) were determined by the rotation speed of the turntable (n = 0.7 rpm), the deposition time, and the process parameters [16].

Carbide inserts SNUN ISO 1832:2012 (WC + 15% TiC + 6% Co) were used as a substrate. Sample preparation consisted of washing the samples in a special solution with ultrasonic assistance, rinsing in distilled water, and drying. No additional impact on the samples (for example, polishing) was carried out. Thus, the technological sequence of sample preparation and coating deposition fully corresponded to the real conditions for the production of metal-cutting tools.

A JEM 2100 (JEOL Company, Tokyo, Japan) transmission electron microscope (TEM) was used to study the structure; the accelerating voltage was 200 kV. The elemental composition was determined by TEM with an EDX INCA Energy (OXFORD Instruments, Abingdon, Oxfordshire, UK) system. Samples (lamellae) were cut using a focused ion beam (FIB) on Strata 205 (FEI, Hillsboro, OR, USA) equipment.

Resistance of samples to wear was determined during the turning of 1045 steel at a CU 500 MRD lathe (ZMM-BULGARIA HOLDING, Sofia, Bulgaria) with a ZMM CU 500 MRD variable-speed drive at dry cutting. Geometric parameters of the cutting process were as follows: $\gamma = -7°$, $\alpha = 7°$, $\lambda = 0$, $r = 0.4$ mm; at cutting mode: $f = 0.1$ rpm, $a_p = 0.5$ mm, and $v_c = 400$ m/min. A tool with the reference coating of Cr-(Cr,Al)N and an uncoated tool were considered for comparison. For each type of cutting tool (coated or uncoated), five cutting tests were carried out. The presented wear graph shows the average wear values for five tests. The corresponding error bars show the spread of values. Four experiments were conducted for each coating, and the obtained values of flank wear were processed to obtain the polynomial functions exhibited on the curve. The limit wear criterion was assumed as $VB_{max} = 0.4$ mm.

Wear resistance was studied when turning steel 1045 on a CU 500 MRD (ZMM-BULGARIA HOLDING, Sofia, Bulgaria) lathe with a ZMM CU 500 MRD variator, under dry cutting conditions. The following cutting geometry was used: $\gamma = -7°$, $\alpha = 7°$, $\lambda = 0$, $r = 0.4$ mm; with cutting parameters: $f = 0.1$ rpm, $a_p = 0.5$ mm, $v_c = 400$ m/min. As objects of comparison, we used a tool with a commercial coating of Cr-(Cr,Al)N and without a coating. For each type of coating, five tests were carried out under similar conditions. The graphs show average wear values, the spread of values is described by error bars. The wear limit criterion was assumed to be $VB_{max} = 0.4$ mm.

To measure the hardness of the coatings, a mechanical tester (CSM Instruments, Needham, MA, USA) was used, the Oliver–Pharr method [63] was used, and the load was 10 mN. Scratch fracture resistance was determined according to ASTM1624-05 [64] on a Nanovea M1 (Micro Scratch, Nanovea, Irvine, CA, USA). The value of the critical load L_{C2}, which characterizes the beginning of the complete destruction of the coating, was determined by the method of acoustic emission. When choosing parameters and research methods, the data presented in [65–74] were taken into account.

3. Results

Research results prove that its wear-resistant layer contains on average (given a change in the proportion of elements over the thickness of each nanolayer) 70 ($\pm$12) at.% Cr, 20 ($\pm$11) at.% Mo, and 10 ($\pm$5) at.% Al (Figure 1a). On the presented diffraction patterns, copper is the peak from the copper mesh on which the lamella is attached. A single phase of c-(Cr,Mo,Al)N is detected in the wear-resistant layer of the coating (Figure 1b). The reference coating of CrAlN contains 89 ($\pm$2) at.% Cr and 11($\pm$2) at.% Al.

The hardness of the Cr,Mo-(Cr,Mo)N-(Cr,Mo,Al)N coating (27.6 $\pm$ 0.9 GPa) is slightly lower than the hardness of the Cr-(Cr,Al)N coating (30.0 $\pm$ 1.1 GPa), with equal values of the critical load of destruction L_{C2} for both coatings (40 N). The study of the structure of the Cr,Mo-(Cr,Mo)N-(Cr,Mo,Al)N coating reveals nanolayer construction with a modulation period $\lambda = 50$ nm (Figure 1c). The nanolayer structure has a complex construction with variable thickness of nanolayers formed during the planetary rotation of the sample in the unit chamber [16].

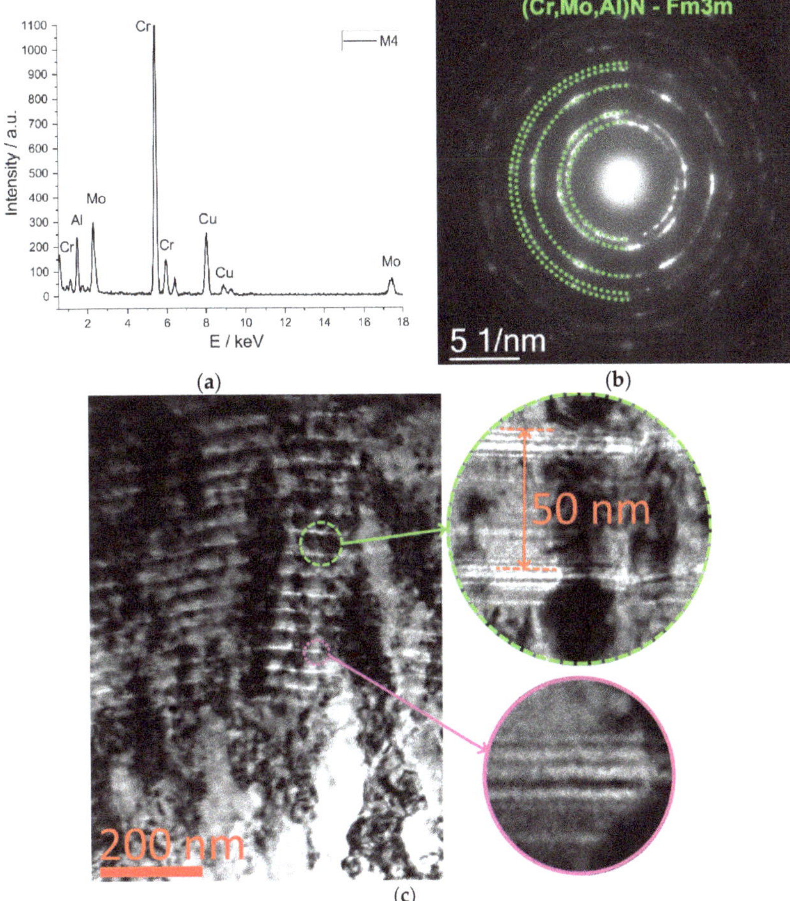

Figure 1. (**a**) Diffraction patterns, (**b**) selected area electron diffraction (SAED) pattern, (**c**) micro- and nanostructure of the Cr,Mo-(Cr,Mo)N-(Cr,Mo,Al)N coating.

The study of wear resistance of the cutting tool finds (Figure 2) that the uncoated tool has a very short tool life at the given cutting modes. The tool with the Cr,Mo-(Cr,Mo)N-(Cr,Mo,Al)N coating demonstrates the best wear resistance both along the rake and flank faces compared to the tool with the Cr-(Cr,Al)N coating and the uncoated tool.

The study of the wear dynamics on the rake face of the tool reveals a noticeably more active formation of wear crater on the tool with the Cr-(Cr,Al)N coating compared to the tool with the Cr,Mo-(Cr,Mo)N-(Cr,Mo,Al)N coating (Figure 3). The formation of a wear crater on the rake face is typical both for the Cr-(Cr,Al)N-coated and Cr,Mo-(Cr,Mo)N-(Cr,Mo,Al)N-coated tools, but the process is more intensive for the Cr-(Cr,Al)N-coated tool. There is a more intensive formation of the wear crater and noticeable differences in its geometry. In contrast, for the tool with the Cr,Mo-(Cr,Mo)N-(Cr,Mo,Al)N coating, the edge of the wear crater, far from the cutting edge, resembles a circular arc, then for the tool with the Cr-(Cr,Al)N coating, an almond-like shape of the wear crater is typical. In turn, this (almond-like) shape of the crater is typical for the high cutting speeds and, accordingly, high temperatures in the cutting zone [3,4,75,76]. Thus, it can be assumed that during cutting with the Cr,Mo-(Cr,Mo)N-(Cr,Mo,Al)N-coated tool, the temperature in the

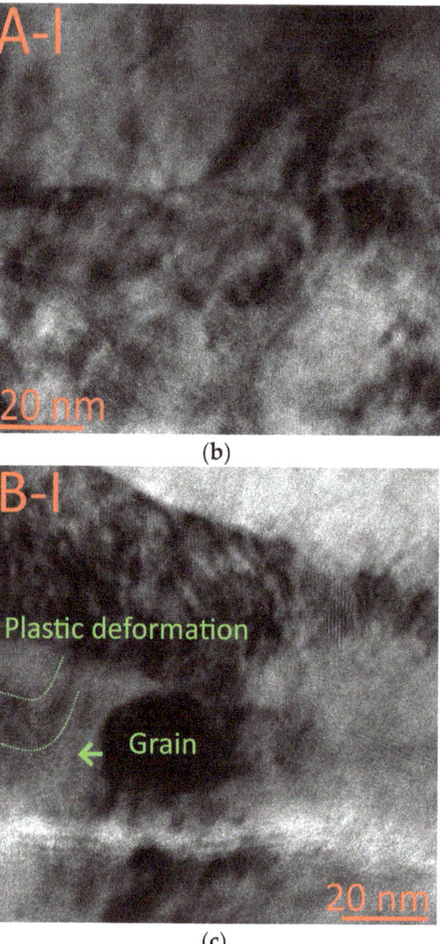

Figure 4. Deformation in the structure of nanolayers of the Cr,Mo-(Cr,Mo)N-(Cr,Mo,Al)N coating under the influence of the cut material flow. (**a**) General view of the investigated coverage area; (**b**) Deformation of the outer layers of the coating under the influence of a moving layer of the cut material; (**c**) Deformation of coating nanolayers under the influence of grains of cut material introduced into the structure.

The structure of the material being machined (steel) can include not only a softer fraction of iron but also harder carbonaceous (for example, martensite) fractions, as well as hard oxide grains [77–79]. These solid particles can affect the structure of the coating (Figure 5). In particular, such particles wedge between the coating nanolayers, resulting in plastic deformation and delamination of the nanolayers.

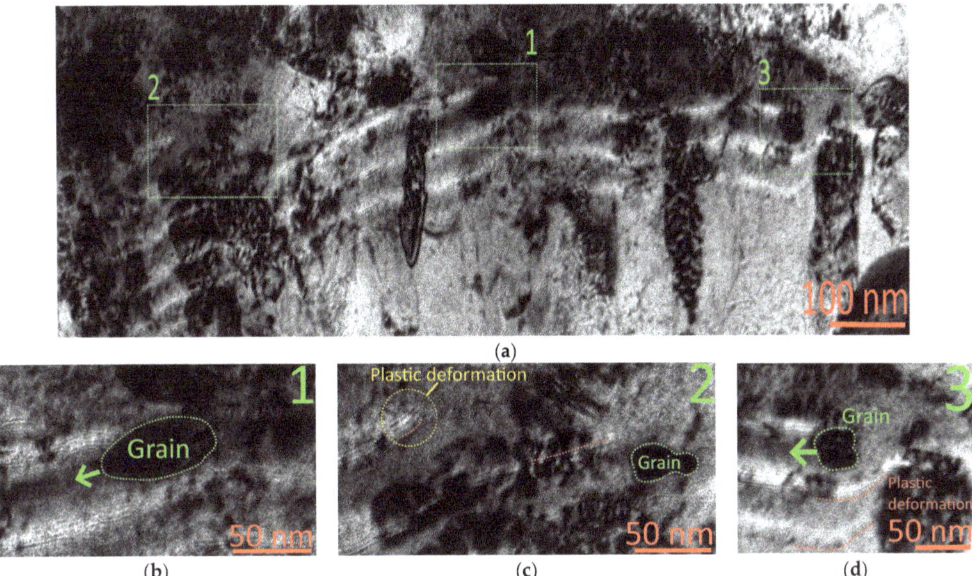

(a)

(b) (c) (d)

Figure 5. (**a**) Specifics of the interaction between the cut material flow and the structure of coating nanolayers, (**b**–**d**) penetration of grains of the machined steel into the structure of the coating.

Figure 6 exhibit an example of the steel penetrating into the coating of the structure. A particle of the machined steel wedges between the coating nanolayers to a depth of no more than 200 nm, breaking the bond between the outer nanolayer, rich in aluminum (light in contrast), and the main structure of the coating. In terms of the interdiffusion processes between the machined steel and the coating (Line L1, Figure 6a), there is the iron diffusion into the coating structure to the depth of no more than 200 nm and the chromium and molybdenum diffusion into the structure of the machined steel to the depth no more than 250 nm (Figure 6d). The Fourier analysis (Figure 6c) of the "coating-material being machined" interface reveals the presence of $Fm\bar{3}m$, the main phase of the (Cr,Mo,Al)N coating, and Im3m, the phase of iron.

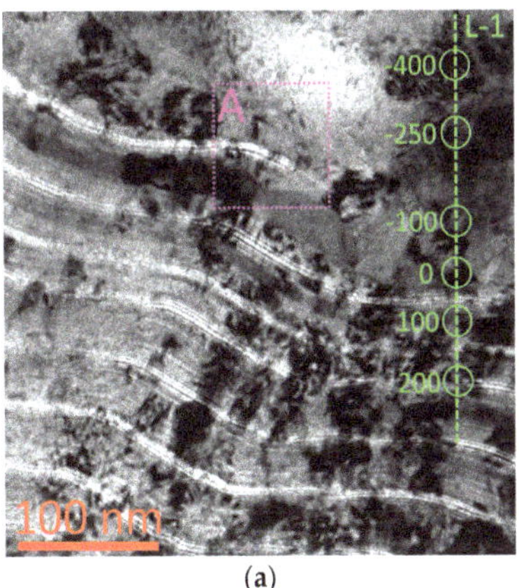

(a)

Figure 6. *Cont.*

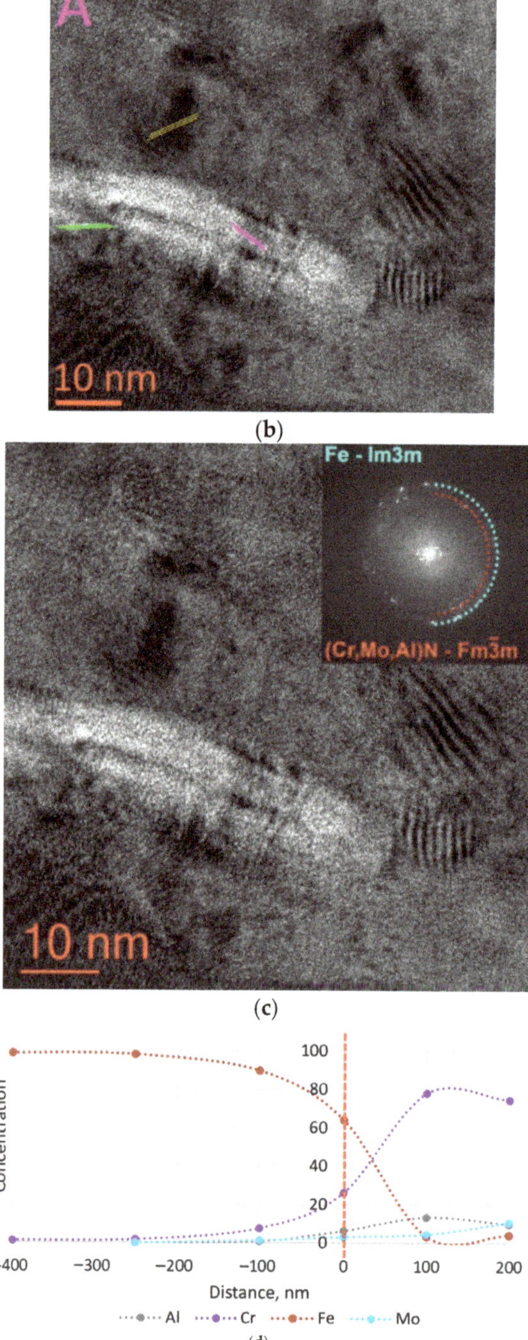

Figure 6. (**a**,**b**) Penetration of a fragment of the machined steel into the structure of the coating, (**c**) Fourier analysis of the area of the "coating-machined steel" interface, (**d**) study of the diffusion processes on the area of the "coating-machined steel" interface (line L1).

The cut material flow has a diverse effect on the structure of the coating. In particular, Figure 7b depict the fracture of a fragment in the coating nanolayer. It can be assumed that later, the fragment could be completely separated from the coating structure and carried away by the cut material flow. Along with the earlier considered mechanism when particles of the machined steel wedge between the nanolayers of the coating, the described mechanism of fracture and separation of fragments from the coating can affect the general pattern of wear of the coating.

Another example of the effect of the cut material flow on the coating structure is the rounding of the ends of the collapsing nanolayers. The described effect results in the formation of a fan-like structure exhibited in Figure 7c.

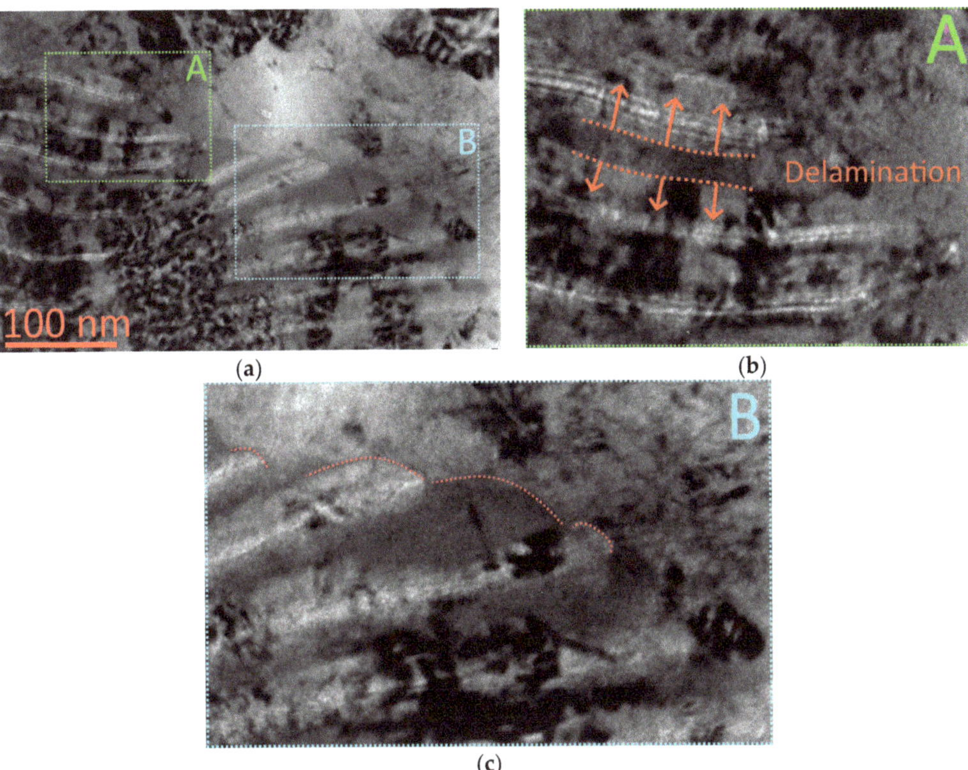

Figure 7. (a–c) Effect of the cut material flow on the nanolayer structure of the coating.

The study of the boundary between the contact of the cut material flow and the coating reveals the presence of the grains of c-(Cr,Mo,Al)N in the coating, while both grains of iron and harder grains of oxide iron (see Figure 8) are detected in the material being machined.

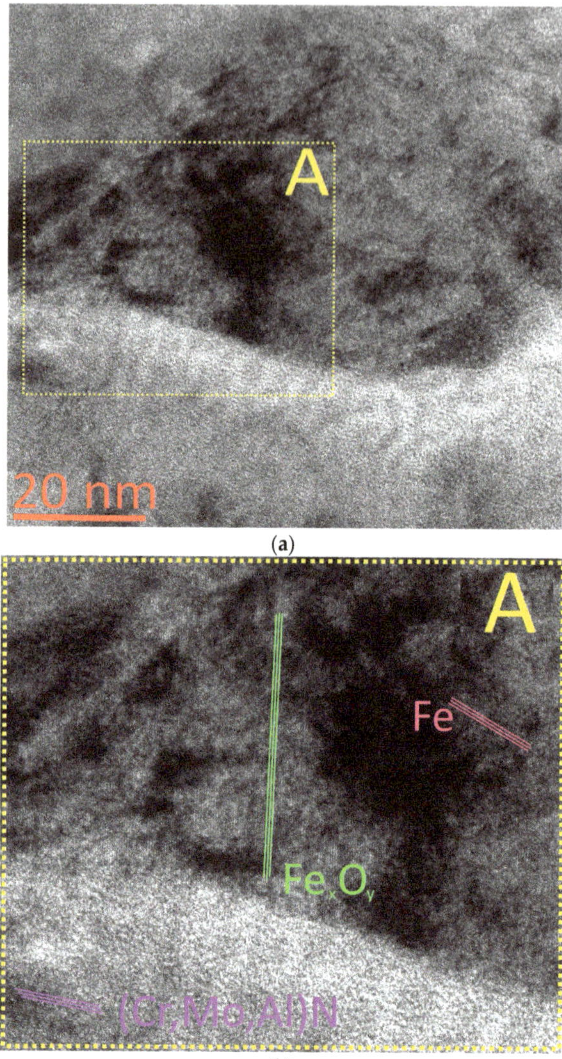

Figure 8. Study of the crystal structure of the area on the boundary between the contact of the cut material layer and the coating. (**a**) General view of the studied coverage area; (**b**) Identification of grains: iron, iron oxide and cubic structure of the coating.

4. Conclusions

- Introduction of 20 at.% Mo into the composition of the (Cr,Al)N coating increases the wear resistance of metal-cutting tools.
- For the coating based on the (Cr,Al)N system, more active growth of wear crater on the rake face is typical compared to the coating based on the (Cr,Mo,Al)N system.
- The analysis of the pattern of wear on the coating with nanolayer structure in contact with the cut material flow reveals the following fracture mechanisms:
- Penetration of particles of the machined steel between nanolayers of the coating, resulting in interlayer delamination;

- Plastic deformation (bending) of nanolayers of the coating under the influence of the moving flow of the machined steel;
- Fracture of fragments from nanolayers of the coating with their further separation and removal by the cut material flow;
- During the cutting, the diffusion of iron into the coating (to the depth not exceeding 200 nm) and the diffusion of coating elements (Cr and Mo) into the machined steel to the depth not exceeding 250 nm occur;
- Along with the particles of iron on the "coating- machined steel" interface, the particles of iron oxides are detected with significantly higher hardness compared to iron, which can thereby more actively affect the wear of the coating.

Author Contributions: Conceptualization, A.V. and A.S.; methodology, A.V. and A.S.; writing—original draft preparation, A.V.; project administration, V.G. All authors have read and agreed to the published version of the manuscript.

Funding: This research was funded by Russian Science Foundation, grant number 20-79-00222.

Institutional Review Board Statement: Not applicable.

Informed Consent Statement: Not applicable.

Acknowledgments: The study used the equipment from the Centre for collective use of Moscow State Technological University STANKIN (agreement No. 075-15-2021-695, 26 July 2021). The coating structure was investigated using the equipment of the Centre for collective use of scientific equipment "Material Science and Metallurgy", purchased with the financial support of the Ministry of Science and Higher Education of the Russian Federation (GK 075-15-2021-696).

Conflicts of Interest: The authors declare no conflict of interest.

References

1. PalDey, S.; Deevi, S.C. Single layer and multilayer wear resistant coatings of (Ti, Al)N: A review. *Mater. Sci. Eng. A* **2003**, *342*, 58–79. [CrossRef]
2. Bobzin, K. High-performance coatings for cutting tools CIRP. *J. Manuf. Sci. Technol.* **2017**, *18*, 1–9. [CrossRef]
3. Ariharan, N.; Sriram, C.G.; Radhika, N.; Aswin, S.; Haridas, S. A comprehensive review of vapour deposited coatings for cutting tools: Properties and recent advances. *Trans. Inst. Met. Finish.* **2022**, *in press*. [CrossRef]
4. Klaus, M.; Genzel, C. Multilayer systems for cutting tools: On the relationship between coating design, surface processing, and residual stress. *Adv. Eng. Mater.* **2011**, *13*, 845–850. [CrossRef]
5. Deng, Y.; Chen, W.; Li, B.; Wang, C.; Kuang, T.; Li, Y. Physical vapor deposition technology for coated cutting tools: A review. *Ceram. Int.* **2020**, *46*, 18373–18390. [CrossRef]
6. Kuzin, V.V.; Grigoriev, S.N.; Volosova, M.A. The role of the thermal factor in the wear mechanism of ceramic tools: Part 1. Macrolevel. *J. Frict. Wear* **2014**, *35*, 505–510. [CrossRef]
7. Grigoriev, S.N.; Gurin, V.D.; Volosova, M.A.; Cherkasova, N.Y. Development of residual cutting tool life prediction algorithm by processing on CNC machine tool. *Mater. Werkst.* **2013**, *44*, 790–796. [CrossRef]
8. Grigoriev, S.; Fominski, V.; Gnedovets, A.; Romanov, R. Experimental and numerical study of the chemical composition of WSex thin films obtained by pulsed laser deposition in vacuum and in a buffer gas atmosphere. *Appl. Surf. Sci.* **2012**, *258*, 7000–7007. [CrossRef]
9. Fominski, V.Y.; Grigoriev, S.N.; Celis, J.P.; Romanov, R.I.; Oshurko, V.B. Structure and mechanical properties of W-Se-C/diamond-like carbon and W-Se/diamond-like carbon bi-layer coatings prepared by pulsed laser deposition. *Thin Solid Film.* **2012**, *520*, 6476–6483. [CrossRef]
10. Grigoriev, S.; Vereschaka, A.; Milovich, F.; Sitnikov, N.; Andreev, N.; Bublikov, J.; Kutina, N. Investigation of the properties of the Cr,Mo-(Cr,Mo,Zr,Nb)N-(Cr,Mo,Zr,Nb,Al)N multilayer composite multicomponent coating with nanostructured wear-resistant layer. *Wear* **2021**, *468–469*, 203597. [CrossRef]
11. Grigoriev, S.; Vereschaka, A.; Milovich, F.; Tabakov, V.; Sitnikov, N.; Andreev, N.; Sviridova, T.; Bublikov, J. Investigation of multicomponent nanolayer coatings based on nitrides of Cr, Mo, Zr, Nb, and Al. *Surf. Coat. Technol.* **2020**, *401*, 126258. [CrossRef]
12. Vereschaka, A.; Grigoriev, S.; Milovich, F.; Sitnikov, N.; Migranov, M.; Andreev, N.; Bublikov, J.; Sotova, C. Investigation of tribological and functional properties of Cr,Mo-(Cr,Mo)N-(Cr,Mo,Al)N multilayer composite coating. *Tribol. Int.* **2021**, *155*, 106804. [CrossRef]
13. Grigoriev, S.; Vereschaka, A.; Milovich, F.; Andreev, N.; Bublikov, J.; Sitnikov, N.; Sotova, C.; Kutina, N. Investigation of wear mechanisms of multilayer nanostructured wear-resistant coatings during turning of steel. Part 2: Diffusion, oxidation processes and cracking in Ti-TiN-(Ti,Cr,Mo,Al)N coating. *Wear* **2021**, *486–487*, 204096. [CrossRef]

14. Mayrhofer, P.H.; Mitterer, C.; Hultman, L.; Clemens, H. Microstructural design of hard coatings. *Prog. Mater. Sci.* **2006**, *51*, 1032–1114. [CrossRef]
15. Yamamoto, K.; Sato, T.; Takahara, K.; Hanaguri, K. Properties of (Ti,Cr,Al)N coatings with high Al content deposited by new plasma enhanced arc-cathode. *Surf. Coat. Technol.* **2003**, *174–175*, 620–626. [CrossRef]
16. Vereschaka, A.; Tabakov, V.; Grigoriev, S.; Sitnikov, N.; Milovich, F.; Andreev, N.; Sotova, C.; Kutina, N. Investigation of the influence of the thickness of nanolayers in wear-resistant layers of Ti-TiN-(Ti,Cr,Al)N coating on destruction in the cutting and wear of carbide cutting tools. *Surf. Coat. Technol.* **2020**, *385*, 125402. [CrossRef]
17. Vereschaka, A.; Milovich, F.; Migranov, M.; Andreev, N.; Alexandrov, I.; Muranov, A.; Mikhailov, M.; Tatarkanov, A. Investigation of the tribological and operational properties of $(Me_x,Mo_y,Al_{1-(x+y)})N$ (Me–Ti, Zr or Cr) coatings. *Tribol. Int.* **2022**, *165*, 107305. [CrossRef]
18. Fox-Rabinovich, G.S.; Yamomoto, K.; Veldhuis, S.C.; Kovalev, A.I.; Dosbaeva, G.K. Tribological adaptability of TiAlCrN PVD coatings under high performance dry machining conditions. *Surf. Coat. Technol.* **2005**, *200*, 1804–1813. [CrossRef]
19. Mayrhofer, P.H.; Willmann, H.; Reiter, A.E. Structure and phase evolution of Cr-Al-N coatings during annealing. *Surf. Coat. Technol.* **2008**, *202*, 4935–4938. [CrossRef]
20. Reiter, A.E.; Derflinger, V.H.; Hanselmann, B.; Bachmann, T.; Sartory, B. Investigation of the properties of Al1-xCrxN coatings prepared by cathodic arc evaporation. *Surf. Coat. Technol.* **2005**, *200*, 2114–2122. [CrossRef]
21. Wistrela, E.; Schmied, I.; Schneider, M.; Gillinger, M.; Mayrhofer, P.M.; Bittner, A.; Schmid, U. Impact of sputter deposition parameters on the microstructural and piezoelectric properties of $Cr_xAl_{1-x}N$ thin films. *Thin Solid Film.* **2018**, *648*, 76–82. [CrossRef]
22. Bobzin, K.; Brögelmann, T.; Kruppe, N.C.; Engels, M. Correlation of the Debye sheath thickness and (Cr,Al)N coating properties for HPPMS, dcMS, CAE and PCAE processes. *Surf. Coat. Technol.* **2017**, *332*, 233–241. [CrossRef]
23. Hu, C.; Xu, Y.X.; Chen, L.; Pei, F.; Zhang, L.J.; Du, Y. Structural, mechanical and thermal properties of CrAlNbN coatings. *Surf. Coat. Technol.* **2018**, *349*, 894–900. [CrossRef]
24. Hu, C.; Chen, L.; Moraes, V. Structure, mechanical properties, thermal stability and oxidation resistance of arc evaporated CrAlBN coatings. *Surf. Coat. Technol.* **2021**, *417*, 127191. [CrossRef]
25. Chen, Y.; Zhang, Z.; Yuan, T.; Mei, F.; Lin, X.; Gao, J.; Chen, W.; Xu, Y. The synergy of V and Si on the microstructure, tribological and oxidation properties of AlCrN based coatings. *Surf. Coat. Technol.* **2021**, *412*, 127082. [CrossRef]
26. Wang, R.; Mei, H.; Li, R.; Zhang, T.; Wang, Q. Influence of V addition on the microstructure, mechanical, oxidation and tribological properties of AlCrSiN coatings. *Surf. Coat. Technol.* **2021**, *407*, 126767. [CrossRef]
27. Hu, C.; Xu, Y.X.; Chen, L.; Pei, F.; Du, Y. Mechanical properties, thermal stability and oxidation resistance of Ta-doped CrAlN coatings. *Surf. Coat. Technol.* **2019**, *368*, 25–32. [CrossRef]
28. Miyake, T.; Kishimoto, A.; Hasegawa, H. Tribological properties and oxidation resistance of (Cr,Al,Y)N and (Cr,Al,Si)N films synthesized by radio-frequency magnetron sputtering method. *Surf. Coat. Technol.* **2010**, *205* (Suppl. 1), S290–S294. [CrossRef]
29. Zhao, Y.; Feng, K.; Yao, C.; Li, Z. Effect of MoO_3 on the microstructure and tribological properties of laser-clad Ni60/nanoCu/h-BN/MoO_3 composite coatings over wide temperature range. *Surf. Coat. Technol.* **2020**, *387*, 125477. [CrossRef]
30. Tao, H.; Tsai, M.T.; Chen, H.W.; Huang, J.C.; Duh, J.G. Improving high-temperature tribological characteristics on nanocomposite CrAlSiN coating by Mo doping. *Surf. Coat. Technol.* **2018**, *349*, 752–756. [CrossRef]
31. Lu, Y.C.; Chen, H.W.; Chang, C.C.; Wu, C.Y.; Duh, J.G. Tribological properties of nanocomposite Cr-Mo-Si-N coatings at elevated temperature through silicon content modification. *Surf. Coat. Technol.* **2018**, *338*, 69–74. [CrossRef]
32. Li, F.; Zhu, S.; Cheng, J.; Qiao, Z.; Yang, J. Tribological properties of Mo and CaF_2 added SiC matrix composites at elevated temperatures. *Tribol. Int.* **2017**, *111*, 46–51. [CrossRef]
33. Qi, D.; Chen, J.; Liu, J.; Lv, W.; Song, J.; Shen, L. Influence of Molybdenum Addition on Oxidation Resistance of CrN Coatings. *Rare Metal Mater. Eng.* **2021**, *50*, 1505–1512.
34. Qi, D.; Lei, H.; Fan, D.; Pei, Z.; Gong, J.; Sun, C. Effect of Mo content on the microstructure and properties of CrMoN composite coatings. *Acta Metall. Sin.* **2015**, *51*, 371–377.
35. Wang, Y.; Tang, Y.; Wan, W.; Zhang, X. High-Temperature Oxidation Resistance of CrMoN Films. *J. Mater. Eng. Perform.* **2020**, *29*, 6412–6416. [CrossRef]
36. Amaya-Roncancio, S.; Arias-Mateus, D.F.; Gómez-Hermida, M.M.; Riaño-Rojas, J.C.; Restrepo-Parra, E. Molecular dynamics simulations of the temperature effect in the hardness on Cr and CrN films. *Appl. Surf. Sci.* **2012**, *258*, 4473–4477. [CrossRef]
37. Cai, Z.; Zhang, P.; Di, Y. Microstructure, hardness and oxidation resistance of CrN and CrAlN coatings synthesized by multi-arc ion plating technology. *Adv. Mater. Res.* **2011**, *168–170*, 2430–2433. [CrossRef]
38. Zeilinger, A.; Daniel, R.; Schöberl, T.; Stefenelli, M.; Sartory, B.; Keckes, J.; Mitterer, C. Resolving depth evolution of microstructure and hardness in sputtered CrN film. *Thin Solid Film.* **2015**, *581*, 75–79. [CrossRef]
39. Ichimura, H.; Ando, I. Mechanical properties of arc-evaporated CrN coatings: Part I—Nanoindentation hardness and elastic modulus. *Surf. Coat. Technol.* **2001**, *145*, 88–93. [CrossRef]
40. Tian, C.X.; Han, B.; Zou, C.W.; Xie, X.; Li, S.Q.; Liang, F.; Tang, X.S.; Wang, Z.S.; Pelenovich, V.O.; Zeng, X.M.; et al. Synthesis of monolayer MoNx and nanomultilayer CrN/Mo2N coatings using arc ion plating. *Surf. Coat. Technol.* **2019**, *370*, 125–129. [CrossRef]

41. Shirazi, M.; Ghasemloo, M.; Etaati, G.R.; Hosseinnejad, M.T.; Toroghinejad, M.R. Plasma focus method for growth of molybdenum nitride thin films: Synthesis and thin film characterization. *J. Alloys Compd.* **2017**, *727*, 978–985. [CrossRef]

42. Wang, J.; Munroe, P.; Zhou, Z.; Xie, Z. Nanostructured molybdenum nitride-based coatings: Effect of nitrogen concentration on microstructure and mechanical properties. *Thin Solid Film.* **2019**, *682*, 82–92. [CrossRef]

43. Wang, T.; Zhang, G.; Ren, S.; Jiang, B. Effect of nitrogen flow rate on structure and properties of MoN_x coatings deposited by facing target sputtering. *J. Alloys Compd.* **2017**, *701*, 1–8. [CrossRef]

44. Klimashin, F.; Koutná, N.; Euchner, H.; Holec, D.; Mayrhofer, P. The impact of nitrogen content and vacancies on structure and mechanical properties of Mo–N thin films. *J. Appl. Phys.* **2016**, *120*, 185301. [CrossRef]

45. Vetter, J.; Eriksson, A.O.; Reiter, A.; Derflinger, V.; Kalss, W. Quo vadis: Alcr-based coatings in industrial applications. *Coatings* **2021**, *11*, 344. [CrossRef]

46. Yoon, C.S.; Kim, K.H.; Kwon, S.H.; Park, I.W. Syntheses and properties of Cr-Al-Mo-N coatings fabricated by using a hybrid coating system. *J. Korean Phys. Soc.* **2009**, *54*, 1237–1241. [CrossRef]

47. Klimashin, F.F.; Mayrhofer, P.H. Ab initio-guided development of super-hard Mo-Al-Cr-N coatings. *Scr. Mater.* **2017**, *140*, 27–30. [CrossRef]

48. Bobzin, K.; Brögelmann, T.; Kalscheuer, C.; Stahl, K.; Lohner, T.; Yilmaz, M. Effects of (Cr,Al)N and (Cr,Al,Mo)N coatings on friction under minimum quantity lubrication. *Surf. Coat. Technol.* **2020**, *402*, 126154. [CrossRef]

49. Wang, Y.; Lou, B. Microstructure and High-Temperature Friction and Wear Properties of CrAlMoN Film. *Oxid. Met.* **2021**, *95*, 239–250. [CrossRef]

50. Iram, S.; Cai, F.; Wang, J.; Zhang, J.; Liang, J.; Ahmad, F.; Zhang, S. Effect of addition of Mo or V on the structure and cutting performance of AlCrN-based coatings. *Coatings* **2020**, *10*, 298. [CrossRef]

51. Bobzin, K.; Brögelmann, T.; Kalscheuer, C. Arc PVD (Cr,Al,Mo)N and (Cr,Al,Cu)N coatings for mobility applications. *Surf. Coat. Technol.* **2020**, *384*, 125046. [CrossRef]

52. Iram, S.; Wang, J.; Cai, F.; Zhang, J.; Ahmad, F.; Liang, J.; Zhang, S. Effect of bilayer number on mechanical and wear behaviours of the AlCrN/AlCrMoN coatings by AIP method. *Surf. Eng.* **2021**, *37*, 536–544. [CrossRef]

53. Grigoriev, S.; Vereschaka, A.; Milovich, F.; Sitnikov, N.; Andreev, N.; Bublikov, J.; Sotova, C.; Oganian, G.; Sadov, I. Investigation of the properties of Ti-TiN-(Ti,Cr,Mo,Al)N multilayered composite coating with wear-resistant layer of nanolayer structure. *Coatings* **2020**, *10*, 1236. [CrossRef]

54. Vereshchaka, A.A.; Vereshchaka, A.S.; Mgaloblishvili, O.; Morgan, M.N.; Batako, A.D. Nano-scale multilayered-composite coatings for the cutting tools. *Int. J. Adv. Manuf. Technol.* **2014**, *72*, 303–317. [CrossRef]

55. Vereschaka, A.; Tabakov, V.; Grigoriev, S.; Aksenenko, A.; Sitnikov, N.; Oganyan, G.; Seleznev, A.; Shevchenko, S. Effect of adhesion and the wear-resistant layer thickness ratio on mechanical and performance properties of ZrN-(Zr,Al,Si)N coatings. *Surf. Coat. Technol.* **2019**, *357*, 218–234. [CrossRef]

56. Metel, A.S.; Grigoriev, S.N.; Melnik, Y.A.; Bolbukov, V.P. Characteristics of a fast neutral atom source with electrons injected into the source through its emissive grid from the vacuum chamber. *Instrum. Exp. Tech.* **2012**, *55*, 288–293. [CrossRef]

57. Metel, A.; Grigoriev, S.; Melnik, Y.; Panin, V.; Prudnikov, V. Cutting Tools Nitriding in Plasma Produced by a Fast Neutral Molecule Beam. *Jpn. J. Appl. Phys.* **2011**, *50*, 08JG04. [CrossRef]

58. Metel, A.S.; Grigoriev, S.N.; Melnik, Y.A.; Bolbukov, V.P. Broad beam sources of fast molecules with segmented cold cathodes and emissive grids. *Instrum. Exp. Tech.* **2012**, *55*, 122–130. [CrossRef]

59. Grigoriev, S.N.; Metel, A.S.; Fedorov, S.V. Modification of the structure and properties of high-speed steel by combined vacuum-plasma treatment. *Met. Sci. Heat Treat.* **2012**, *54*, 8–12. [CrossRef]

60. Fominski, V.Y.; Grigoriev, S.N.; Gnedovets, A.G.; Romanov, R.I. Pulsed laser deposition of composite Mo–Se–Ni–C coatings using standard and shadow mask configuration. *Surf. Coat. Technol.* **2012**, *206*, 5046–5054. [CrossRef]

61. Grigoriev, S.N.; Melnik, Y.A.; Metel, A.S.; Panin, V.V.; Prudnikov, V.V. A compact vapor source of conductive target material sputtered by 3-keV ions at 0.05-Pa pressure. *Instrum. Exp. Tech.* **2009**, *52*, 731–737. [CrossRef]

62. Grigoriev, S.; Vereschaka, A.; Zelenkov, V.; Sitnikov, N.; Bublikov, J.; Milovich, F.; Andreev, N.; Mustafaev, E. Specific features of the structure and properties of arc-PVD coatings depending on the spatial arrangement of the sample in the chamber. *Vacuum* **2022**, *200*, 111047. [CrossRef]

63. Oliver, W.C.; Pharr, G.M.J. An improved technique for determining hardness and elastic modulus using load and displacement sensing indentation. *J. Mater. Res.* **1992**, *7*, 1564–1583. [CrossRef]

64. *ASTM C1624-05(2015)*; Standard Test Method for Adhesion Strength and Mechanical Failure Modes of Ceramic Coatings by Quantitative Single Point Scratch Testing. ASTM International: West Conshohocken, PA, USA, 2010. [CrossRef]

65. Grigoriev, S.; Vereschaka, A.; Zelenkov, V.; Sitnikov, N.; Bublikov, J.; Milovich, F.; Andreev, N.; Sotova, C. Investigation of the influence of the features of the deposition process on the structural features of microparticles in PVD coatings. *Vacuum* **2022**, *202*, 111144. [CrossRef]

66. Mohamed, Y.S.; El-Gamal, H.; Zaghloul, M.M.Y. Micro-hardness behavior of fiber reinforced thermosetting composites embedded with cellulose nanocrystals. *Alex. Eng. J.* **2018**, *57*, 4113–4119. [CrossRef]

67. Zaghloul, M.M.Y.; Zaghloul, M.Y.M.; Zaghloul, M.M.Y. Experimental and modeling analysis of mechanical-electrical behaviors of polypropylene composites filled with graphite and MWCNT fillers. *Polym. Test.* **2017**, *63*, 467–474. [CrossRef]

68. Zaghloul, M.M.Y.; Steel, K.; Veidt, M.; Heitzmann, M.T. Wear behaviour of polymeric materials reinforced with man-made fibres: A comprehensive review about fibre volume fraction influence on wear performance. *J. Reinf. Plast. Compos.* **2022**, *41*, 215–241. [CrossRef]

69. Mahmoud Zaghloul, M.Y.; Yousry Zaghloul, M.M.; Yousry Zaghloul, M.M. Developments in polyester composite materials—An in-depth review on natural fibres and nano fillers. *Compos. Struct.* **2021**, *278*, 114698. [CrossRef]

70. Zaghloul, M.M.Y.M. Mechanical properties of linear low-density polyethylene fire-retarded with melamine polyphosphate. *J. Appl. Polym. Sci.* **2018**, *135*, 46770. [CrossRef]

71. Zaghloul, M.M.Y.; Zaghloul, M.M.Y. Influence of flame retardant magnesium hydroxide on the mechanical properties of high density polyethylene composites. *J. Reinf. Plast. Compos.* **2017**, *36*, 1802–1816. [CrossRef]

72. Varga, G.; Ferencsik, V. Analysis of Cylindricity Error of High and Low Temperature Storage Tested Alternator Stators. *Int. J. Automot. Technol.* **2020**, *21*, 1519–1526. [CrossRef]

73. Kundrák, J.; Pálmai, Z.; Varga, G. Analysis of Tool Life Functions in Hard Turning. *Teh. Vjesn.* **2020**, *27*, 166–173.

74. Karpuschewski, B.; Kundrák, J.; Felhő, C.; Varga, G.; Borysenko, D. Effects of the tool edge design on the roughness of face milled surfaces. *Iop Conf. Ser. Mater. Sci. Eng.* **2018**, *448*, 012056. [CrossRef]

75. Astakhov, V.P. Tribology of Cutting Tools, Chapter 1. In *Tribology in Manufacturing Technology*; Springer: Dordrecht, The Netherlands, 2013; pp. 1–66.

76. Gates, J.D. Two-body and three-body abrasion: A critical discussion. *Wear* **1998**, *214*, 139–146. [CrossRef]

77. Moore, M.A. Abrasive wear. *Mater. Eng. Appl.* **1978**, *1*, 97–111. [CrossRef]

78. Torrance, A.A. An explanation of the hardness differential needed for abrasion. *Wear* **1981**, *68*, 263–266. [CrossRef]

79. Coronado, J.J.; Rodríguez, S.A.; Sinatora, A. Effect of particle hardness on mild-severe wear transition of hard second phase materials. *Wear* **2013**, *301*, 82–88. [CrossRef]

Article

Investigation of the Properties of Multilayer Nanostructured Coating Based on the (Ti,Y,Al)N System with High Content of Yttrium

Sergey Grigoriev [1], Alexey Vereschaka [2,*], Filipp Milovich [3], Nikolay Sitnikov [4], Jury Bublikov [2], Anton Seleznev [1], Catherine Sotova [1] and Alexander Rykunov [5]

[1] Department of High-Efficiency Processing Technologies (VTO), Moscow State University of Technology, STANKIN, Vadkovsky per. 1, 127055 Moscow, Russia

[2] Institute of Design and Technological Informatics, Russian Academy of Sciences (IDTI RAS), 127055 Moscow, Russia

[3] Materials Science and Metallurgy Shared Use Research and Development Center, National University of Science and Technology MISiS, Leninsky Prospect 4, 119049 Moscow, Russia

[4] State Scientific Center, Russian Federation "Keldysh Research Center", 8, Onezhskaya Str., 125438 Moscow, Russia

[5] Federal State-Financed Educational Institution, High Professional Education P. A. Solovyov Rybinsk State Aviation Technical University, St. Pushkina, 53, 152934 Rybinsk, Russia

* Correspondence: dr.a.veres@yandex.ru

Abstract: The studies are focused on the properties of the multilayer composite coating based on the (Ti,Y,Al)N system with high content of yttrium (about 40 at.%) of yttrium (Y). The hardness and elastic modulus were defined, and the resistance to fracture was studied during the scratch testing. Two cubic solid solutions (fcc phases), including c-(Ti,Y,Al)N and c-(Y,Ti,Al)N, are formed in the coating. The investigation of the wear resistance of the (Ti,Y,Al)N-coated tools during the turning of steel in comparison with the wear resistance of the tools with the based on the (Ti,Cr,Al)N system coating and the uncoated tools found a noticeable increase (by 250%–270%) in rake wear resistance. Active oxidation processes are observed in the (Ti,Y,Al)N coating during wear. It can be assumed that yttrium oxide is predominantly formed with a possible insignificant formation of titanium and aluminum oxides. At the same time, complete oxidation of c-(Y,Ti,Al)N nanolayers is not observed. Some hypotheses explaining the rather high performance of a coating with a high yttrium content are considered.

Keywords: nanostructured coatings; yttrium nitride; tool wear; oxidation; metal cutting

Citation: Grigoriev, S.; Vereschaka, A.; Milovich, F.; Sitnikov, N.; Bublikov, J.; Seleznev, A.; Sotova, C.; Rykunov, A. Investigation of the Properties of Multilayer Nanostructured Coating Based on the (Ti,Y,Al)N System with High Content of Yttrium. *Coatings* **2023**, *13*, 335. https://doi.org/10.3390/coatings13020335

Academic Editor: Gianni Barucca

Received: 9 December 2022
Revised: 24 January 2023
Accepted: 30 January 2023
Published: 1 February 2023

1. Introduction

The increasing requirements for tool materials in general and for coatings in particular predetermine the need to find ways for further improvement in their properties [1–5]. Coatings with a nanolayer structure, which combines nanolayers with different properties, are widely used. In particular, it has been found that plastic flow in a (Cr,Al)N nanolayer coating with alternating layers dominated by CrN and AlN is not associated with the movement of classical dislocations, but occurs as a result of rotation of nanosized grains and sliding along the boundary for larger grains [6]. Changing the parameters of the nanolayer structure of the (Ti,Cr,Al)N coating makes it possible to control its properties [7,8].

It is known that YN forms a cubic crystal fcc lattice with the parameter of a = 4.88 Å [9]. In the Y-containing coatings, yttrium can either form its own nitride phase of YN or enter into the composition of other phases (for example, as an element of a substitutional solid solution). For example, no single (Y,Al)N phase is detected in the system of (Y,Al)N, and AlN reacts with Y to form the phases of YN and YAl$_2$ [10,11]. When the coating, which includes the (Al,Y)N phase, is heated to 600–1000 °C upon contact with the atmosphere, oxides Al_2O_3 and Y_2O_3 are formed on the surface [12]. The introduction of Y in the composition of nitride coating increases its resistance to oxidation. The temperature of the onset of active oxidation of the (Nb,Y)N coating, which was 500 °C for a system without the addition of yttrium, increases to 760 °C at a Y content of 12.1 at.% [13]. An increase in the yttrium content in the nitride coating leads to the formation of an amorphous microstructure with a finer and denser morphology and a smoother surface [14].

When Y is added to the (Ti,Cr,Al)N system, its heat resistance increased to 1200 °C [15]. Upon heating up to 1000 °C, the number of oxides formed in the Y-containing coating was half the size in comparison with the Y-free coating. When Y is added to the (Cr,Al)N system, the cohesive energy increases (the energy of formation decreases) due to the substitution of Cr atoms for Y atoms in the crystal lattice, and, as a result, the phase stability increases [16]. The described effects contribute to a noticeable growth of the wear resistance of the coating, especially when the coating is exposed to high temperatures [17].

The addition of Y to the coating composition leads to a decrease in the grain size [18]. The grain size decreases from 76 to 21 nm as the Y content increases from 0 to 5.8 at.%. Upon heating to the temperature of 1200 °C, the (Cr,Al)N coating demonstrates the growth of grains from the initial size of 30–40 nm to 100 nm. In the (Cr,Al,Y)N coating, the grain sizes remain at the level of 30–40 nm at the temperature of 1200 °C. The great hardness of the (Cr,Al,Y)N coating is retained at temperatures up to 1100 °C [13]. The addition of Y to the composition of nitride coatings also noticeably slows down the diffusion processes [13,19].

Thus, the introduction of yttrium into the composition of the coatings can significantly enhance the cutting properties of the tools due to the improvement in the tribological properties and resistance to oxidation and diffusion, which may be explained by the formation of complex oxide films in the cutting zone [10,15,20–25].

Several studies consider the properties of the (Ti,Al,Y)N coating. It has been found that the introduction of yttrium into the (Al,Ti)N coating leads to a decrease in the average grain size and compaction of the microstructure, as well as an increase in the oxidation resistance of the coating [26,27]. The introduction of yttrium also increases heat resistance and hardness at high temperatures (above 1000 °C) [28]. Yttrium slows down the decomposition of the $Ti_{1-x-y}Al_xY_yN$ solid solution and contributes to the formation in the coating of c-TiN, c-YN, and w-AlN phases. The introduction of Y reduces the compressive stress, increases the hardness, and also enhances the predominant formation of dense Al_2O_3 oxide films instead of porous TiO_2 oxide [12]. In general, the introduction of yttrium increases the resistance to oxidation [29,30]. It is found that the wear resistance of the (Ti,Al,Y)N coating grows with an increase in the Y content [31]. At the same time, the average rates of abrasive wear of the (Ti,Al,Y)N coating were 3–5 times lower in comparison with the (Ti,Al)N coating and 10 times lower in comparison with the TiN coating. In [31], such properties of the (Ti,Al,Y)N coating are associated with the formation of a dense structure with nanosized grains and a decrease in surface roughness and the level of internal stresses.

It should be noted that almost all of the above studies consider the coatings with the low (1–2 at%) content of yttrium. Such choice may be explained by the fact that yttrium nitride (YN) hydrolyzes upon contact with water and forms Y_2O_3 oxide upon heating in an oxygen-containing atmosphere [32–35]. The hardness of yttrium oxide is significantly lower in comparison with the hardness of YN or TiN (hardness of Y_2O_3 stays in a range from 3 to 8 GPa [36]). At the same time, Y_2O_3 is characterized by a very high thermal stability (up to 2100 °C) and rather low thermal conductivity, which makes it possible to effectively use it as a barrier to heat fluxes in the coating [37]. A possible useful property of YN and Y_2O_3 is their relatively high fracture toughness [38,39].

Of particular interest could be an investigation focused on the properties of a coating in which the layers with great hardness and wear resistance (for example, the layers based on the (Ti,Al)N system) would be alternated with the layers with good barrier properties combined with high fracture toughness (for example, the layers based on YN). It has been established that yttrium and yttrium nitride increase the fracture toughness when they are introduced into the composition of alloys and nitride phases [40–44].

Fracture toughness increases significantly (from 0.1 to 1.0 MPa.m$^{1/2}$) with an increase in the yttrium content from 0 to 20 at.% in the TiN-based coating [39].

The alternation of the mentioned nanolayers will protect YN-based layers from rapid oxidation, while allowing the formation of tribologically active Y_2O_3 oxides that have a positive effect on the conditions in the cutting zone [22–24].

It has been found that the introduction of up to 7.8 at.% Y into TiN contributes to the formation of the dominant fcc phase of the c (Ti,Y)N solid solution, with the formation of a mixture of c-TiN and c-YN fcc phases upon an increase in the yttrium content [39,45]. The hardness increases from 21 GPa for the yttrium-free coating to 26 GPa for the coating with 10.2 at.% yttrium. The fracture toughness of the coating increases with a growth of the yttrium content.

An additional positive effect is provided by the application of a multilayer coating architecture, which includes an adhesive layer with a substrate, a transition layer, and a wear-resistant layer [46–50]. In this case, the wear-resistant layer has a nanolayer structure with alternating nanolayers characterized by different mechanical properties [50–52].

Therefore, the Ti-TiN-(Ti,Y,Al)N coating was chosen for the study. This coating has a three-layer architecture [46,53], including the Ti adhesive layer, the TiN transition layer, and the (Ti,Y,Al)N wear-resistant layer. The (Ti,Y,Al)N wear-resistant layer has a nanolayer structure with the modulation period λ of about 60 nm. Due to the yttrium content in the wear-resistant layer (about 40 at.%), the formation of two phases (c-(Ti,Al)N and c-YN) can be predicted.

The object of comparison was the Ti-TiN-(Ti,Cr,Al)N coating with similar architecture parameters. This composition was chosen due to the wide application of the coatings based on the (Ti,Cr,Al)N system in the manufacturing of metal-cutting tools [10,54–58].

For convenience, the coatings under comparison will be further designated only by their wear-resistant layer: (Ti,Y,Al)N and (Ti,Cr,Al)N, respectively.

2. Materials and Methods

2.1. Sample Preparation

The deposition of the coatings under comparison was carried out on the specialized PVD unit of VIT-2 [46,48–50,59,60] (IDTI RAS–MSTU STANKIN, Moscow, Russia), using the filtered cathodic vacuum arc deposition (FCVAD) system (IDTI RAS–MSTU STANKIN, Moscow, Russia) [48–50] for the aluminum cathode and the Controlled Accelerated Arc system (CAA-PVD) (IDTI RAS–MSTU STANKIN, Moscow, Russia) [61,62] –for other cathodes. A detailed circuit diagram and layout of the VIT-2 unit is presented in [46].

The following cathodes were used for coating deposition: Al (99.80%), Cr (99.90%), Ti (99.60%), and Y (99.98%). Due to the tendency of yttrium to hydrolyze at high temperatures, the yttrium cathode had a copper mounting part with a cooling system to prevent any contact of yttrium with water. Chemically pure nitrogen of high purity (grade 6.0) was used. Volume fraction of nitrogen, % not less than 99.99990.

The substrate used was carbide inserts SNUN ISO 1832:2012, 12.00 mm × 12.00 mm × 4.75 mm (WC + 15% TiC + 6% Co) (KZTS, Kirovograd, Russia). Each type of coating was deposited on 8 samples.

Before coating deposition, preparation procedures were carried out, including:

- Pre-washing when exposed to ultrasound in a special solution;
- Washing in purified water;
- Drying and wiping with pure medical alcohol;
- After that, the samples were placed on the special fixture and loaded into the chamber of the unit;
- Before coating deposition, the samples are finely cleaned and thermally activated in a gas and metal plasma flow, with the following process parameters: gas (Ar) pressure = 2.0 Pa, cathode arc current = 110 A, voltage on substrate U = 110 V;
- After the above stages of preparation, coatings were deposited on the samples;
- Key characteristics of the coating process are presented in Table 1.

Table 1. Main parameters of the coating deposition process.

Nitrogen Pressure (Pa)	Voltage on Substrate U (V)	Cathode Arc Current (A)			
		Al	Ti	Cr	Y
0.42	−150 DC	160	110	75	85

The rotation speed of the turntable was 0.7 rpm [7,48].

The surface temperature of the substrate during the deposition of the coating was 650–700 °C.

In the deposition of adhesive and transition layers, only the Ti cathode was used; deposition took place in an argon and nitrogen atmosphere, respectively. When depositing (Ti,Y,Al)N wear-resistant layer, three cathodes were used: Ti, Y, and Al.

2.2. Characterization

After coating deposition, characteristics such as hardness, modulus of elasticity, and fracture strength of the coatings during the scratch testing are investigated.

A CB-500 tester (Nanovea, Irvine, CA, USA) with a nanomodulus was used to measure the elastic modulus and hardness of the samples. A precision piezoelectric drive and a high sensitivity load cell were used. Instrumental indentation with a Berkovich indenter was used at a load of 50 mN. Since the study of the hardness of the coating as a thin object presents a certain difficulty, 20 measurements were carried out for each sample, then the average values were determined and the deviations of the values were indicated.

The scratch resistance was determined on a Nanovea instrument (Nanovea, Irvine, CA, USA) according to the ASTM C1624-05 method [63]. The indenter was a Rockwell C Diamond with a tip radius R = 100 μm and a taper angle of 120 degrees. The tests were carried out with a linearly increasing load from 0.2 to 40 N at a loading rate of 5 N/min. After microfurrow formation, dimensions were measured using an optical SX45 stereo microscope (Vision Engineering Ltd., Woking, UK). The moment of complete destruction of the coating corresponds to the load L_{c2}.

Then the structural characteristics and composition (elemental and phase) of the coatings were studied. To study the nanostructure of the coatings, a transmission electron microscope (TEM) JEM 2100 (JEOL, Tokyo, Japan) at an accelerating voltage of 200 kV was used. The elemental composition was determined with the EDX (energy-dispersive X-ray) system of INCAEnergy (OXFORD Instruments, Oxford, UK). Samples (lamellas) were cut out using Strata focused ion beam (FIB) 205 (FEI, Hillsboro, OR, USA).

A scanning electron microscope (SEM) Quanta 600 FEG (FEI, Hillsboro, OR, USA) was also used. Preparation of samples for metallographic studies was carried out on equipment for the production of metallographic sections Isomet 1000 (Bühler AG, Uzwil, Switzerland), automatic hydraulic press for hot fitting Simplimet 1000 (Bühler AG, Uzwil, Switzerland), automatic grinding and polishing machine EcoMet 250 and AutoMet 250 (Bühler AG, Uzwil, Switzerland).

Then, the wear resistance of the coated cutting tool during turning is investigated. The wear resistance of the samples under consideration was studied during the turning of Inconel 718 at a CU 500 MRD lathe (Sliven, Bulgaria) with a ZMM CU 500 MRD variable-speed drive, without cutting fluid (dry cutting), with the cutting parameters as follows: $\gamma = -7°$, $\alpha = 7°$, $\lambda = 0$, r = 0.4 mm; at cutting mode: f = 0.25 mm/rev, a_p = 1.0 mm, and v_c = 300 m/min. The flank wear VB_{max} = 0.3 mm was assumed as the criterion for the limit wear of the tools. Each test was repeated 5 times, and error bars are depicted in the graph.

To consider the wear process on the (Ti,Y,Al)N coating, a transverse section was made passing through the cutting edge, in parallel to the minor flank face (Figure 1). A flank wear land can be noticed on the transverse section, and there is no wear crater on the rake face.

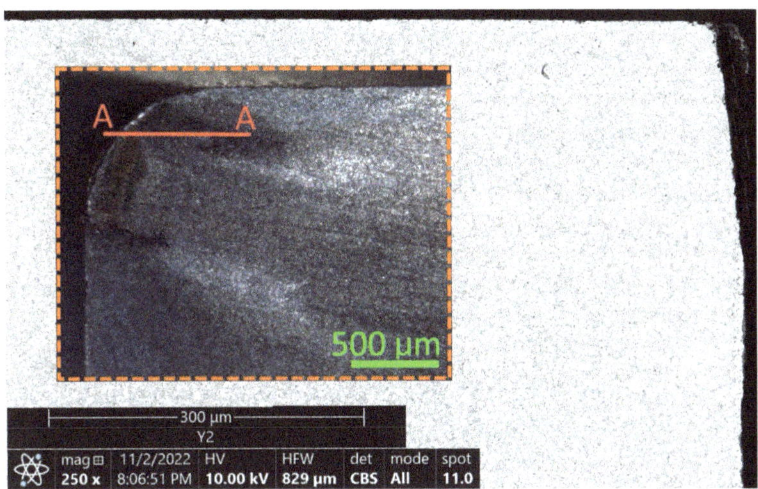

Figure 1. Transverse section of a worn insert (SEM); (center) view of a worn insert coated with (Ti,Y,Al)N and the designation of the A-A transverse section plane.

3. Results and Discussion

At the elastic modulus of 356 ± 24 GPa, the hardness of the (Ti,Y,Al)N coating is considerably high (HV 2758 ± 78). The reference coating of (Ti,Cr,Al)N has the hardness of HV 3182 ± 53 and the elastic modulus of 438 ± 32 GPa. The compositions of the (Ti,Y,Al)N coating and the (Ti,Cr,Al)N reference coating are exhibited in Table 2. It can be noted that the compositions of the coatings under comparison in terms of the Ti and Al content are quite close (taking into account the gradient distribution of elements over the coating thickness). The study of the fracture strength of the coatings during the scratch testing reveals that both coatings have good adhesion to the substrate and that the point of complete coatings failure of L_{C2} was not reached at the load of 50 N.

Table 2. Compositions of the coatings under comparison.

Coating	Element Content, at.%			
	Ti	Cr	Al	Y
(Ti,Cr,Al)N	63.75	23.98	12.27	-
(Ti,Y,Al)N	51.82	-	7.69	40.49

Figure 2a,b depict the nanolayer structure of the coatings under comparison. The (Ti,Y,Al)N coating exhibits noticeable differences in the structural features of its nanolayers, while the structure of the (Ti,Cr,Al)N coating is much more homogeneous. The value of the modulation period λ is 65 nm for the (Ti,Y,Al)N coating and 53 nm for the (Ti,Cr,Al)N coating. These values are quite close and are in the range of values that provide the optimal properties of the coating [7,48]. In terms of the phase composition (Figure 2c,d), there is a noticeable difference between the coatings under comparison. While a single fcc phase of the c-(Ti,Cr,Al)N solid solution is formed in the (Ti,Cr,Al)N coating, then two cubic solid solutions (fcc phases of c-(Ti,Y,Al)N and c-(Y,Ti,Al)N)) are formed in the (Ti,Y,Al)N coating. This two-phase structure of the (Ti,Y,Al)N coating correlates to some extent with its significantly differentiated nanolayer structure.

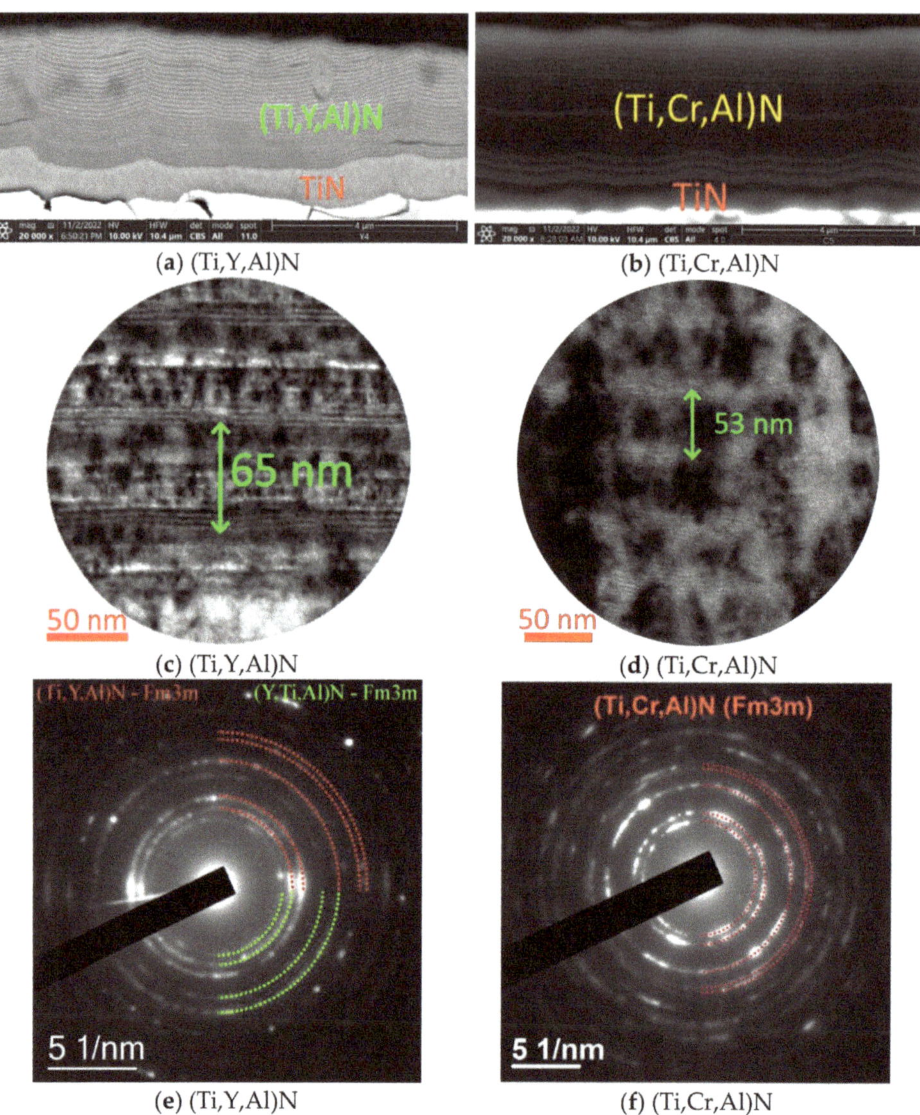

(**a**) (Ti,Y,Al)N   (**b**) (Ti,Cr,Al)N

(**c**) (Ti,Y,Al)N   (**d**) (Ti,Cr,Al)N

(**e**) (Ti,Y,Al)N   (**f**) (Ti,Cr,Al)N

Figure 2. (**a**,**b**) View of transverse sections of the studied coatings (SEM), comparison of (**c**,**d**) the nanolayer structures (TEM) and (**e**,**f**) phase compositions, defined by the selected area electron diffraction pattern (SAED) method.

As a result of the cutting tests, the relationship between the tool flank wear VB and the cutting time was investigated (Figure 3). The tools with the coatings of (Ti,Y,Al)N and (Ti,Cr,Al)N, as well as uncoated tools were compared. The coated tools demonstrated significantly higher wear resistance in comparison with the uncoated tools. While the wear rates of the coated tools were fairly close, the tool with the reference coating of (Ti,Cr,Al)N demonstrated slightly higher wear resistance.

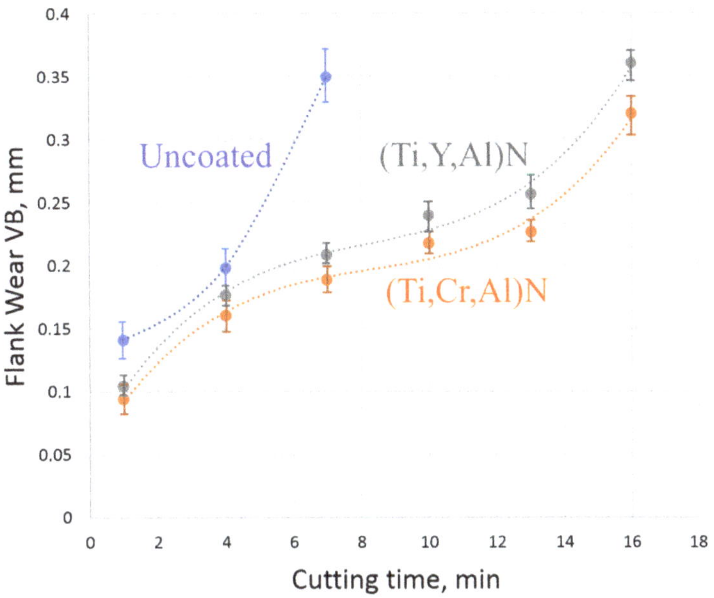

Figure 3. Wear dynamics along the tool flank faces of the tools with the coatings under comparison and the uncoated tools (after 16 min of cutting at f = 0.25 mm/rev, a_p = 1.0 mm, and v_c = 300 m/min).

The studies of the wear pattern of the coating conducted with a scanning electron microscope (SEM) (Figure 4) reveal the retained TiN transition layer in the worn area on the rake face. The (Ti,Y,Al)N wear-resistant layer has been preserved in fragments (see the area lighter in contrast). No delamination of the coating from the substrate is observed, which indicates good adhesion between the coating and the substrate. This confirms the data obtained earlier as a result of the scratch testing.

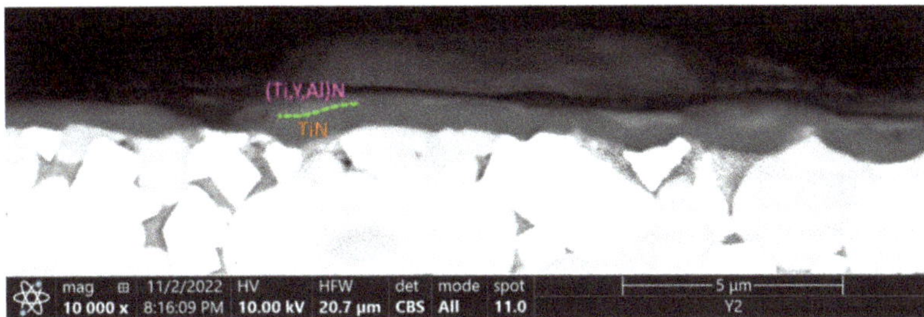

Figure 4. General wear pattern of the coating (SEM, after 16 min of cutting at f = 0.25 mm/rev, a_p = 1.0 mm, and v_c = 300 m/min).

The studies conducted with a transmission electron microscope (TEM) provide more information about the wear pattern on the coating. Figure 5 exhibits a general view of a lamella cut out from the area of the wear boundary of the coating. Areas A, B, and C are highlighted for further research. The TEM image correlates well with the previously obtained SEM. The well-preserved TiN transition layer can be clearly seen, with the structure, coarse-grained relative to the grains of the overlying (Ti,Y,Al)N coating. There are also preserved fragments of the (Ti,Y,Al)N wear-resistant layer, which has a nanolayer tructure.

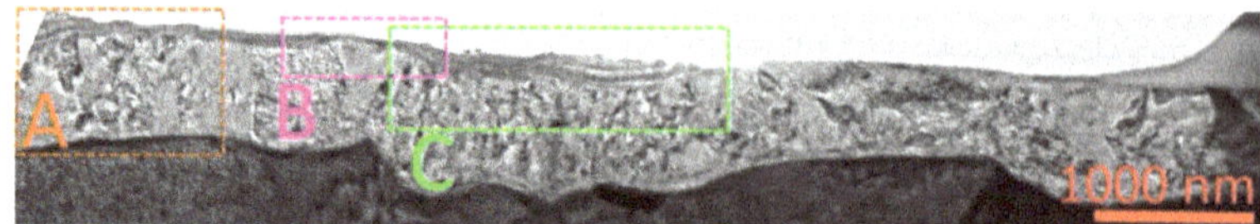

Figure 5. General view of the lamella and localization of Areas A, B, and C (after 16 min of cutting at f = 0.25 mm/rev, a_p = 1.0 mm, and v_c = 300 m/min) (TEM).

The examination of Areas A, B, and C in Figure 6 reveals a significant differentiation between the separate nanolayers of the coating. This differentiation is slightly enhanced compared to the coating after deposition (Figure 1). The described effect may be associated with the formation of yttrium oxide (Y_2O_3), which partially or completely replaces yttrium nitride (YN). The difference in the structure of the nanolayers is particularly noticeable in Image C2 (Figure 6d). The c-(Ti,Y,Al)N layer (darker in contrast) has the layered structure, while the (Y,Ti,Al)N layer (lighter in contrast) does not have layered structure in the region under consideration. The deeper layers of (Y,Ti,Al)N have the layered structure (Figure 6a,b), which may indicate the oxidation of the (Y,Ti,Al)N layers coming to the surface (and, accordingly, contacting oxygen in air), with the corresponding transition of YN → Y_2O_3.

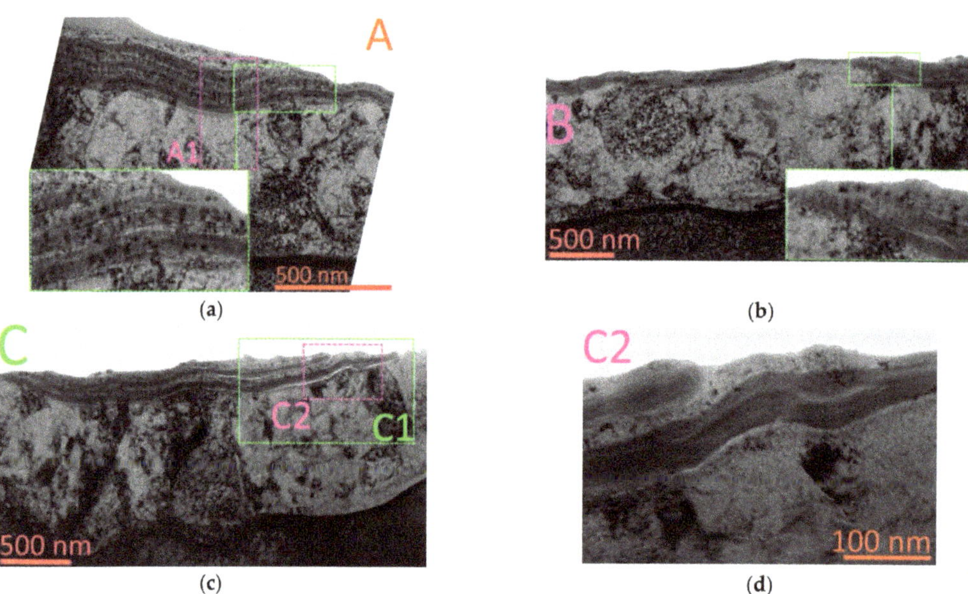

Figure 6. Investigation of the failure of the nanolayer structure of the coating. (**a,b**) (Y,Ti,Al)N nanolayers retain their layered structure, (**c,d**) (Y,Ti,Al)N nanolayers have lost their layered structure (possible YN → Y_2O_3 transformation) (TEM).

Due to the small area of the studied regions, no reliable SAED analysis can be conducted, but the analysis of the elemental composition makes it possible to identify zones of possible oxidation (Figure 7). The surface of the coating includes a layer up to 75 nm thick, in which an insignificant content of Y is combined with the high content of Ti and Al (see Point 1, Area A1 in Figure 7a). The content of oxygen in this layer is also high. It can be assumed that the oxides of Al_2O_3 and TiO_2 are formed as a result of oxidation and spinodal decomposition [64–69]. The next two Points 2 and 3 exhibit an increased content of yttrium with the high content of oxygen retained and a decrease in the content of Ti and Al. It can be assumed that yttrium oxide (Y_2O_3) is predominantly formed in this region. The insignificant presence of iron and chromium in the outer layers of the coating can be explained by the diffusion of these elements from the material being machined. The analysis of the distribution of elements in the outer layers of the worn coating (Area C1 in Figure 7b) finds some increase in oxygen content at Point 2, in which an increased content of yttrium is also detected. Taking into account that Point 2 is located farther from the coating surface compared to Point 1, and the oxygen content at Point 2 is higher, it can be assumed that the high oxygen content is associated with the active transformation of $YN \rightarrow Y_2O_3$. This conclusion is also confirmed by the results of the analysis of Regions 6 and 7. Whereas Region 6 (with the dominant yttrium content of 36.03 at.%) has a high oxygen concentration (45.63 at.%), then in Region 7 with the dominance of titanium (49.14 at.%) the content of oxygen is noticeably lower (26.20 at.%). This may also indicate the predominant formation of yttrium oxide with a significantly less intense formation of titanium and aluminum oxides, or the absence of these oxides in the considered region.

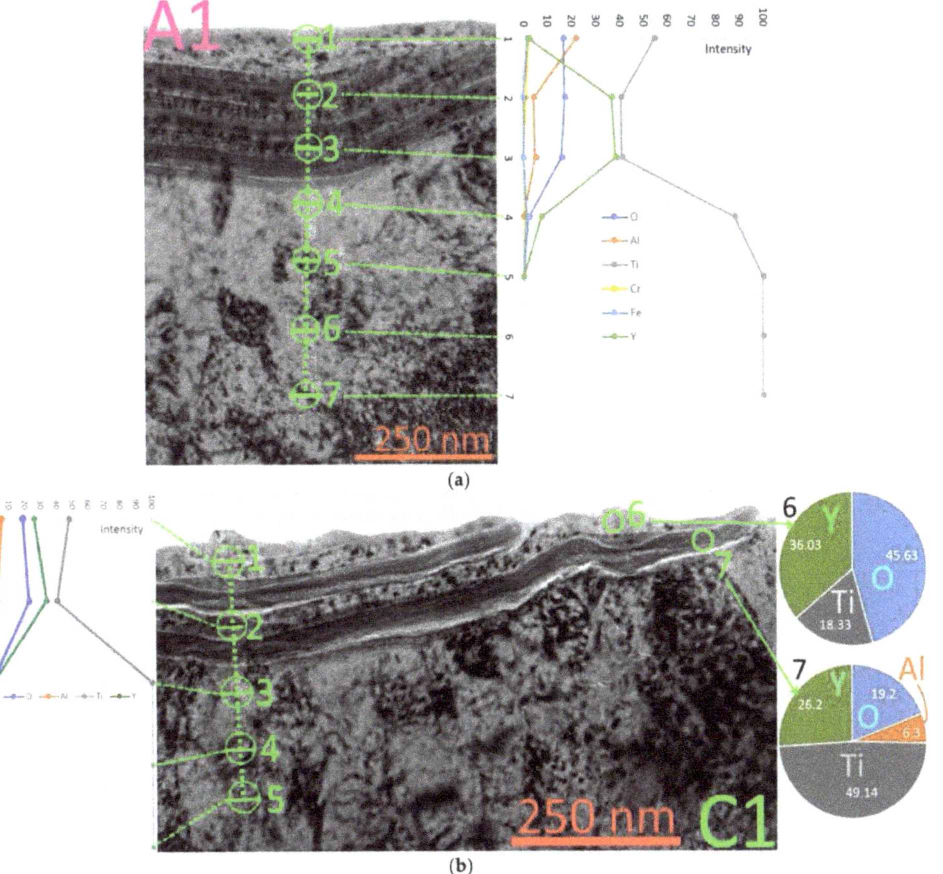

Figure 7. Analysis of the distribution of elements in the outer layers of the worn coating. (a) Area A1, (b) Area C1. The localization of Areas A1 and C1 is exhibited in Figure 6 (TEM).

Despite the active oxidation processes in the surface layers, the (Ti,Y,Al)N coating with a high yttrium content showed a rather high efficiency. As a direction for further research, it would be interesting to study the oxidation processes in the coating at different yttrium contents. Thus, it is possible to establish both the yttrium content, which is optimal for the tool life, and the ratio of the negative (coating wear) and positive (tribological conditions optimization) influence of oxidative processes on the general properties of the coating.

4. Conclusions

The studies were focused on the properties of the Ti-TiN-(Ti,Y,Al)N multilayer composite coating with the high content (about 40 at.%) of yttrium in its wear-resistant layer.

- The Ti-TiN-(Ti,Y,Al)N coating is characterized by the considerably high hardness (HV 2758 $\pm$ 78) with the elastic modulus of 356 $\pm$ 24 GPa;
- Two cubic solid solutions (fcc phases)–c-(Ti,Y,Al)N and c-(Y,Ti,Al)N–are formed in the coating;
- The study of the wear resistance of the Ti-TiN-(Ti,Y,Al)N-coated tools during the turning of steel in comparison with the wear resistance of the tools with the reference coating of Ti-TiN-(Ti,Cr,Al)N and the uncoated tools detects a noticeable increase in the wear resistance on the rake face (by 250%–270%) for the tools with both coatings. With the wear rates of the coated tools being fairly close, the tool life of the tool with the reference coating of (Ti,Cr,Al)N was slightly longer (by 10%–15%);
- During the process of wear, active oxidation processes take place in the layers of the Ti-TiN-(Ti,Y,Al)N coating that are in contact with the cut material flow. The mentioned processes consist in the dominant formation of yttrium oxide of Y_2O_3 with a possible slight formation of oxides of Al_2O_3 and TiO_2. Thus, for the described cutting conditions, the mechanisms of oxidative wear dominate in the coating.

Therefore, despite the tendency of YN to hydrolyze upon contact with water (including that contained in the atmosphere) and oxidize upon heating in an oxygen-containing environment, the Ti-TiN-(Ti,Y,Al)N coating demonstrated fairly good wear resistance. After 16 min of cutting the worn area on the rake face still demonstrates undamaged fragments of not only the transition layer, but also the wear-resistant layer of the coating. A possible reason for the lack of complete oxidation or noticeable hydrolysis of (Y,Ti,Al)N may be the protective functions of the layers with the dominance of (Ti,Al)N, which protect the underlying layers of (Y,Ti,Al)N from early oxidative damage. Another reason for the described phenomenon may be the substitution of the YN phase (which is prone to hydrolysis and thermal oxidation) for the solid solution phase of (Y,Ti,Al)N, the properties of which may be different. At the same time, the oxide layers formed on the surface of the coating improve the tribological conditions in the cutting zone and thus slow down the tool wear rate.

Author Contributions: Conceptualization, A.V.; methodology, A.V., F.M., N.S. and A.S.; investigation, F.M., J.B., N.S., C.S. and A.S.; resources, S.G.; data curation, C.S. and A.R.; writing—original draft preparation, A.V.; writing—review and editing, A.V.; project administration, S.G.; funding acquisition, S.G. All authors have read and agreed to the published version of the manuscript.

Funding: This work was supported financially by the Ministry of Science and Higher Education of the Russian Federation (project No FSFS-2021-0006).

Institutional Review Board Statement: Not applicable.

Informed Consent Statement: Not applicable.

Data Availability Statement: Not applicable.

Acknowledgments: The study used the equipment from the Centre for collective use of Moscow State Technological University STANKIN (agreement No. 075-15-2021-695, 26/07/2021). The coating structure was investigated using the equipment of the Centre for collective use of scientific equipment "Material Science and Metallurgy", purchased with the financial support of the Ministry of Science and Higher Education of the Russian Federation (GK 075-15-2021-696).

Conflicts of Interest: The authors declare no conflict of interest.

References

1. Krella, A. Resistance of PVD coatings to erosive and wear processes: A review. *Coatings* **2020**, *10*, 21. [CrossRef]
2. Fox-Rabinovich, G.S.; Gershman, I.S.; Veldhuis, S. Thin-film PVD coating metamaterials exhibiting similarities to natural processes under extreme tribological conditions. *Nanomaterials* **2020**, *10*, 1720. [CrossRef] [PubMed]
3. Baptista, A.; Silva, F.; Porteiro, J.; Míguez, J.; Pinto, G. Sputtering physical vapour deposition (PVD) coatings: A critical review on process improvement and market trend demands. *Coatings* **2018**, *8*, 402. [CrossRef]
4. Mehran, Q.M.; Fazal, M.A.; Bushroa, A.R.; Rubaiee, S. A Critical Review on Physical Vapor Deposition Coatings Applied on Different Engine Components. *Crit. Rev. Solid State Mater. Sci.* **2018**, *43*, 158–175. [CrossRef]
5. Bobzin, K. High-performance coatings for cutting tools. *CIRP J. Manuf. Sci. Technol.* **2017**, *18*, 1–9. [CrossRef]
6. Bobzin, K.; Brögelmann, T.; Kruppe, N.C.; Arghavani, M.; Mayer, J.; Weirich, T.E. Plastic deformation behavior of nanostructured CrN/AlN multilayer coatings deposited by hybrid dcMS/HPPMS. *Surf. Coat. Technol.* **2017**, *332*, 253–261. [CrossRef]
7. Vereschaka, A.; Tabakov, V.; Grigoriev, S.; Sitnikov, N.; Milovich, F.; Andreev, N.; Sotova, C.; Kutina, N. Investigation of the influence of the thickness of nanolayers in wear-resistant layers of Ti-TiN-(Ti,Cr,Al)N coating on destruction in the cutting and wear of carbide cutting tools. *Surf. Coat. Technol.* **2020**, *385*, 125402. [CrossRef]
8. Teppernegg, T.; Czettl, C.; Michotte, C.; Mitterer, C. Arc evaporated Ti-Al-N/Cr-Al-N multilayer coating systems for cutting applications. *Int. J. Refract. Hard Met.* **2018**, *72*, 83–88. [CrossRef]
9. Ramírez-Montes, L.; López-Pérez, W.; González-García, A.; González-Hernández, R. Structural, optoelectronic, and thermodynamic properties of YxAl1-xN semiconducting alloys. *J. Mater. Sci.* **2016**, *51*, 2817–2829. [CrossRef]
10. Schuster, J.C.; Bauer, J. The ternary systems ScAlN and YAlN. *J. Less-Common Met.* **1985**, *109*, 345–350. [CrossRef]
11. Ben Sedrine, N.; Zukauskaite, A.; Birch, J.; Jensen, J.; Hultman, L.; Schöche, S.; Schubert, M.; Darakchieva, V. Infrared dielectric functions and optical phonons of wurtzite $Y_xAl_{1-x}N$ (0 < x < 0.22). *J. Phys. D* **2015**, *48*, 415102.
12. Miyake, T.; Kishimoto, A.; Hasegawa, H. Tribological properties and oxidation resistance of (Cr,Al,Y)N and (Cr,Al,Si)N films synthesized by radio-frequency magnetron sputtering method. *Surf. Coat. Technol.* **2010**, *205* (Suppl. 1), S290–S294. [CrossRef]
13. Ju, H.; Jia, P. Microstructure, Oxidation Resistance and Mechanical Properties of Nb–Y–N Films by Reactive Magnetron Sputtering. *Prot. Met. Phys. Chem. Surf.* **2020**, *56*, 328–332. [CrossRef]
14. Scheerer, H.; Berger, C. Wear mechanisms of (Cr,Al,Y)N PVD coatings at elevated temperatures. *Plasma Process Polym.* **2009**, *6* (Suppl. 1), S157–S161. [CrossRef]
15. Yamamoto, K.; Kujime, S.; Fox-Rabinovich, G. Effect of alloying element (Si,Y) on properties of AIP deposited (Ti,Cr,Al)N coating. *Surf. Coat. Technol.* **2008**, *203*, 579–583. [CrossRef]
16. Rovere, F.; Music, D.; Schneider, J.M.; Mayrhofer, P.H. Experimental and computational study on the effect of yttrium on the phase stability of sputtered Cr-Al-Y-N hard coatings. *Acta Mater.* **2010**, *58*, 2708–2715. [CrossRef]
17. Aninat, R.; Valle, N.; Chemin, J.-B.; Duday, D.; Michotte, C.; Penoy, M.; Bourgeois, L.; Choquet, P. Addition of Ta and Y in a hard Ti-Al-N PVD coating: Individual and conjugated effect on the oxidation and wear properties. *Corros. Sci.* **2019**, *156*, 171–180. [CrossRef]
18. Wu, Z.; Qi, Z.; Zhang, D.; Wang, Z. Evolution of the microstructure and oxidation resistance in co-sputtered Zr-Y-N coatings. *Appl. Surf. Sci.* **2014**, *321*, 268–274. [CrossRef]
19. Rovere, F.; Mayrhofer, P.H. Thermal stability and thermo-mechanical properties of magnetron sputtered Cr-Al-Y-N coatings. *J. Vac. Sci. Technol. A* **2008**, *26*, 29–35. [CrossRef]
20. Dabees, S.; Mirzaei, S.; Kaspar, P.; Holcman, V.; Sobola, D. Characterization and Evaluation of Engineered Coating Techniques for Different Cutting Tools—Review. *Materials* **2022**, *15*, 5633. [CrossRef]
21. Kovalev, A.; Wainstein, D.; Rashkovskiy, A.; Fox-Rabinovich, G.; Veldhuis, S.; Agguire, M.; Yamamoto, K. Investigation of electronic and atomic structure of tribofilms on the surface of cutting tools with TiAlCrSiYN and multilayer TiAlCrSiYN/TiAlCrN coatings during machining of hardened steels. *Surf. Interface Anal.* **2010**, *42*, 1368–1372. [CrossRef]
22. PalDey, S.; Deevi, S.C. Single layer and multilayer wear resistant coatings of (Ti,Al)N: A review. *Mater. Sci. Eng. A* **2003**, *342*, 58–79. [CrossRef]
23. Smith, I.J.; Münz, W.D.; Donohue, L.A.; Petrov, I.; Greene, J.E. Improved $Ti_{1-x}Al_xN$ PVD coatings for dry high speed cutting operations Greene. *Surf. Eng.* **1998**, *14*, 37–41. [CrossRef]
24. Dosbaeva, G.K.; Veldhuis, S.C.; Yamamoto, K.; Wilkinson, D.S.; Beake, B.D.; Jenkins, N.; Elfizy, A.; Fox-Rabinovich, G.S. Oxide scales formation in nano-crystalline TiAlCrSiYN PVD coatings at elevated temperature. *Int. J. Refract. Hard Met.* **2010**, *28*, 133–141. [CrossRef]
25. Vetter, J.; Eriksson, A.O.; Reiter, A.; Derflinger, V.; Kalss, W. Quo vadis: Alcr-based coatings in industrial applications. *Coatings* **2021**, *11*, 344. [CrossRef]
26. Mo, J.; Wu, Z.; Yao, Y.; Zhang, Q.; Wang, Q. Influence of Y-addition and multilayer modulation on microstructure, oxidation resistance and corrosion behavior of $Al_{0.67}Ti_{0.33}N$ coatings. *Surf. Coat. Technol.* **2018**, *342*, 129–136. [CrossRef]
27. Zhu, L.; Zhang, Y.; Ni, W.; Liu, Y. The effect of yttrium on cathodic arc evaporated $Ti_{0.45}Al_{0.55}N$ coating. *Surf. Coat. Technol.* **2013**, *214*, 53–58. [CrossRef]
28. Moser, M.; Kiener, D.; Scheu, C.; Mayrhofer, P.H. Influence of yttrium on the thermal stability of Ti-Al-N thin films. *Materials* **2010**, *3*, 1573–1592. [CrossRef]

29. Li, M.; Wang, F.; Wu, W. High temperature oxidation of (Ti,Al)N and (Ti,Al,Y)N coatings on a steel prepared by arc ion plating (AIP). *Mat. Sci. Forum* **2004**, *461–464*, 351–358. [CrossRef]

30. Fan, Y.; Zhang, S.; Tu, J.; Sun, X.; Liu, F.; Li, M. Influence of doping with Si and Y on structure and properties of (Ti,Al)N coating. *Acta Metall. Sin.* **2012**, *48*, 99–106. [CrossRef]

31. Belous, V.; Vasyliev, V.; Luchaninov, A.; Marinin, V.; Reshetnyak, E.; Strel'nitskij, V.; Goltvyanytsya, S.; Goltvyanytsya, V.; Goltvyanytsya, S.; Goltvyanytsya, V. Cavitation and abrasion resistance of Ti-Al-Y-N coatings prepared by the PIII&D technique from filtered vacuum-arc plasma. *Surf. Coat. Technol.* **2013**, *223*, 68–74.

32. Singh, S.; Chauhan, R.; Gour, A. TBI calculations of the elastic properties and structural phase transformation in novel materials: Yttrium nitride. *Acta Phys. Pol.* **2011**, *120*, 1021–1025. [CrossRef]

33. Zerroug, S.; Ali Sahraoui, F.; Bouarissa, N. Ab initio calculations of yttrium nitride: Structural and electronic properties. *Appl. Phys. A* **2009**, *97*, 345–350. [CrossRef]

34. Saha, B.; Sands, T.D.; Waghmare, U.V. Electronic structure, vibrational spectrum, and thermal properties of yttrium nitride: A first-principles study. *J. Appl. Phys.* **2011**, *109*, 73720. [CrossRef]

35. Yang, J.W.; An, L. Ab initio calculation of the electronic, mechanical, and thermodynamic properties of yttrium nitride with the rocksalt structure. *Phys. Status Solidi B Basic Res.* **2014**, *251*, 792–802. [CrossRef]

36. Barve, S.A.; Jagannath Mithal, N.; Deo, M.N.; Biswas, A.; Mishra, R.; Kishore, R.; Bhanage, B.M.; Gantayet, L.M.; Patil, D.S. Effects of precursor evaporation temperature on the properties of the yttrium oxide thin films deposited by microwave electron cyclotron resonance plasma assisted metal organic chemical vapor deposition. *Thin Solid Film.* **2011**, *519*, 3011–3020. [CrossRef]

37. Cho, M.-H.; Ko, D.-H.; Jeong, K.; Whangbo, S.W.; Whang, C.N.; Choi, S.C.; Cho, S.J. Structural transition of crystalline Y_2O_3 film on Si(111) with substrate temperature. *Thin Solid Film.* **1999**, *349*, 266–269. [CrossRef]

38. Fantozzi, G.; Orange, G.; Liang, K.; Gautier, M.; Duraud, J.-P.; Maire, P.; Le Gressus, C.; Gillet, E. Effect of Nonstoichiometry on Fracture Toughness and Hardness of Yttrium Oxide Ceramics. *J. Am. Ceram. Soc.* **1989**, *72*, 1562–1563. [CrossRef]

39. Ju, H.; Yu, L.; He, S.; Asempah, I.; Xu, J.; Hou, Y. The enhancement of fracture toughness and tribological properties of the titanium nitride films by doping yttrium. *Surf. Coat. Technol.* **2017**, *321*, 57–63. [CrossRef]

40. Zhang, T.G.; Zhuang, H.F.; Zhang, Q.; Yao, B.; Yang, F. Influence of Y_2O_3 on the microstructure and tribological properties of Ti-based wear-resistant laser-clad layers on TC4 alloy. *Ceram. Int.* **2020**, *46*, 13711–13723. [CrossRef]

41. Tabatchikov, A.S. Appearance of fracture and impact toughness of weld metal alloyed with yttrium. *Weld. Int.* **1988**, *2*, 40–41. [CrossRef]

42. Jun, L.; Huiping, W.; Manping, L.; Zhishui, Y. Effect of yttrium on microstructure and mechanical properties of laser clad coatings reinforced by in situ synthesized TiB and TiC. *J. Rare Earths* **2011**, *29*, 477–483.

43. Li, K.W.; Wang, X.B.; Li, S.M.; Zhong, H.; Xue, Y.L.; Fu, H.Z. Microstructure and fracture toughness of Cr/Cr_2Nb alloys with trace Y addition. *Mater. Sci. Technol.* **2016**, *32*, 195–199. [CrossRef]

44. Sun, G.; Jia, L.; Ye, C.; Jin, Z.; Wang, Y.; Li, H.; Zhang, H. Balancing the fracture toughness and tensile strength by multiple additions of Zr and Y in Nb–Si based alloys. *Intermetallics* **2021**, *133*, 107172. [CrossRef]

45. Barshilia, H.C.; Acharya, S.; Ghosh, M.; Suresh, T.N.; Rajam, K.S.; Konchady Manohar, S.; Pai, D.M.; Sankar, J. Performance evaluation of TiAlCrYN nanocomposite coatings deposited using four-cathode reactive unbalanced pulsed direct current magnetron sputtering system. *Vacuum* **2010**, *85*, 411–420. [CrossRef]

46. Vereschaka, A.; Tabakov, V.; Grigoriev, S.; Aksenenko, A.; Sitnikov, N.; Oganyan, G.; Seleznev, A.; Shevchenko, S. Effect of adhesion and the wear-resistant layer thickness ratio on mechanical and performance properties of ZrN-(Zr,Al,Si)N coatings. *Surf. Coat. Technol.* **2019**, *357*, 218–234. [CrossRef]

47. Liu, Y.; Yu, S.; Shi, Q.; Ge, X.; Wang, W. Multilayer Coatings for Tribology: A Mini Review. *Nanomaterials* **2022**, *12*, 1388. [CrossRef]

48. Vereschaka, A.; Grigoriev, S.; Tabakov, V.; Migranov, M.; Sitnikov, N.; Milovich, F.; Andreev, N. Influence of the Nanostructure of Ti-TiN-(Ti,Al,Cr)N Multilayer Composite Coating on Tribological Properties and Cutting Tool Life. *Tribol. Int.* **2020**, *150*, 106388. [CrossRef]

49. Grigoriev, S.; Vereschaka, A.; Milovich, F.; Tabakov, V.; Sitnikov, N.; Andreev, N.; Sviridova, T.; Bublikov, J. Investigation of multicomponent nanolayer coatings based on nitrides of Cr, Mo, Zr, Nb, and Al. *Surf. Coat. Technol.* **2020**, *401*, 126258. [CrossRef]

50. Vereshchaka, A.A.; Vereshchaka, A.S.; Mgaloblishvili, O.; Morgan, M.N.; Batako, A.D. Nano-scale multilayered-composite coatings for the cutting tools. *Int. J. Adv. Manuf. Technol.* **2014**, *72*, 303–317. [CrossRef]

51. Beake, B.D. Nano-and Micro-Scale Impact Testing of Hard Coatings: A Review. *Coatings* **2022**, *12*, 793. [CrossRef]

52. Pogrebnjak, A.D.; Beresnev, V.M.; Bondar, O.V.; Abadias, G.; Chartier, P.; Postol'nyi, B.A.; Andreev, A.A.; Sobol', O.V. The effect of nanolayer thickness on the structure and properties of multilayer TiN/MoN coatings. *Tech. Phys. Lett.* **2014**, *40*, 215–218. [CrossRef]

53. Vereschaka, A.; Tabakov, V.; Grigoriev, S.; Sitnikov, N.; Milovich, F.; Andreev, N.; Bublikov, J. Investigation of wear mechanisms for the rake face of a cutting tool with a multilayer composite nanostructured Cr–CrN-(Ti,Cr,Al,Si)N coating in high-speed steel turning. *Wear* **2019**, *438–439*, 203069.

54. Yamamoto, K.; Kujime, S.; Takahara, K. Structural and mechanical property of Si incorporated (Ti,Cr,Al)N coatings deposited by arc ion plating process. *Surf. Coat. Technol.* **2005**, *200*, 1383–1390.

55. Ichijo, K.; Hasegawa, H.; Suzuki, T. Microstructures of (Ti,Cr,Al,Si)N films synthesized by cathodic arc method. *Surf. Coat. Technol.* **2007**, *201*, 5477–5480. [CrossRef]

56. Zhang, J.; Lv, H.; Cui, G.; Jing, Z.; Wang, C. Effects of bias voltage on the microstructure and mechanical properties of (Ti,Al,Cr)N hard films with N-gradient distributions. *Thin Solid Film.* **2011**, *519*, 4818–4823. [CrossRef]

57. Bobzin, K.; Brögelmann, T.; Kruppe, N.C.; Carlet, M. Wear behavior and thermal stability of HPPMS (Al,Ti,Cr,Si)ON, (Al,Ti,Cr,Si)N and (Ti,Al,Cr,Si)N coatings for cutting tools. *Surf. Coat. Technol.* **2020**, *385*, 125370. [CrossRef]

58. Fukumoto, N.; Ezura, H.; Yamamoto, K.; Hotta, A.; Suzuki, T. Effects of bilayer thickness and post-deposition annealing on the mechanical and structural properties of (Ti,Cr,Al)N/(Al,Si)N multilayer coatings. *Surf. Coat. Technol.* **2009**, *203*, 1343–1348.

59. Grigoriev, S.N.; Volosova, M.A.; Vereschaka, A.A.; Sitnikov, N.N.; Milovich, F.; Bublikov, J.I.; Fyodorov, S.V.; Seleznev, A.E. Properties of (Cr,Al,Si)N-(DLC-Si) composite coatings deposited on a cutting ceramic substrate. *Ceram. Int.* **2020**, *46*, 18241–18255.

60. Vereschaka, A.A.; Volosova, M.A.; Grigoriev, S.N.; Vereschaka, A.S. Development of wear-resistant complex for high-speed steel tool when using process of combined cathodic vacuum arc deposition. *Procedia CIRP* **2013**, *9*, 8–12. [CrossRef]

61. Grigoriev, S.; Vereschaka, A.; Zelenkov, V.; Sitnikov, N.; Bublikov, J.; Milovich, F.; Andreev, N.; Mustafaev, E. Specific features of the structure and properties of arc-PVD coatings depending on the spatial arrangement of the sample in the chamber. *Vacuum* **2022**, *200*, 111047. [CrossRef]

62. Grigoriev, S.; Vereschaka, A.; Zelenkov, V.; Sitnikov, N.; Bublikov, J.; Milovich, F.; Andreev, N.; Sotova, C. Investigation of the influence of the features of the deposition process on the structural features of microparticles in PVD coatings. *Vacuum* **2022**, *202*, 111144. [CrossRef]

63. *ASTM C1624-05*; Standard Test Method for Adhesion Strength and Mechanical Failure Modes. ASTM: West Conshohocken, PA, USA, 2010.

64. Povstugar, I.; Choi, P.-P.; Tytko, D.; Ahn, J.-P.; Raabe, D. Interface-directed spinodal decomposition in TiAlN/CrN multilayer hard coatings studied by atom probe tomography. *Acta Mater.* **2013**, *61*, 7534–7542. [CrossRef]

65. Rogström, L.; Ullbrand, J.; Almer, J.; Hultman, L.; Jansson, B.; Odén, M. Strain evolution during spinodal decomposition of TiAlN thin films. *Thin Solid Film.* **2012**, *520*, 5542–5549. [CrossRef]

66. Endrino, J.L.; Rhammar, C.; Gutiérrez, A.; Gago, R.; Horwat, D.; Soriano, L.; Fox-Rabinovich, G.; Martín, Y.; Marero, D.; Guo, J.; et al. Spectral evidence of spinodal decomposition, phase transformation and molecular nitrogen formation in supersaturated TiAlN films upon annealing. *Acta Mater.* **2011**, *59*, 6287–6296. [CrossRef]

67. Knutsson, A.; Schramm, I.C.; Asp Grönhagen, K.; Mücklich, F.; Odén, M. Surface directed spinodal decomposition at TiAlN/TiN interfaces. *J. Appl. Phys.* **2013**, *113*, 114305. [CrossRef]

68. Volosova, M.A.; Grigor'ev, S.N.; Kuzin, V.V. Effect of Titanium Nitride Coating on Stress Structural Inhomogeneity in Oxide-Carbide Ceramic. Part 4. Actionof Heat Flow. *Refract. Ind. Ceram.* **2015**, *56*, 91–96. [CrossRef]

69. Vereschaka, A.A.; Vereschaka, A.S.; Grigoriev, S.N.; Kirillov, A.K.; Khaustova, O. Development and research of environmentally friendly dry technological machining system with compensation of physical function of cutting fluids. *Procedia CIRP* **2013**, *7*, 311–316. [CrossRef]

Article

Influence of Surface Layer Condition of Al$_2$O$_3$+TiC Ceramic Inserts on Quality of Deposited Coatings and Reliability during Hardened Steel Milling

Marina A. Volosova *, Mikhail M. Stebulyanin, Vladimir D. Gurin and Yury A. Melnik

Department of High-Efficiency Processing Technologies, Moscow State University of Technology STANKIN, Vadkovskiy per. 3A, 127055 Moscow, Russia
* Correspondence: m.volosova@stankin.ru; Tel.: +7-916-308-49-00

Abstract: The specific features of the destruction of tool ceramics, associated with structural heterogeneity and defects formed during diamond grinding, largely determine their reduced reliability (dispersion of resistance). This is most pronounced at increased heat and power loads on the contact surfaces and limits the industrial application of ceramic cutting tools. The surface layer of industrially produced Al$_2$O$_3$+TiC cutting inserts contains numerous defects, such as deep grooves and torn grains. During the milling of hardened steels of the 100CrMn type with increased cutting parameters, the "wear–cutting time" curves have a fan-shaped character with different wear rates. The resistance of the tool that was taken from one batch before reaching the accepted failure criterion has a significant variation in values (VarT is 30%). The study is aimed to evaluate the influence of the condition of the surface layer of Al$_2$O$_3$+TiC inserts processed by various types of abrasive treatments, such as diamond grinding, lapping and polishing, on the quality of the (TiAl)N and (TiZr)N coatings and the reliability of prefabricated end mills. The obtained "wear–cutting time" curves are characterized as closely intertwined bundles. The coefficient of resistance variation (the tool's reliability) decreases by more than two times (14%). This can be used further in coating development to improve the performance of CCT.

Keywords: ceramic inserts; surface layer; diamond grinding defects; lapping and polishing; vacuum arc coatings; hardened steel milling; dispersion of resistance; tool reliability

Citation: Volosova, M.A.; Stebulyanin, M.M.; Gurin, V.D.; Melnik, Y.A. Influence of Surface Layer Condition of Al$_2$O$_3$+TiC Ceramic Inserts on Quality of Deposited Coatings and Reliability during Hardened Steel Milling. *Coatings* **2022**, *12*, 1801. https://doi.org/10.3390/coatings12121801

Academic Editor: Jinyang Xu

Received: 30 October 2022
Accepted: 17 November 2022
Published: 23 November 2022

Publisher's Note: MDPI stays neutral with regard to jurisdictional claims in published maps and institutional affiliations.

1. Introduction

A favorable combination of the most critical properties of tool ceramics (increased hardness, heat resistance, low affinity with most processed materials) allows using ceramic cutting tools (CCT) at extremely high cutting speeds, unattainable for carbide tools, while ensuring the high surface quality of the machined parts. Due to this, using CCT allows us to multiply the machining productivity concerning the level achieved by using a carbide tool [1,2]. The finish hard turning technologies for machining hardened structural steel with CCT within a hardness of over 50 HRC were developed more than 20 years ago. They found application in some industries (primarily in bearings) as an excellent alternative to grinding. The surface layer of the machined parts is characterized by an unfavorable stress–strain condition, which often negatively affects the performance of parts [3–5].

However, with all the advantages of CCT, the real share of its industrial use in the total global market for bladed tools is at a low level and does not show significant growth; if the share of tool ceramics in the world market for replaceable multifaceted inserts in 1991 was about 4% [6], then it increased by more than two times after almost 30 years. According to Ref. [7], the CCTs used in machining technologies accounted for about 9% of the total global market volume in 2018. Currently, the share of CCT use in the world market of cutting tools has increased slightly and is about 11%–12%.

The broader spread of CCT in the industry is limited by the low operational efficiency of such a tool, characterized by reduced reliability (significant variation in resistance) [8,9]. This disadvantage is especially pronounced when the ceramic tool is used under a combination of increased mechanical and thermal loads and cyclic loads during milling [10–12]. When working at increased cutting speeds and relatively large cross-sections of the cut layer, an accelerated (in some cases, sudden) destruction of the CCT contact surfaces is often observed. This is due to a few reasons, such as structural heterogeneity of the ceramics and defects of a technological character present in the volumetric structure and tool surface formed at various stages of the tool life cycle (sintering and diamond grinding).

There are three possible mechanisms of destruction of the surface layer of a ceramic material under the action of external loads:

(1) The mechanism of intragranular fracture of the CCT surface layer, according to which the external acting loads lead to the formation of microcracks of subcritical size in areas. The process is repeated many times, and it is the most favorable option from the point of view of the wear process and looks like a gradual abrasion of micro-sections on the CCT's contact surfaces.

(2) The mechanism of intergranular fracture with the separation of a single grain under the influence of a complex of thermal and force loads and the formation of unfavorable local areas in the CCT's surface layer. A crack at the stable growth stage does not encounter obstacles during its development along the intergranular phase. The presence of multiple defects accelerates the development of a crack, and the critical growth stage begins with an almost instantaneous exit to the surface.

(3) The mechanism of mixed destruction with separating a conglomerate of grains. This mechanism is the most unfavorable and unpredictable option for the CCT's surface layer destruction. It is observed in the case of a critical combination of increased mechanical and thermal loads and numerous defects present in the volume and surface layer.

An analysis of the mechanisms of destruction of the CCT surface layer suggests that technological defects will be additional stress concentrators, which will lead to accelerated destruction of the tool's contact surfaces. Unfortunately, even when using the most advanced sintering processes, the subsequent CCT diamond grinding (sharpening) negatively contributes to the surface layer's condition and reduces the tool's operating efficiency.

With a complex thermomechanical effect on a ceramic workpiece in the process of diamond grinding, the removal of the surface layer to the required depth occurs when stresses are created in it, the level of which exceeds the fracture stress of the material. The surface of the ceramic parts has a specific relief due to the impact of diamond grains and wheel binder friction on the sintered ceramic surface and the local plastic deformation, which occurs during high-speed heating of the ceramic surface areas and their rapid cooling [13]. Numerous technological defects reduce the efficiency of CCT operation and hinder the industrial use of the tool [14–16]. Therefore, the leading industrial application of CCT is high-speed lapping continuous machining with small cut thicknesses.

Today, a separate scientific direction, "ceramic surface engineering", has been formed to minimize CCT surface layer defects. Within it, various technologies of ion-plasma, beam, mechanical, combined and other types of exposure are used to modify the characteristics of the surface layer, including for "healing" surface defects and improving wear resistance. Among them, the most common and cost-effective approach is the deposition of functional coatings, which have proven to increase the resistance of carbide tools—(TiAl)N, (TiZr)N, (CrAlSi)N, etc. [17–21]. Some tool manufacturers (e.g., Iscar and Sandvik) are now producing coated CCTs. Taking into account the fact that CCT is a more expensive tool and is operated at cutting speeds significantly higher than the corresponding values for a carbide tool, an increase in resistance even by 1.5–2 times due to the use of functional coatings should be regarded as a significant result [22]. Another important feature of the coatings formed on the CCT is their rational thickness. Whereas for high-speed steel and carbide, the optimal thickness is usually around 6.0 μm, for CCT, the maximum thickness of the applied coatings should not exceed 4.0 μm. With an increase in this value, it is impossible to ensure the high strength of the adhesive bond between coatings and ceramic substrates, and their delamination is observed even under insignificant external loads [23–27].

The well-known works of scientists [28–32] are aimed at the research of the effect of various coatings of Al_2O_3+TiC ceramics on the wear pattern of CCT working surfaces, evaluating the coating effect on the average tool resistance and on the roughness of the workpiece surface layer when turning workpieces made of high-hard cast irons, hardened bearing and tool steels. According to various studies, by applying coatings of TiN, (TiAl)N (AlTi)N and (TiAlSi)N during turning, an increase in the average tool resistance of cutting inserts by 1.2–2.5 times compared to uncoated CCT is achieved. The authors of Ref. [22] increased the average tool resistance of ceramic cutting inserts based on Al_2O_3+TiC by 1.8 times compared to an uncoated tool when milling hardened bearing steel by applying (TiZr)N coating. At the same time, the analysis of the results of previous studies shows that coatings with a thickness of up to 4.0 μm are not able to "heal" numerous defects in the form of deep grooves and torn grains, which are visible already on the surface of coatings when studying their microstructure, but only reduce the depth of the defective layer [32–34].

However, even with the specific features of the coating deposition on a defective surface described above, their application to CCT certainly changes the conditions for the physicochemical interaction of the tool contact surfaces with the workpiece and the shearing chips during cutting, which increases the average cutting resistance. However, when evaluating the efficiency of CCT operation under conditions of increased cross-sections of the cut layer and high cutting speeds, it is impossible to focus only on the average tool resistance (Tav). This performance indicator is often not informative. The dependence of wear along the back surface of Al_2O_3+TiC-based ceramic cutting inserts on cutting time is a random variable, and a group of wear curves can have a pronounced fan-like character [9]. This nature of wear development over time is difficult to predict, and the resistance of CCT (time to reach the accepted failure criterion) has a significant variation in values. Under the conditions of machine-building production, when making decisions about the appropriateness of using a cutting tool, along with average resistance, reliability is evaluated by variation (dispersion) of resistance (VarT). So far, this critical performance characteristic has yet to be the focus of attention of researchers and specialists involved in applying various coatings on CCT.

With a particular value of previously obtained experimental results in improving the wear resistance of CCT by applying functional coatings, they are united by a narrow formulation of the tasks to be solved. The understudied issues regard assessing the effect of the condition of the CCT surface layer on the characteristics of the formed coatings and their performance indicators during cutting under conditions of intense mechanical and thermal loads.

The purpose of this work is to study the influence of the condition of the surface layer of Al_2O_3+TiC ceramic cutting inserts processed by diamond grinding and polishing on the quality of the formed coatings (for example, (TiAl)N and (TiZr)N) and the reliability of prefabricated face mills when cutting bearing steel 100CrMn6 with increased cross-sections of the cut layer.

The novelty of the work lies in evaluating the coated CCT reliability by variation (dispersion) of resistance (VarT), a critical performance characteristic that has not been researched so far. The study is conducted with sample Al_2O_3+TiC ceramic cutting inserts processed by diamond grinding and polishing, on which surface (TiAl)N and (TiZr)N coatings were deposed, in conditions of 100CrMn6 bearing steel cutting with increased cross-sections of the cut layer.

The practical significance of the work lies in evaluating the (TiAl)N and (TiZr)N coatings' effect on industrially diamond grinded and additionally polished Al_2O_3+TiC ceramic cutting inserts on the reliability of the CCT under increased mechanical and thermal cyclic contact loads.

2. Materials and Methods

2.1. Cutting Tool, Processed Material and Reliability Assessment Methodology

As a cutting tool for the research, prefabricated 160×50 mm face mills with a mechanical fastening of 10 square ceramic cutting inserts of 12.7×4.76 mm were used (Figure 1). The tool material was ceramics based on Al_2O_3+TiC, in which the content of the main phases, revealed by X-ray diffraction analysis and processing of the results using the PANalytical X'Pert HighScore Plus software by PANalytical B.V. (version 3.0) and the ICCD PDF–2 database (version 2023), was 70 vol.% Al_2O_3 and 30 vol.% TiC.

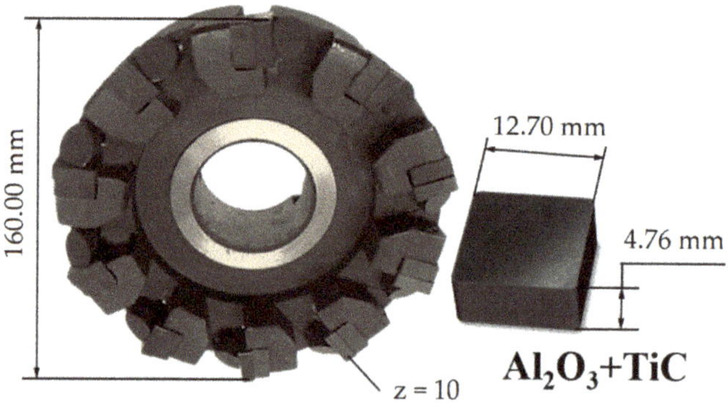

Figure 1. Construction of a prefabricated face mill with mechanical fastening of square cutting ceramic inserts used in testing.

The material to be machined was 100CrMn6 bearing steel, which is common in the industry, and the chemical composition of which is given in Table 1.

Table 1. Chemical composition of 100CrMn6 bearing steel used in testing.

Element	Fe	Cr	Mn	C	Si	Ni	Cu	S	P
Content (%)	95.3	1.56	1.1	1.0	0.5	0.3	0.2	0.02	0.02

The experiments were carried out during the processing of prismatic workpieces made of steel 100CrMn6 (hardness 62–63 HRC) on a vertical milling machine BM127 (JSC Votkinsk Machine Building Plant, Votkinsk, Russia) at a cutting mode that provides high mechanical and thermal loads on the contact surfaces of ceramic inserts: cutting speed $V = 380$ m/min, feed $S = 0.15$ mm/tooth and depth $t = 1$ mm. To construct a group of curves, "tool wear–cutting time", two faces of each insert (tooth) of the prefabricated cutter were tested. The size of the wear area along the back surface was measured every 2 min of operation during cutting on a Stereo Discovery V12 Zeiss optical microscope (Carl Zeiss AG, Oberkochen, Germany), and the wear of 400 μm was the limiting value. The standard calculation formula [35] was used to estimate the reliability indicator of the tool, i.e., the variation in the tool resistance VarT, which characterizes the dissipation of the tool's operating time until the wear limit is reached. Subsequently, VarT is defined as the ratio of the values of the mean square deviation and the arithmetic mean value of resistance. As a rule, for a batch of tools used under production conditions for critical machining operations on CNC machines, VarT should not exceed 10%–15%. It should be noted that ceramic tools are mainly used for machining critical engineering products on high-precision CNC machines. When the tool life dispersion exceeds 10%–15%, there is a high probability that tool failure will occur directly during the production cycle, and an expensive part will be culled. In such a situation, one has to either significantly reduce the cutting conditions, which is not economically feasible, or carry out a forced tool replacement (underestimating the failure criterion) to avoid accidents associated with premature tool failures, i.e., a part of the fully functional tool will be recognized as unusable, which leads to additional production costs. Therefore, the generally accepted practice in high-tech production is the requirement that the dispersion of resistance of 10%–15% and high dispersion of tool resistance values indicate low reliability of the tool and force technologists to underestimate the cutting modes [36–40].

2.2. Preparation of Cutting Ceramic Inserts with a Different Condition of the Surface Layer

To form experimental groups of ceramic inserts with different surface layer conditions, commercially available square-shaped Al_2O_3+TiC CC650 inserts manufactured by Sandvik AB, Sandviken, Sweden (Figure 2) were subjected to the additional diamond abrasive processing operations—lapping and polishing. Additional processing of ceramic inserts was carried out on a lapping and polishing machine (Lapmaster Wolters, Mt Prospect, IL, USA) with unique lapping and polishing wheels using various diamond suspensions (with a grain size of 50/40, 40/28 for lapping, and 10/7, 5/3 for polishing) at a cutting speed of 3 m/s. Thus, two groups of Al_2O_3+TiC ceramic inserts were prepared for research: after diamond grinding (I), after diamond grinding, lapping and polishing (II). Lapping and polishing as additional operations for the abrasive processing of ceramic inserts were chosen based on the fact that these processes can significantly minimize the degree of defectiveness of the surface layer formed during diamond grinding [41–43].

(**a**)   (**b**)

Figure 2. SEM images of the general view (**a**) and the cutting part (**b**) of commercially available CC650 ceramic inserts used in testing.

2.3. Coating of Ceramic Inserts

The deposition of (TiAl)N and (TiZr)N PVD coatings with a thickness of ~3.7 µm on two groups of Al_2O_3+TiC ceramic inserts was carried out on a pilot technological unit developed at the Moscow State University of Technology "STANKIN", equipped with devices for plasma generation [44–48]. The schematic diagram and the main components of the unit are shown in Figure 3. The vacuum-arc coating deposition method was chosen as a method that provides high productivity for the needs of tool production [49–52]. The technological process of applying coatings to ceramic inserts included three stages: purification in gas plasma, bombardment with metal ions and deposition of a coating of the required composition by the vacuum-arc method [53–56]. The values of the technological modes under which the coating was applied are given in Table 2.

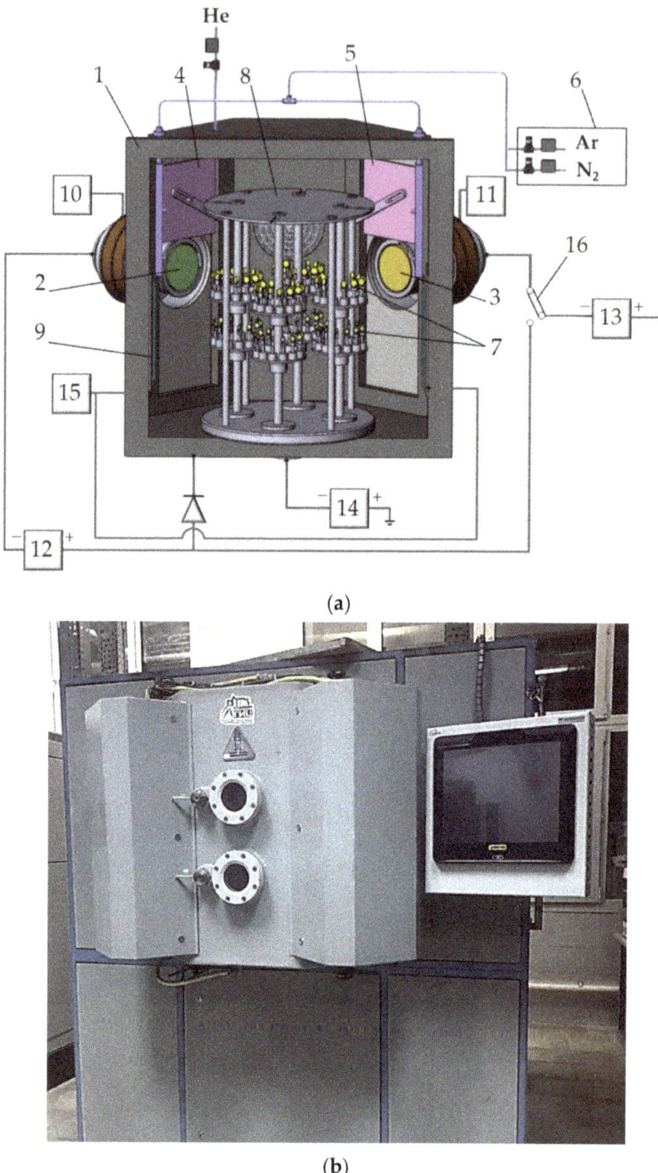

Figure 3. (**a**) Schematic diagram of a technological unit for coating ceramic inserts: 1—vacuum chamber; 2, 3—cathodes; 4, 5—cathode shutters; 6—gas supply system; 7—ceramic inserts; 8—planetary rotation device; 9—heating element; 10, 11—power sources of cathode coils; 12, 13—cathode current sources; 14—reference voltage source; 15—heating element power source; 16—switch; (**b**) installation appearance.

Table 2. Technological modes of coating on Al_2O_3+TiC ceramic inserts.

Stage of the Process	Technological Modes of the Process	Value of the Modes When Applying Two Options of Coatings	
		(TiZr)N	(TiAl)N
Purification in gas plasma with argon ions	Arc currents at the cathode, A	100	
	Bias voltage, V	600 … 800	
	Argon pressure, Pa	1.0×10^{-1}	
	Purification time, min	10	
Metal ion bombardment	Number and material of cathodes	1 Ti + 1 Zr	1 Ti + 1 Al
	Arc currents at the cathode, A	110 (Ti) 100 (Zr)	110 (Ti) 80 (Al)
	Bias voltage, V	950	
	Ion bombardment time, min	8	
Coating deposition	Number and material of cathodes	1 Ti + 1 Zr	1 Ti + 1 Al
	Arc currents at the cathode, A	110 (Ti) 100 (Zr)	110 (Ti) 80 (Al)
	Bias voltage, V	300	250
	Reaction gas pressure N_2/Ar, Pa	4.0×10^{-1}	4.5×10^{-1}
	Deposition time (for h = 3.7 μm), min	50	40

2.4. Investigation of the Properties of the Surface Layer of Ceramic Inserts after Abrasive Treatment and Coating

To construct the surface profilograms of two groups of ceramic inserts without coatings and after coating, a Dektak XT stylus profilometer (Bruker, Billerica, MA, USA) was used, which performs a set of electromechanical measurements using contact scanning of the required surface area with a highly sensitive diamond tip at a given motion speed. Based on the measurement results, a specialized software processes the information, visualizes and constructs the necessary profilograms. The specified equipment was used to evaluate the following characteristics of the surface layer in accordance with the ISO 4287:1997 standard: Ra is the arithmetic average of the absolute values of microroughness within the length of the sample in the direction of the X-axis of the ceramic insert; Rt is the total height of the profile, estimated as the sum of the most significant height of the profile peak and the most significant depth of the profile cavity within the length of the evaluated section of the ceramic insert. The values of these parameters were determined by the results of the analysis of ten inserts of each group.

For the microstructure study of the surface of ceramic inserts subjected to diamond processing and coating, the method of scanning electron microscopy on VEGA3 LMH (Tescan, Brno, Czech Republic) equipment was used.

The crack resistance and microhardness of ceramic inserts after various abrasive processing technologies were determined on a QnessQ10A universal microhardness tester (Qness GmbH, Mammelzen, Germany) by Vickers pyramid indentation. When evaluating microhardness, the load on the indenter was 2 kg, and when assessing crack resistance, it was 5 kg. After indentation, the values of the diagonals of the indentations and the lengths of cracks propagating from the corners of the indentations were measured, on the basis of which the crack resistance (Kc) and microhardness HV were determined from the known dependences [57].

To study the influence of the condition of the surface layer and applied coatings of ceramic inserts on the abrasion resistance under abrasive conditions, tests were carried out on a Calowear (CSM Instruments, Peseux, Switzerland) device under pressure on samples with a force of 0.2 N of a rotating sphere of hardened steel, where a water-based abrasive suspension was fed into the contact zone. Optical analysis of the geometric dimensions of the wear holes, as well as their measurement on a stylus profilometer, made it possible to quantify and qualitatively assess the volumetric wear of the samples [58].

The evaluation of the change in the friction coefficient of ceramic inserts after various options for abrasive treatment and coating over time was carried out on a THT-S-AX0000 (CSEM, Neuchatel, Switzerland) tribometer during rotation of the ceramic inserts relative to a fixed ceramic ball with a diameter of 6 mm at a load of 1 N, a sliding speed of 10 cm/s and a test temperature of 800 °C [59].

To obtain data on the nanohardness and modulus of elasticity of coatings formed on ceramic inserts with different surface layer conditions, we used the method of nanoindentation with a Berkovich diamond indenter on Nano Hardness Tester (CSEM, Neuchatel, Switzerland) equipped with specialized software based on the Oliver W.C. and Pharr G.M. algorithm [60]. The measurements were carried out at a load of 2.0 mN, and the duration of the load–unload cycle was 50 s. Based on the obtained experimental data, the nanohardness (H) and modulus of elasticity (E) of the coatings were calculated. In addition, the H/E ratio, which was called the index of plasticity, was evaluated, which can be used to judge the viscosity of coatings and their ability to resist possible deformation and destruction under external loads [61]. To reduce possible measurement errors, data were obtained by evaluating 10 inserts from each group. An assessment was performed of the adhesive bond strength of the formed coatings with ceramic substrates by sclerometry (scratch testing) with the fixation of the spectrum of acoustic emission signals on the NANOVEA M1 scratch tester (Irvine, CA, USA). A Rockwell indenter in the form of a diamond cone with a radius at the apex R = 100 μm and a taper angle of 120 degrees was used as an indenter. Three scratches 5 mm long were applied to each ceramic insert. The tests were performed at a linearly increasing load of up to 50 N and a loading speed of 5 N/min. During the trial, acoustic emission spectra and the corresponding forces were recorded. According to the results of three measurements, the normal load was identified, which corresponded to the moment of delamination of the coatings [62–64].

3. Results and Discussion

3.1. Influence of Various Types of Abrasive Treatment on the Condition and Characteristics of the Surface Layer of Ceramic Inserts

The characteristic 3D profilograms and SEM images of the microstructure of the surface layer of Al_2O_3+TiC ceramic inserts of the two groups under study are shown in Figure 4, and Table 3 presents the generalized data on the characteristics of the surface layer of ceramic samples after various options for abrasive treatment.

The obtained experimental data clearly demonstrate the pronounced changes that occur on the surface layer of industrially produced ceramic inserts as a result of the use of additional abrasive processing—lapping and polishing. The surface layer of Al_2O_3+TiC inserts present on the market after diamond grinding (Figure 4a) is full of numerous defects; deep grooves are observed, profiled by diamond grains of the grinding wheel, as well as tearing of ceramic material grains occurring under the influence of force loads. It should be noted that despite the pronounced defectiveness, the roughness parameter Ra of ceramic inserts after diamond grinding does not exceed the restrictions imposed by the current standards for the manufacture of ceramic tools, in particular, those specified in GOST 25003-81: "Ceramic indexable throw-away inserts for cutting tools. Specifications". The average value of Ra, according to the results of evaluation of 10 samples, does not exceed 0.3 μm (Table 3), and the maximum value of Ra is 0.33 μm (Figure 4a). Thus, from the point of view of assessing the defectiveness of the surface layer of ceramic inserts, the parameter Ra is not informative, since the value of the Ra parameter is not significantly affected by the grooves and sites of local destruction. The most informative parameter that could be used in practice to assess the defectiveness of the surface layer of ceramic inserts after grinding is the Rt parameter, the measurement of which makes it possible to take into account all kinds of grooves, micropits and other defects. Thus, the Rt parameter can in fact be regarded as the depth of the defective layer. The average value of Rt from the evaluation results of 10 samples after diamond grinding was 3.32 μm (Table 3), and the maximum set value was 4.0 μm (Figure 4a).

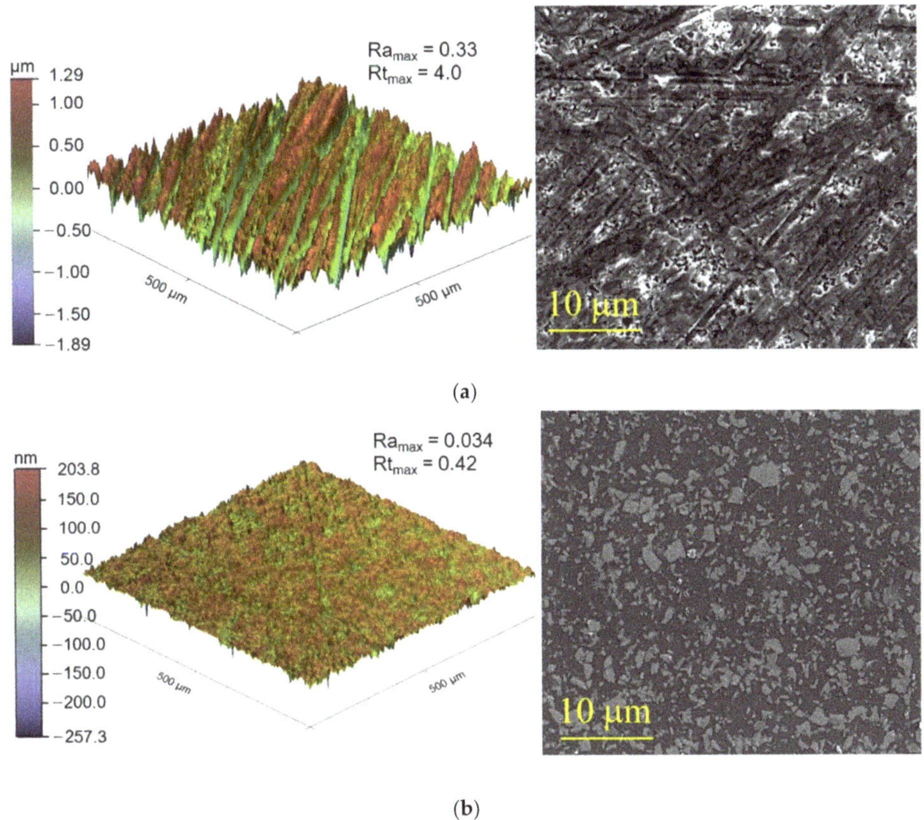

Figure 4. 3D profilograms (**left**) and SEM images (**right**) of the microstructure of the surface layer of Al_2O_3+TiC ceramic inserts after diamond grinding (**a**) and after diamond grinding, lapping and polishing (**b**).

Table 3. Characteristics of the surface layer of Al_2O_3+TiC ceramic inserts after various types of abrasive treatments.

No	Option of Abrasive Treatment of Ceramic Inserts Al_2O_3+TiC	Average Values of the Characteristics of the Surface Layer			
		Crack Resistance Kc $(MPa·m^{1/2})$	Microhardness HV (GPa)	Roughness Ra (μm)	Defect Layer Depth Rt (μm)
1	Diamond grinding (I)	3.68	14.82	0.3	3.32
2	Diamond grinding, lapping and polishing (II)	3.9	15.1	0.02	0.31

The use of additional lapping and polishing minimizes the defects formed during the diamond grinding of ceramic inserts, as illustrated by the profilogram and SEM image of the microstructure of the surface layer (Figure 4b). The average value of Rt, according to the evaluation results of 10 samples after diamond grinding, lapping and polishing, was

0.33 μm (Table 3), and the maximum value was 0.42 μm (Figure 4a). Thus, the depth of the defective layer with the use of additional abrasive treatment was reduced by ten times.

The results of studies of the crack resistance (Kc) of the surface layer of Al_2O_3+TiC ceramic inserts that underwent various types of abrasive treatments showed a certain relationship between this indicator and the presence of defects. In quantitative terms, the increase in Kc of ceramic inserts after diamond grinding (I group) and after minimizing defects through additional lapping and polishing (II group) was ~6%—from 3.68 to 3.9 MPa·m$^{1/2}$, respectively (Table 3). At the same time, there were no significant differences in the microhardness (HV) of the surface ceramic inserts that underwent various types of abrasive treatments using the assessment method.

The dependences of the volume of worn material on the test time presented in Figure 5 give a particular idea of the wear kinetics of ceramic inserts with different surface layer conditions and their resistance to abrasive wear. It is noticeable that the presence of diamond grinding defects in the surface layer of ceramic inserts made of Al_2O_3+TiC (group I) significantly reduces the ability of ceramics to resist abrasive wear. Inserts with minimal surface layer defects, which underwent additional lapping and polishing (group II) throughout the entire test distance, showed significantly lower wear values; after 20 min, their volumetric wear of the samples was two times less than that of the samples of group I. Taking into account that the microhardness of the contact surfaces of ceramic inserts after additional abrasive treatment increases very slightly, the increase in abrasion resistance under abrasive conditions for samples with minimal surface layer defects can be explained by the minimization of stress concentrators [12,65], which lead to accelerated destruction of the contact surfaces of the tool, including the mixed mechanism described above.

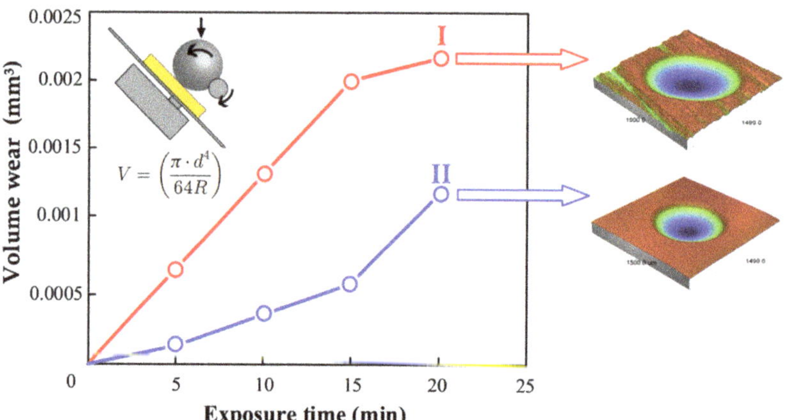

Figure 5. Dependences of the volume wear of Al_2O_3+TiC ceramic inserts with the different conditions of the surface layer (I—diamond grinding; II—diamond grinding, lapping and polishing) on the time of abrasive exposure.

The contact surfaces of ceramic inserts during the cutting process are subjected to high thermal loads, so it is important to study the condition of the surface layer for tribological characteristics during high-temperature heating. Figure 6 shows the dependences of the coefficient of friction (COF) during high-temperature heating of Al_2O_3+TiC ceramic inserts that underwent diamond grinding (I) and after minimizing defects using lapping and polishing (II). The experimental curves show that samples with the lowest level of defects somewhat reduce the average COF value compared with samples of inserts subjected to diamond grinding. A significant result is a change in the nature of COF development over time (Figure 6). For samples in group I with high defectiveness, the COF changes abruptly; first, its value increases, reaching maximum values, then sharply decreases, increases again and only stabilizes with time. Such a non-monotonic change in COF is apparently the result of alternating processes of the adhesive setting of the contacting surfaces and destruction of the "bridges" of adhesive bonds. For ceramic samples in group II, the change in COF over the entire friction path is monotonic, which indicates more favorable conditions for frictional interaction with the counterbody. It can be assumed that the noted changes in the nature of the contact interaction under friction-sliding conditions are also largely associated with a significant decrease in the roughness of the contact surfaces after the application of lapping and polishing, the data on which are given in Table 3.

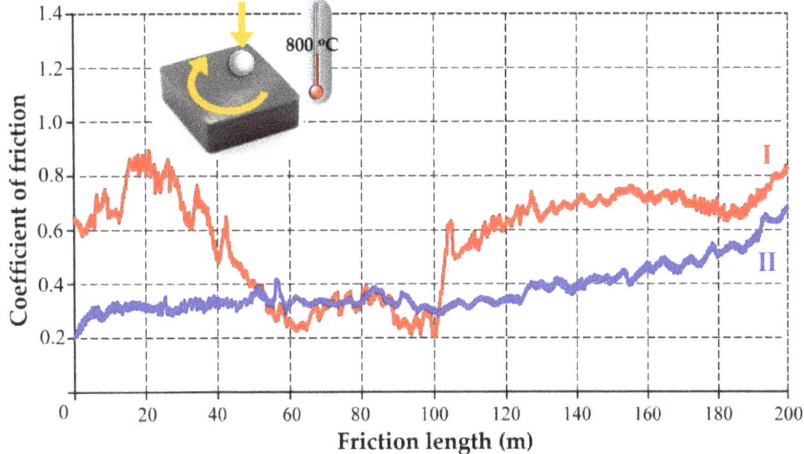

Figure 6. Dependences of the coefficient of friction during high-temperature heating of Al_2O_3+TiC ceramic inserts with the different conditions of the surface layer (I—diamond grinding; II—diamond grinding, lapping and polishing) on the length of the friction path.

3.2. Influence of the Condition of the Surface Layer of Ceramic Inserts on the Quality of the Formed Coatings

The results of experimental studies carried out in this work on the study of the characteristics of tool coatings with a total thickness of ~3.7 μm based on (TiAl)N and (TiZr)N nitrides deposited on ceramic inserts by vacuum-arc evaporation show that the microstructure, morphology and physicomechanical properties of the coatings are strongly dependent on the defectiveness of the surface layer of the ceramic substrate. Figures 7a and 8a show SEM images and 3D profilograms of the microstructure of (TiAl)N and (TiZr)N coatings formed on ceramic substrates after diamond grinding with high imperfection. It can be seen that the morphological pattern of the coatings largely copies the characteristic defects present in the surface layer of ceramic inserts, which were discussed above (Figure 4). The microstructure of the deposited coatings includes various pores and discontinuities, and numerous defects in the form of deep grooves and torn grains, which are present on industrially produced ceramic inserts, are clearly visible on them. At the same time, a quantitative assessment of the samples by the Rt parameter made it possible to find out that the (TiAl)N and (TiZr)N coatings reduce this indicator by ~30%–35%.

To differentiate the defects of the formed coatings into those related to the condition of the ceramic substrates and the technological features of the condensation processes, it is necessary to consider the microstructures of coatings deposited under identical conditions on ceramic substrates after additional lapping and polishing, which have minimal defects (Figures 7b and 8b). A comparison of the presented SEM images of microstructures and 3D profilograms with the data of Figures 7a and 8a shows that the coatings deposited on "defect-free" ceramic substrates only have well-known defects and features associated with the processes of coating synthesis, crystallite growth and micro-drops formation [66,67]. It should be noted that it is conditionally considered that if the SEM analysis (a quite fine study) does not reveal defects, then such a surface can be regarded as "defect-free" in comparison with the gross defects of industrial samples.

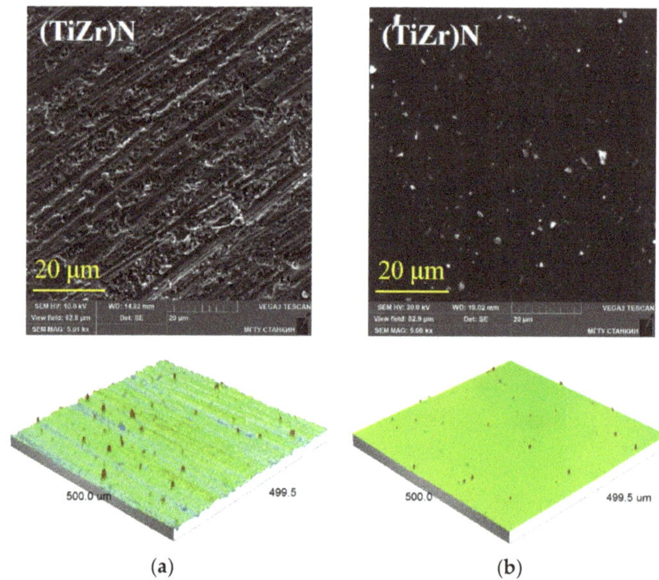

(a) **(b)**

Figure 7. SEM images of the microstructure and 3D profilogram of (TiZr)N coatings deposited on Al$_2$O$_3$+TiC ceramic inserts with different surface layer conditions after diamond grinding (**a**) and after diamond grinding, lapping and polishing (**b**).

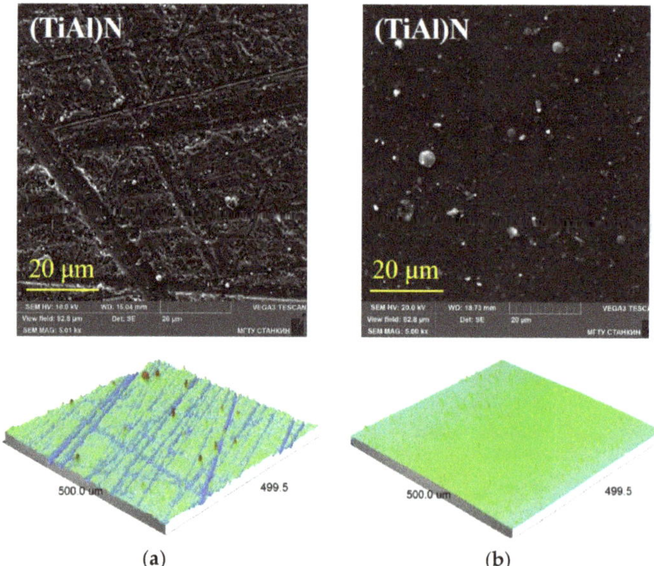

(a) **(b)**

Figure 8. SEM images of the microstructure and 3D profilogram of (TiAl)N coatings deposited on Al$_2$O$_3$+TiC ceramic inserts with different surface layer conditions after diamond grinding (**a**) and after diamond grinding, lapping and polishing (**b**).

As the data shown in Table 4 demonstrate, the degree of defectiveness of the surface layer on which the coatings are deposited significantly affects their physical and mechanical characteristics. The nanoindentation curves calculated from the analysis of (TiAl)N and (TiZr)N coatings formed on ceramic inserts with different conditions of the surface layer made it possible to determine their nanohardness (H) and modulus of elasticity (E). Table 4 also provides reference data on the characteristics of commercially available Al_2O_3+TiC inserts with TiN coatings. It should be noted that minimizing the degree of defectiveness slightly increases the nanohardness of (TiZr)N coatings (from 29 to 31 GPa) and also reduces the spread of this index for the two coatings under study. In addition, for the formed coatings based on (TiAl)N and (TiZr)N, a slight (~5%) decrease in the modulus of elasticity and a decrease in the spread of this value were noted. The ratio (H/E), called the index of plasticity and by which it is possible to approximate the fracture toughness of a coating and its ability to resist possible deformation and fracture during cutting [68], also varies depending on the condition of the ceramic substrate. As can be seen from the presented data (Table 4), the maximum H/E ratio at the level of 0.1 had coatings (TiAl)N and (TiZr)N formed on ceramic inserts after additional lapping and polishing, while this indicator for coatings formed on non-defective inserts after diamond grinding was 0.09 and 0.08, respectively.

Table 4. Physical and mechanical characteristics of coatings deposited on Al_2O_3+TiC ceramic inserts with different conditions of the surface layer.

No	Composition of the Coating on the Ceramic Insert	Nanohardness H (GPa)	Modulus of Elasticity E (GPa)	Index of Plasticity H/E	Friction Coefficient when Heated to 800 °C at a Distance of 200 m		Breaking Load When Assessing Adhesion (H)
					Min Value	Max Value	
1			Al_2O_3+TiC after diamond grinding				
1.1	TiN (industrial)	24 ± 3	305 ± 10	0.07	0.59	1.1	28
1.2	(TiAl)N	33 ± 4	342 ± 10	0.09	0.54	0.92	31
1.3	(TiZr)N	29 ± 3	329 ± 12	0.08	0.42	0.78	32
2			Al_2O_3+TiC after diamond grinding, lapping and polishing				
2.1	(TiAl)N	33 ± 1	326 ± 6	0.1	0.52	0.8	41
2.2	(TiZr)N	31 ± 1	310 ± 4	0.1	0.4	0.65	43

A significant influence of the condition of the surface layer of ceramic inserts on the coefficient of friction of coatings during high-temperature heating was found (Table 4). The (TiAl)N and (TiZr)N coatings deposited on "defect-free" substrates had lower average COF values—0.6 and 0.5 relative to 0.7 and 0.6 for coatings deposited on samples after diamond grinding, and a decrease in COF differences along the entire test distance in friction with a counterbody.

The assessment of the adhesive bond strength of the formed coatings with ceramic substrates, carried out within the framework of the studies, showed that the coatings (TiAl)N and (TiZr)N formed on defective substrates delaminate at relatively low loads—31 N and 32 N, respectively (Table 4). Similarly low adhesion values were obtained by the authors of Ref. [25] in a study on various nitride coatings. This is an extremely important observation, as understanding the reasons behind it can largely explain the insufficient effectiveness of coatings when applied to ceramic tools with numerous defects in the surface layer. For example, the authors of Ref. [34], who studied the effect of deposition of various PVD coatings on the characteristics of tool ceramics under the influence of external loads, place particular emphasis on the fact that the performance properties of a ceramic substrate with coatings depend mainly on their adhesion to the substrate.

It is known [69,70] that the adhesive bond strength of PVD coatings with a substrate depends on many factors and is one of the key characteristics of a coated tool. In particular, the presence and number of bonds between the contacting bodies and the area of actual contact between the coating and the substrate are of great importance, which decreases in the presence of numerous grooves and torn grains on the surface of the ceramic substrate. In addition, the presence of various defects on the ceramic insert during the deposition of thin vacuum-plasma coatings contributes to the formation of common defects in their growth in the form of porosity and deformation of crystallites. Microdefects on a ceramic substrate can lead to misorientation of the axes of growing crystallites, and a high density of microroughness leads to the formation of a large volume of porosity near the coating–substrate interface [71,72]. Thus, the increased defectiveness of the surface layer of the ceramic substrate, formed during diamond grinding, contributes to the formation of high internal stresses in the coating, which, when exposed to heat and power loads during the cutting process, can lead to both small and significant delamination centers. In other words, a defective ceramic substrate contributes to the formation of defective coatings with insufficient adhesive bond strength. The data given in Table 4 show that the (TiAl)N and (TiZr)N coatings formed on "defect-free" substrates subjected to additional lapping and polishing had significantly higher values of loads at which the coating delaminates—41 and 43 N, respectively, which is ~30% higher than the loads established for coatings formed on ceramic inserts after diamond grinding.

The dependences of the volume of worn material on the test time presented in Figure 9 make it possible to judge the resistance of (TiAl)N and (TiZr)N coatings formed on ceramic inserts with different surface layer conditions to abrasion. The coatings noticeably reduce the volume wear of Al_2O_3+TiC ceramics, both in the case of their deposition on defective substrates and in the case of formation on "defect-free" inserts—(TiAl)N by 1.4–1.5 times and (TiZr)N by 2.0 times. That is to say, even with the above-described negative effect of a defective surface layer on the quality of the formed coatings (Figures 7a and 8a) and the failure to provide adequate adhesive strength, their application to ceramic inserts certainly changes the conditions for the interaction of the contact surfaces with the counter body and inhibits the development of wear holes. Since the intensity of abrasive wear is strongly dependent on the hardness of the surface layer of the contacting pair, these changes, first of all, should be associated with a higher hardness of nitride coatings (29 GPa or more, according to Table 4) in comparison with the hardness of the original ceramics based on Al_2O_3+TiC, which is 14.82 GPa for diamond grinding and 15.1 GPa for additional lapping and polishing (Table 3). However, these changes may not be sufficient to increase the wear resistance of ceramic inserts during the cutting of hardened steels, when the contact surfaces experience not only the mechanical pressure of the highly hard material being processed but are also subjected to intense thermal loads. Therefore, for a comprehensive assessment of the contribution of the condition of the surface layer of the ceramic substrate to the performance of ceramic inserts, it is necessary to conduct full-scale cutting tests. For these purposes, the (TiZr)N coating was chosen, as it showed the best results in friction tests and under abrasive conditions compared with (TiAl)N coating.

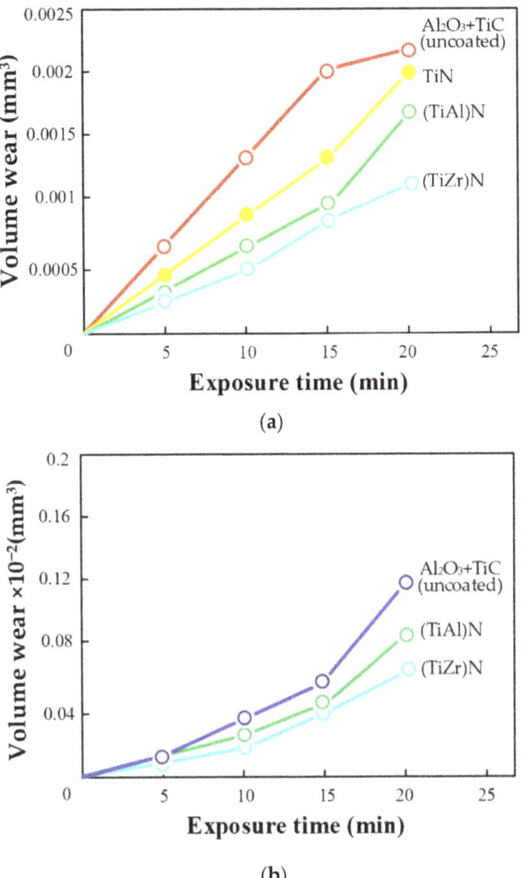

(a)

(b)

Figure 9. Dependences of volume wear of ceramic inserts made of Al_2O_3+TiC with coatings deposited on substrates after diamond grinding (**a**) and after diamond grinding, lapping and polishing (**b**) on the abrasive exposure time.

3.3. Influence of the Condition of the Surface Layer of Ceramic Inserts with Coatings on Reliability when Milling Hardened Steel

To determine the contribution of the surface layer condition of ceramic inserts to CCT performance, the following four options of face mills equipped with Al_2O_3+TiC square inserts were tested when machining prismatic 100CrMn6 hardened steel workpieces:

(1) industrially produced Al_2O_3+TiC ceramic inserts subjected to diamond grinding at the final stage of manufacturing;

(2) industrially produced Al_2O_3+TiC ceramic inserts subjected to diamond grinding and (TiZr)N coating with a thickness of ~3.7 μm;

(3) industrially produced Al_2O_3+TiC ceramic inserts subjected to diamond grinding and additional lapping and polishing;

(4) industrially produced Al_2O_3+TiC ceramic inserts subjected to diamond grinding and additional lapping and polishing and (TiZr)N coating with a thickness of ~3.7 μm.

Figure 10a,b show groups of "wear–cutting time" curves constructed based on the test results of 20 faces of industrially produced cutting inserts made of Al_2O_3+TiC without coating and after coating with (TiZr)N at cutting speed V = 380 m/min, feed S = 0.15 mm/tooth and depth t = 1 mm.

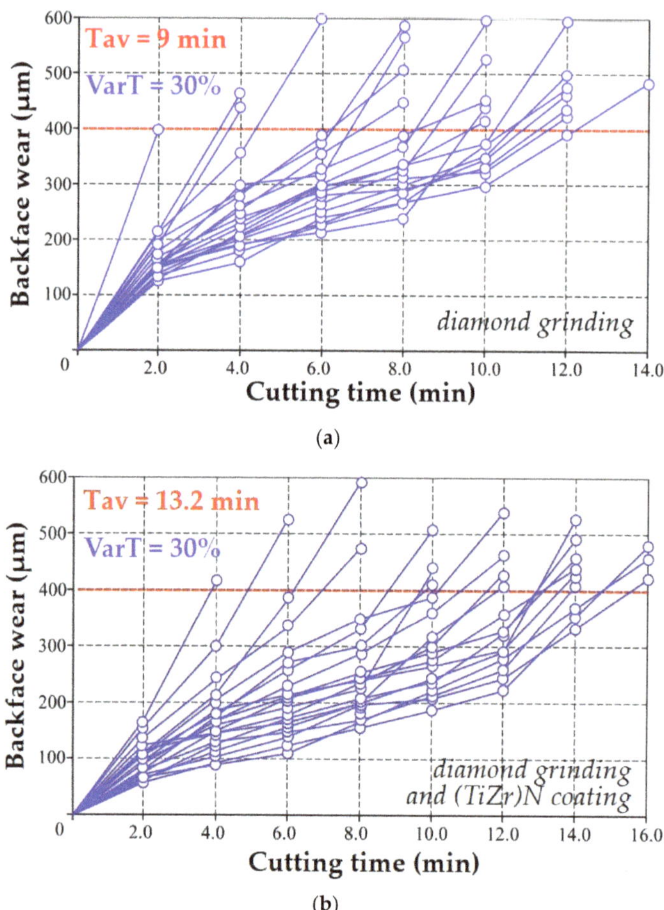

Figure 10. Group of "wear–cutting time" curves of cutting faces of ceramic inserts made of Al_2O_3+TiC after diamond grinding when milling hardened steel 100CrMn6 (V = 380 m/min, S = 0.15 mm/tooth, t = 1 mm) uncoated (**a**) and coated with (TiZr)N (**b**).

The experimental data clearly illustrate (Figure 10a) the disadvantages associated with the low reliability of CCT and limiting its industrial application. It can be seen that the curves of the "wear–cutting time" implementations have a pronounced fan-shaped character with different wear rates of the cutting faces after the completion of the run-in stage. Attention is drawn to the extremely large variation in the values of the resistance of ceramic inserts before reaching the accepted failure criterion (400 µm). The average resistance (Tav) of a batch of ceramic inserts under the selected cutting modes is 9 min, and the resistance variation (VarT) is 30%.

Figure 10b shows the "wear–cutting time" dependences of industrially produced ceramic inserts with (TiZr)N coatings. It can be seen that the formation of coatings practically does not affect the nature of the development of wear over the cutting time, and the curves have a difficult-to-predict fan-shaped character with a large dispersion of resistance (VarT is ~30%). At the same time, the coating increases the average resistance by 1.4 times (Tav is 13.2 min) of ceramic inserts made of Al_2O_3+TiC when milling hardened steel 100CrMn6 compared to an uncoated tool.

Figure 11a,b show the experimentally obtained "wear–cutting time" curves constructed based on the test results of 20 faces of industrially produced cutting inserts made of Al_2O_3+TiC, subjected to additional lapping and polishing, uncoated and after deposition of (TiZr)N coating. A comparison of the experimental data presented in Figures 10a and 11a demonstrates the pronounced changes that occur when milling hardened steel with "defect-free" ceramic inserts; the group of "wear–cutting time" curves has the form of a fairly closely intertwined bundle of curves with relatively small dispersion in resistance values. For ceramic inserts that underwent additional lapping and polishing, the Tav is 10.8 min, which, in comparison with industrially produced inserts, cannot be called a significant result (an increase of only 1.2 times). The two-fold reduction in the dispersion of resistance should be considered the most critical result; when milling hardened steel with "defect-free" ceramic inserts, the VarT is 14%. A comparison of test results of "defect-free" ceramic inserts with (TiZr)N coating (Figure 11b) with the test results of ceramics after diamond grinding (Figure 10a) shows similar changes; the obtained dependences of "wear–cutting time" have the form of an intertwining bundle of curves with a significantly smaller dispersion of resistance. VarT was reduced to 15%, which is 2 times less than that of industrially produced inserts. Simultaneously, with the increase in reliability, the use of "defect-free" ceramic inserts made of Al_2O_3+TiC with (TiZr)N coating when milling hardened steel 100CrMn6 demonstrates an increase in the average resistance by 1.7 times compared to ceramic inserts present on the market with pronounced defects in the surface layer.

To quantify the contribution of surface layer defects to the performance of coated CCTs when milling 100CrMn6-type hardened steel, Figure 12 shows the averaged curves of development of wear focused over time along the back surface of the cutting faces of Al_2O_3+TiC ceramic inserts with (TiZr)N coatings formed on substrates after diamond grinding (I) and diamond grinding, lapping and polishing (II). Additionally, Figure 12 shows data on the range of variations in the average resistance (ΔT) during the testing of 20 faces of ceramic inserts with (TiZr)N coatings, with different conditions of the surface layer (in Figure 12, the ΔT range areas are highlighted with the corresponding color). These data summarize the above results shown in Figures 10b and 11b.

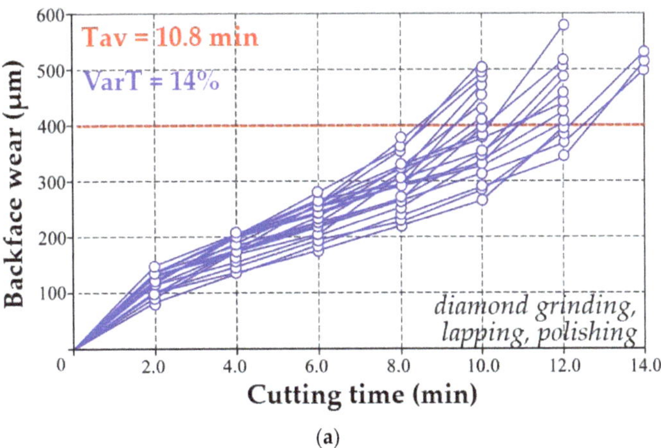

(a)

Figure 11. *Cont.*

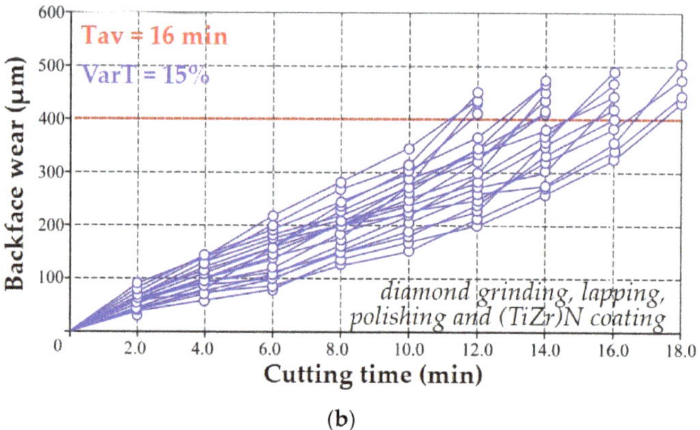

(**b**)

Figure 11. Group of "wear–cutting time" curves of cutting faces of ceramic inserts made of Al$_2$O$_3$+TiC after diamond grinding, lapping and polishing when milling hardened steel 100CrMn6 (V = 380 m/min, S = 0.15 mm/tooth, t = 1 mm) uncoated (**a**) and coated with (TiZr)N (**b**).

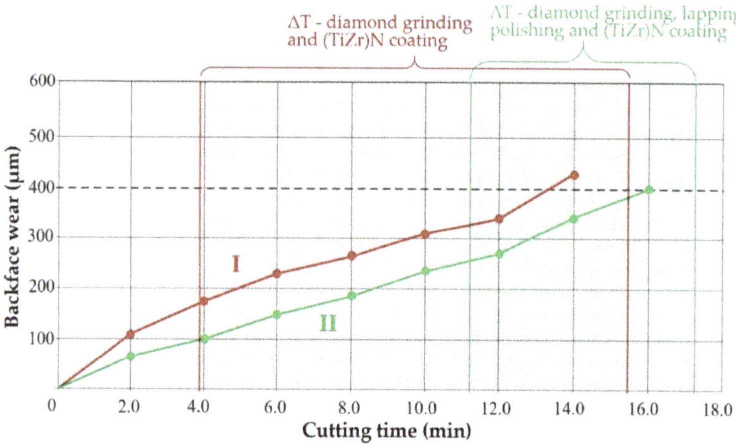

Figure 12. Average curves of wear development over time along the back surface and the range of changes in resistance (ΔT) of cutting edges of Al$_2$O$_3$+TiC ceramic inserts with (TiZr)N coatings formed on substrates after diamond grinding (I) and diamond grinding, lapping and polishing (II) when milling hardened steel 100CrMn6 (V = 380 m/min, S = 0.15 mm/tooth, t = 1 mm).

In the case of the formation of (TiZr)N coatings on "defect-free" ceramic substrates, a slight increase in the average resistance is provided—by 1.2 times in comparison with the same coatings formed on the inserts after diamond grinding. The most important result of minimizing the degree of defectiveness of the ceramic substrate on which the coating is deposited is a significant decrease in the ΔT range (by ~1.9 times) and the spread of VarT. The change in VarT in accordance with the performed calculations was 15%, which is two times less than that of ceramic inserts after diamond grinding and (TiZr)N coating.

4. Discussion

The specific features of the destruction of tool ceramics, associated with structural heterogeneity and defects present on the surface layer formed during diamond grinding, largely determine their reduced reliability (large dispersion of resistance), which is most pronounced at increased heat and power loads on the contact surfaces, including the ones of cyclic nature, which significantly limits the industrial application of CCT. At the same time, premature destruction of the cutting part can occur at various stages of tool operation—both during the run-in period and at the stage of regular wear [8,11,22].

The surface layer of industrially produced Al_2O_3+TiC ceramic inserts after diamond grinding is full of numerous defects and contains deep grooves profiled by diamond grains of the grinding wheel and the tearing of ceramic material grains occurring under the action of force loads [10,16]. The "wear–cutting time" curves have a pronounced fan-shaped character with different wear rates of the cutting edges during the use of these ceramic inserts in the conditions of milling hardened steels of 100CrMn type with increased cutting speeds and increased sections of the cut layer [9]. The resistance of a tool from one batch before reaching the accepted failure criterion has a significant variation in values (VarT is 30%), which does not provide high reliability. The use of widespread instrumental vacuum-plasma coatings, such as TiN, (TiAl)N and (TiZr)N, does not provide significant "healing" of surface layer defects formed during diamond grinding of ceramic inserts but is only able to minimize the depth of the defective layer [20,23,25]. The morphological pattern of the formed coatings largely copies the characteristic defects present on the surface layer of the tool [32–34].

The increased defectiveness of the surface layer of the ceramic substrate contributes to the formation of defective coatings (porous and discontinuous), characterized by reduced adhesive bond strength, which significantly reduces their effectiveness when applied to CCT [22,23,33]. Despite the negative impact of the defective surface layer of the ceramic substrate on the quality of the formed coatings, their deposition on the CCT changes the conditions of the contact interaction between the working surfaces of the tool and the material to be processed.

During the operation of industrially produced Al_2O_3+TiC ceramic inserts, on which the (TiZr)N coating was deposed, an increase in the average resistance (Tav) by 1.4 times is noted compared to the base-coated tool in the conditions of milling hardened steels of the 100CrMn type with increased cutting speeds and increased sections of the cut layer (Figure 10). At the same time, the coating does not solve the main problem of CCT associated with low reliability; the tool resistance has a significant variation (VarT is 30%).

The use of additional lapping and polishing of Al_2O_3+TiC ceramic inserts significantly increases resistance to abrasive wear (by 2 times, Figure 5), increases the crack resistance (by ~6%) (Table 3), determined by indentation, and also favorably affects (stabilizes) the conditions of frictional interaction with the counter body during high-temperature heating. When using the ceramic inserts in the conditions of milling hardened steels of the 100CrMn type with increased cutting speeds and increased sections of the cut layer, the "wear–cutting time" curves have the form of fairly closely intertwined bundles of curves, and the coefficient of resistance variation, which characterizes the reliability of the tool, decreases by more than two times (VarT is 14%, Figure 11a). When (TiAl)N and (TiZr)N coatings are deposited on "defect-free" ceramic substrates that underwent additional lapping and polishing, the microstructure of the coatings is determined by the features of the processes of coating synthesis, the growth of crystallites of coating elements and micro-drops formation (Figures 7b, 8b and 11b).

The index of plasticity (H/E), which characterizes the ability to resist possible deformations and fracture during cutting for coatings deposited on "defect-free" substrates, increases by an average of 10% for (TiAl)N coatings and by 14% for (TiZr)N coatings concerning similar coatings formed on ceramic samples that underwent diamond grinding and have numerous defects on the surface layer (Table 4). The use of "defect-free" Al_2O_3+TiC ceramic inserts with (TiZr)N coating during the milling of hardened steels of the 100CrMn type with increased cutting speeds and increased sections of the cut layer demonstrates an increase in average resistance (Tav) by 1.7 times compared to the ceramic inserts present on the market with noticeable defects on the surface layer (Figures 10a and 11b). Compared to ceramic inserts after diamond grinding and (TiZr)N coating, the increase in average resistance is less significant and amounts to 1.2 times (Figures 10b and 11b).

5. Conclusions

The analytical and experimental studies carried out made it possible to obtain original results, indicating a strong influence of the condition of the surface layer of ceramic inserts on their performance in milling hardened steels:

(1) During the operation of industrially produced Al_2O_3+TiC ceramic inserts with the (TiZr)N coating in milling hardened steels of the 100CrMn type, an increase in the average resistance (Tav) by 1.4 times compared to the base-coated tool was noted. However, the coating does not solve the main problem of CCT of low reliability, since the tool resistance has a significant variation (VarT is 30%).

(2) Minimizing the defectiveness of Al_2O_3+TiC ceramic inserts through the use of additional lapping and polishing increases the resistance to abrasive wear by two times and the crack resistance by ~6%. The coefficient of resistance variation, which characterizes the tool's reliability, decreases by more than two times (VarT is 14%).

(3) The use of "defect-free" Al_2O_3+TiC ceramic inserts with (TiZr)N coating demonstrates an increase in the average resistance (Tav) by 1.7 times compared to the ceramic inserts present on the market. The increase in average resistance is less significant and amounts to 1.2 times compared to ceramic inserts after diamond grinding and (TiZr)N coating.

(4) The most important result of minimizing the degree of defectiveness is a considerable decrease in the range of change in the average resistance ΔT (by ~1.9 times) and the variation in the dispersion of resistance VarT up to 15%, which is two times less than that of ceramic inserts with (TiZr)N coating.

Author Contributions: Conceptualization, M.A.V.; methodology, M.A.V. and M.M.S.; software, M.M.S. and Y.A.M.; validation, V.D.G.; formal analysis, Y.A.M.; investigation, M.A.V., V.D.G. and Y.A.M.; resources, M.M.S. and Y.A.M.; data curation, M.M.S.; writing—original draft preparation, M.A.V.; writing—review and editing, M.A.V., V.D.G. and Y.A.M.; visualization, Y.A.M.; supervision, M.A.V. and M.M.S.; project administration, M.A.V.; funding acquisition, M.A.V. All authors have read and agreed to the published version of the manuscript.

Funding: This work is funded by the state assignment of the Ministry of Science and Higher Education of the Russian Federation, Project No. 0707-2020-0025.

Institutional Review Board Statement: Not applicable.

Informed Consent Statement: Not applicable.

Data Availability Statement: Data sharing is not applicable to this article.

Acknowledgments: The study was carried out on the equipment of the Center of collective use of MSUT "STANKIN" supported by the Ministry of Higher Education of the Russian Federation (project No. 075-15-2021-695 from 26.07.2021, unique identifier RF 2296.61321X0013).

Conflicts of Interest: The authors declare no conflict of interest.

References

1. Xing, Y.; Deng, J.; Zhao, J.; Zhang, G.; Zhang, K. Cutting performance and wear mechanism of nanoscale and microscale textured Al_2O_3/TiC ceramic tools in dry cutting of hardened steel. *Int. J. Refract. Met. Hard Mater.* **2014**, *43*, 46–58. [CrossRef]
2. Lee, W.K.; Ratnam, M.M.; Ahmad, Z.A. In-process detection of chipping in ceramic cutting tools during turning of difficult-to-cut material using vision-based approach. *Int. J. Adv. Manuf. Technol.* **2016**, *85*, 1275–1290. [CrossRef]
3. Benga, G.C.; Abrao, A.M. Turning of hardened 100Cr6 bearing steel with ceramic and PCBN cutting tools. *J. Mater. Process. Technol.* **2003**, *143*, 237–241. [CrossRef]
4. Zhou, J.M.; Andersson, M.; Ståhl, J.E. Identification of cutting errors in precision hard turning process. *J. Mater. Process. Technol.* **2004**, *153–154*, 746–750. [CrossRef]
5. Grzesik, W.; Rech, J.; Wanat, T. Surface finish on hardened bearing steel parts produced by superhard and abrasive tools. *Int. J. Mach. Tools Manuf.* **2007**, *47*, 255–262. [CrossRef]
6. Brook, R.J. *Concise Encyclopedia of Advanced Ceramic Materials*, 1st ed.; Pergamon Press: Oxford, UK, 1991; p. 604. [CrossRef]
7. Rizzo, A.; Goel, S.; Luisa Grilli, M.; Iglesias, R.; Jaworska, L.; Lapkovskis, V.; Novak, P.; Postolnyi, B.O.; Valerini, D. The Critical Raw Materials in Cutting Tools for Machining Applications: A Review. *Materials* **2020**, *13*, 1377. [CrossRef] [PubMed]

8. Kuzin, V.V.; Grigor'ev, S.N.; Volosova, M.A. Effect of a TiC Coating on the Stress-Strain State of a Plate of a High-Density Nitride Ceramic Under Nonsteady Thermoelastic Conditions. *Refract. Ind. Ceram.* **2014**, *54*, 376–380. [CrossRef]

9. Kuzin, V.V.; Grigor'ev, S.N.; Fedorov, S.Y. Evaluation of Ceramic Tool Reliability with a Limited Number of Tests Based on Established Wear Criteria. *Refract. Ind. Ceram.* **2018**, *59*, 386–390. [CrossRef]

10. Grguras, D.; Kern, M.; Pusavec, F. Suitability of the full body ceramic end milling tools for high speed machining of nickel based alloy Inconel 718. *Procedia CIRP* **2018**, *77*, 630–633. [CrossRef]

11. Jiang, C.P.; Wu, X.F.; Li, J.; Song, F.; Shao, Y.F.; Xu, X.H.; Yan, P. A study of the mechanism of formation and numerical simulations of crack patterns in ceramics subjected to thermal shock. *Acta Mater.* **2012**, *60*, 4540–4550. [CrossRef]

12. Kuzin, V.V.; Grigoriev, S.N.; Volosova, M.A. The role of the thermal factor in the wear mechanism of ceramic tools: Part 1. Macrolevel. *J. Frict. Wear* **2014**, *35*, 505–510. [CrossRef]

13. Kuzin, V.V.; Grigor'ev, S.N.; Volosova, M.A. Microstructural Model of the Surface Layer of Ceramics After Diamond Grinding Taking into Account Its Real Structure and the Conditions of Contact Interaction with Elastic Body. *Refract. Ind. Ceram.* **2020**, *61*, 303–308. [CrossRef]

14. Vigneau, J.; Bordel, P.; Geslot, R. Reliability of ceramic cutting tools. *CIRP Ann.* **1988**, *37*, 101–104. [CrossRef]

15. Vereschaka, A.A.; Batako, A.D.; Krapostin, A.A.; Sitnikov, N.N.; Oganyan, G.V. Improvement in Reliability of Ceramic Cutting Tool using a Damping System and Nano-structured Multi-layered Composite Coatings. *Procedia CIRP* **2017**, *63*, 563–568. [CrossRef]

16. Wachtman, J.B.; Cannon, W.R.; Matthewson, M.J. *Mechanical Properties of Ceramics*, 2nd ed.; Wiley: Hoboken, NJ, USA, 2009; Volume XVI, p. 479, ISBN 978-0-471-73581-6.

17. Vereschaka, A.; Tabakov, V.; Grigoriev, S.; Sitnikov, N.; Andreev, N.; Milovich, F. Investigation of wear and diffusion processes on rake faces of carbide inserts with Ti-TiN-(Ti,Al,Si)N composite nanostructured coating. *Wear* **2018**, *416–417*, 72–80. [CrossRef]

18. Sobol, O.V.; Andreev, A.A.; Grigoriev, S.N.; Gorban, V.F.; Volosova, M.A.; Aleshin, S.V.; Stolbovoi, V.A. Effect of high-voltage pulses on the structure and properties of titanium nitride vacuum-arc coatings. *Met. Sci. Heat Treat.* **2012**, *54*, 195–203. [CrossRef]

19. Vereschaka, A.A.; Grigoriev, S.N.; Sitnikov, N.N.; Oganyan, G.V.; Batako, A. Working efficiency of cutting tools with multilayer nano-structured Ti-TiCN-(Ti,Al)CN and Ti-TiCN-(Ti,Al,Cr)CN coatings: Analysis of cutting properties, wear mechanism and diffusion processes. *Surf. Coat. Technol.* **2017**, *332*, 198–213. [CrossRef]

20. Long, Y.; Zeng, J.; Yu, D.; Wu, S. Microstructure of TiAlN and CrAlN coatings and cutting performance of coated silicon nitride inserts in cast iron turning. *Ceram. Int.* **2014**, *40*, 9889–9894. [CrossRef]

21. Sobol, O.V.; Andreev, A.A.; Grigoriev, S.N.; Volosova, M.A.; Gorban, V.F. Vacuum-arc multilayer nanostructured TiN/Ti coatings: Structure, stress state, properties. *Met. Sci. Heat Treat.* **2012**, *54*, 28–33. [CrossRef]

22. Volosova, M.; Grigoriev, S.; Metel, A.; Shein, A. The Role of Thin-Film Vacuum-Plasma Coatings and Their Influence on the Efficiency of Ceramic Cutting Inserts. *Coatings* **2018**, *8*, 287. [CrossRef]

23. Wang, H.; Zhang, J.; Yi, M.; Xiao, G.; Chen, Z.; Sheng, C.; Zhang, P.; Xu, C. Simulation study on influence of coating thickness on cutting performance of coated ceramic cutting tools. *J. Phys. Conf. Ser.* **2020**, *1549*, 032140. [CrossRef]

24. Grigoriev, S.N.; Vereschaka, A.A.; Fyodorov, S.V.; Sitnikov, N.N.; Batako, A.D. Comparative analysis of cutting properties and nature of wear of carbide cutting tools with multi-layered nano-structured and gradient coatings produced by using of various deposition methods. *Int. J. Adv. Manuf. Technol.* **2017**, *90*, 3421–3435. [CrossRef]

25. Liu, W.; Chu, Q.; Zeng, J.; He, R.; Wu, H.; Wu, Z.; Wu, S. PVD-CrAlN and TiAlN coated Si3N4 ceramic cutting tools −1. Microstructure, turning performance and wear mechanism. *Ceram. Int.* **2017**, *43*, 8999–9004. [CrossRef]

26. Long, Y.; Zeng, J.; Wu, S. Cutting performance and wear mechanism of Ti–Al–N/Al–Cr–O coated silicon nitride ceramic cutting inserts. *Ceram. Int.* **2014**, *40*, 9615–9620. [CrossRef]

27. Vereschaka, A.A.; Volosova, M.A.; Grigoriev, S.N.; Vereschaka, A.S. Development of wear-resistant complex for high-speed steel tool when using process of combined cathodic vacuum arc deposition. *Procedia CIRP* **2013**, *9*, 8–12. [CrossRef]

28. Das, S.R.; Dhupal, D.; Kumar, A. Experimental investigation into machinability of hardened AISI 4140 steel using TiN coated ceramic tool. *Measurement* **2015**, *62*, 108–126. [CrossRef]

29. Aslantas, K.; Ucun, İ.; Çicek, A. Tool life and wear mechanism of coated and uncoated Al_2O_3/TiCN mixed ceramic tools in turning hardened alloy steel. *Wear* **2012**, *274–275*, 442–451. [CrossRef]

30. Bensouilah, H.; Aouici, H.; Meddour, I.; Yallese, M.A.; Mabrouki, T.; Girardin, F. Performance of coated and uncoated mixed ceramic tools in hard turning process. *Measurement* **2016**, *82*, 1–18. [CrossRef]

31. Dobranski, L.A.; Mikula, J. Structure and properties of PVD and CVD coated Al_2O_3+TiC mixed oxide tool ceramics for dry on high speed cutting processes. *J. Mater. Process. Technol.* **2005**, *164–165*, 822–831. [CrossRef]

32. Kumar, C.S.; Patel, S.K. Experimental and numerical investigations on the effect of varying AlTiN coating thickness on hard machining performance of Al_2O_3-TiCN mixed ceramic inserts. *Surf. Coat. Technol.* **2017**, *309*, 266–281. [CrossRef]

33. Souza, J.V.C.; Macedo, O.M.; Nono, M.C.A.; Machado, J.P.B.; Pimenta, M.; Ribeiro, M.V. Si3N4 ceramic cutting tool sintered with CeO_2 and Al_2O_3 additives with AlCrN coating. *Mat. Res.* **2011**, *14*, 514–518. [CrossRef]

34. Staszuk, M.; Pakuła, D.; Chladek, G.; Pawlyta, M.; Pancielejko, M.; Czaja, P. Investigation of the structure and properties of PVD coatings and ALD + PVD hybrid coatings deposited on sialon tool ceramics. *Vacuum* **2018**, *154*, 272–284. [CrossRef]

35. Nairy, K.S.; Rao, K.A. Tests of coefficient of variation of normal population. *Commun. Stat. Simul. Comput.* **2003**, *32*, 641–661. [CrossRef]

36. Grigoriev, S.N.; Teleshevskii, V.I. Measurement problems in technological shaping processes. *Meas. Tech.* **2011**, *54*, 744–749. [CrossRef]
37. Albertelli, P.; Mussi, V.; Monno, M. Development of generalized tool life model for constant and variable speed turning. *Int. J. Adv. Manuf. Technol.* **2022**, *118*, 1885–1901. [CrossRef]
38. Grigoriev, S.N.; Sinopalnikov, V.A.; Tereshin, M.V.; Gurin, V.D. Control of parameters of the cutting process on the basis of diagnostics of the machine tool and workpiece. *Meas. Tech.* **2012**, *55*, 555–558. [CrossRef]
39. Broderick, M.; Turner, S.; Ridgway, K. Correlation between tool life and cutting force coefficient as the basis for a novel method in accelerated MWF performance assessment. *Procedia CIRP* **2021**, *101*, 366–369. [CrossRef]
40. Grigoriev, S.N.; Gurin, V.D.; Volosova, M.A.; Cherkasova, N.Y. Development of residual cutting tool life prediction algorithm by processing on CNC machine tool. *Materwiss. Werksttech.* **2013**, *44*, 790–796. [CrossRef]
41. Arai, S.A.; Wilson, S.A.; Corbett, J.; Whatmore, R.W. Ultra-precision grinding of PZT ceramics—Surface integrity control and tooling design. *Int. J. Mach. Tools Manuf.* **2009**, *49*, 998–1007. [CrossRef]
42. Canneto, J.J.; Cattani-Lorente, M.; Durual, S.; Wiskott, A.H.W.; Scherrer, S.S. Grinding damage assessment on four high-strength ceramics. *Dent. Mater.* **2016**, *32*, 171–182. [CrossRef]
43. Zhang, C.; Liu, H.; Zhao, Q.; Guo, B.; Wang, J.; Zhang, J. Mechanisms of ductile mode machining for AlON ceramics. *Ceram. Int.* **2020**, *46*, 1844–1853. [CrossRef]
44. Grigoriev, S.N.; Metel, A.S.; Fedorov, S.V. Modification of the structure and properties of high-speed steel by combined vacuum-plasma treatment. *Met. Sci. Heat Treat.* **2012**, *54*, 12. [CrossRef]
45. Metel, A.S.; Grigoriev, S.N.; Melnik, Y.A.; Bolbukov, V.P. Broad beam sources of fast molecules with segmented cold cathodes and emissive grids. *Instrum. Exp. Tech.* **2012**, *55*, 122–130. [CrossRef]
46. Grigoriev, S.; Metel, A. Plasma- and Beam-Assisted Deposition Methods. In *Nanostructured Thin Films and Nanodispersion Strengthened Coatings*; NATO Science Series II: Mathematics, Physics and Chemistry; Voevodin, A.A., Shtansky, D.V., Levashov, E.A., Moore, J.J., Eds.; Springer: Berlin/Heidelberg, Germany, 2004; Volume 155, pp. 147–154. ISBN 9781402022227. [CrossRef]
47. Metel, A.; Bolbukov, V.; Volosova, M.; Grigoriev, S.; Melnik, Y. Source of metal atoms and fast gas molecules for coating deposition on complex shaped dielectric products. *Surf. Coat. Technol.* **2013**, *225*, 34–39. [CrossRef]
48. Grigoriev, S.N.; Melnik, Y.A.; Metel, A.S.; Panin, V.V.; Prudnikov, V.V. A compact vapor source of conductive target material sputtered by 3-keV ions at 0.05-Pa pressure. *Instrum. Exp. Tech.* **2009**, *52*, 731–737. [CrossRef]
49. Pat, S.; Ekem, N.; Akan, T.; Küsmüs, Ö.; Demirkol, S.; Vladoiu, R.; Lungu, C.P.; Musa, G. Study on Termionic Vacuum Arc—A Novel and Advanced Technology for Surface Coating. *J. Optoelectron. Adv. Mater.* **2005**, *7*, 2495–2499.
50. Vladoiu, R.; Tichý, M.; Mandes, A.; Dinca, V.; Kudrna, P. Thermionic Vacuum Arc—A Versatile Technology for Thin Film Deposition and Its Applications. *Coatings* **2020**, *10*, 211. [CrossRef]
51. Vetter, J.; Eriksson, A.O.; Reiter, A.; Derflinger, V.; Kalss, W. Quo Vadis: AlCr-Based Coatings in Industrial Applications. *Coatings* **2021**, *11*, 344. [CrossRef]
52. Zhang, Z.; Zhang, L.; Yuan, H.; Qiu, M.; Zhang, X.; Liao, B.; Zhang, F.; Ouyang, X. Tribological Behaviors of Super-Hard TiAlN Coatings Deposited by Filtered Cathode Vacuum Arc Deposition. *Materials* **2022**, *15*, 2236. [CrossRef]
53. Metel, A.S.; Grigoriev, S.N.; Melnik, Y.A.; Prudnikov, V.V. Glow discharge with electrostatic confinement of electrons in a chamber bombarded by fast electrons. *Plasma Phys. Rep.* **2011**, *37*, 628–637. [CrossRef]
54. Grigoriev, S.N.; Fominski, V.Y.; Gnedovets, A.G.; Romanov, R.I. Experimental and numerical study of the chemical composition of WSex thin films obtained by pulsed laser deposition in vacuum and in a buffer gas atmosphere. *Appl. Surf. Sci.* **2012**, *258*, 7000–7007. [CrossRef]
55. Grigoriev, S.; Melnik, Y.; Metel, A. Broad fast neutral molecule beam sources for industrial scale beam-assisted deposition. *Surf. Coat. Technol.* **2002**, *156*, 44–49. [CrossRef]
56. Metel, A.; Bolbukov, V.; Volosova, M.; Grigoriev, S.; Melnik, Y. Equipment for deposition of thin metallic films bombarded by fast argon atoms. *Instrum. Exp. Tech.* **2014**, *57*, 345–351. [CrossRef]
57. Mei, Z.; Lu, Y.; Lou, Y.; Yu, P.; Sun, M.; Tan, X.; Zhang, J.; Yue, L.; Yu, H. Determination of Hardness and Fracture Toughness of Y-TZP Manufactured by Digital Light Processing through the Indentation Technique. *Biomed. Res. Int.* **2021**, *2021*, 6612840. [CrossRef]
58. Franco Steier, V.; Ashiuchi, E.S.; Reißig, L.; Araújo, J.A. Effect of a Deep Cryogenic Treatment on Wear and Microstructure of a 6101 Aluminum Alloy. *Adv. Mater. Sci. Eng.* **2016**, *2016*, 1582490. [CrossRef]
59. Shulepov, I.A.; Kashkarov, E.B.; Stepanov, I.B.; Syrtanov, M.S.; Sutygina, A.N.; Shanenkov, I.; Obrosov, A.; Weiß, S. The Formation of Composite Ti-Al-N Coatings Using Filtered Vacuum Arc Deposition with Separate Cathodes. *Metals* **2017**, *7*, 497. [CrossRef]
60. Oliver, W.C.; Pharr, G.M. An improved technique for determining hardness and elastic modulus using load and displacement sensing indentation experiments. *J. Mater. Res.* **1992**, *7*, 1564–1583. [CrossRef]
61. Kim, Y.S.; Park, H.J.; Lim, K.S.; Hong, S.H.; Kim, K.B. Structural and Mechanical Properties of AlCoCrNi High Entropy Nitride Films: Influence of Process Pressure. *Coatings* **2020**, *10*, 10. [CrossRef]
62. Choudhary, R.K.; Mishra, P. Use of Acoustic Emission During Scratch Testing for Understanding Adhesion Behavior of Aluminum Nitride Coatings. *J. Mater. Eng. Perform.* **2016**, *25*, 2454–2461. [CrossRef]
63. Grigoriev, S.N.; Kozochkin, M.P.; Sabirov, F.S.; Kutin, A.A. Diagnostic Systems as Basis for Technological Improvement. *Procedia CIRP* **2012**, *1*, 599–604. [CrossRef]

64. Kazlauskas, D.; Jankauskas, V.; Tučkutė, S. Research on Tribological Characteristics of Hard Metal WC-Co Tools with TiAlN and CrN PVD Coatings for Processing Solid Oak Wood. *Coatings* **2020**, *10*, 632. [CrossRef]
65. Wang, H.; Huang, Z.; Qi, J.; Wang, J. A new methodology to obtain the fracture toughness of YAG transparent ceramics. *J. Adv. Ceram.* **2019**, *8*, 418–426. [CrossRef]
66. Muboyadzhyan, S.A.; Budinovskii, S.A.; Gorlov, D.S.; Doronin, O.N. Structure and Microporosity of Ion-Plasma Condensed Coatings Deposited from a Two-Phase Vacuum-Arc Discharge Plasma Flow Containing Evaporated Material Microdroplets. *Russ. Metall.* **2019**, *2019*, 52–62. [CrossRef]
67. Ryabchikov, A.I.; Sivin, D.O.; Bumagina, A.I. Physical mechanisms of macroparticles number density decreasing on a substrate immersed in vacuum arc plasma at negative high-frequency short-pulsed biasing. *Appl. Surf. Sci.* **2014**, *305*, 487–491. [CrossRef]
68. Sobol, O.V.; Pogrebnyak, A.D.; Beresnev, V.M. Effect of the preparation conditions on the phase composition, structure, and mechanical characteristics of vacuum-Arc Zr-Ti-Si-N coatings. *Phys. Met. Metallogr.* **2011**, *112*, 188. [CrossRef]
69. Sveen, S.; Andersson, J.M.; M'Saoubi, R.; Olsson, M. Scratch adhesion characteristics of PVD TiAlN deposited on high speed steel, cemented carbide and PCBN substrates. *Wear* **2013**, *308*, 133–141. [CrossRef]
70. Khlifi, K.; Ben Cheikh Larbi, A. Investigation of adhesion of PVD coatings using various approaches. *Surf. Eng.* **2013**, *29*, 555–560. [CrossRef]
71. Dobrzański, L.A.; Żukowska, L.W. Gradient PVD coatings deposited on the sintered tool materials. *Arch. Mater. Sci. Eng.* **2011**, *48*, 103–111.
72. Tillmann, W.; Stangier, D.; Denkena, B.; Grove, T.; Lucas, H. Influence of PVD-coating technology and pretreatments on residual stresses for sheet-bulk metal forming tools. *Prod. Eng. Res. Devel.* **2016**, *10*, 17–24. [CrossRef]

 *coatings*

Article

High Temperature Low Friction Behavior of h-BN Coatings against ZrO$_2$

Qunfeng Zeng

Key Laboratory of Education Ministry for Modern Design and Rotor-Bearing System, Xi'an Jiaotong University, Xi'an 710049, China; zengqf1949@gmail.com

Abstract: This paper presents high temperature low friction behaviors of the h-BN coatings, which were deposited on high-speed tool steel by radio frequency magnetron sputtering. A tribometer was used to investigate high temperature tribological properties of h-BN coatings against ZrO$_2$ from 500 °C to 800 °C. The surface morphology, mechanical properties and chemical states of the worn surface of the friction pair were characterized and investigated systemically. The experimental results show that h-BN coatings are of significant importance to improve high temperature tribological properties of steel. Moreover, it is found that high temperature super low friction of the friction pairs is successfully achieved due to tribochemistry, which plays a key role in forming the in-situ generated Fe$_2$O$_3$/h-BN composites on the worn surface of h-BN coatings. CoFs of the friction pair are as super low as about 0.02 at 800 °C and around 0.03 at 600 °C at the stable stage. The high temperature super low friction mechanism of the friction pair is discussed in detail. The present study opens a new strategy to achieve high temperature super low friction of the friction system during sliding.

Keywords: hexagonal boron nitride coatings; in-situ generated composite; tribochemistry; high temperature low friction

Citation: Zeng, Q. High Temperature Low Friction Behavior of h-BN Coatings against ZrO$_2$. *Coatings* **2022**, *12*, 1772. https://doi.org/10.3390/coatings12111772

Academic Editor: Sergey N. Grigoriev

Received: 17 October 2022
Accepted: 16 November 2022
Published: 19 November 2022

Publisher's Note: MDPI stays neutral with regard to jurisdictional claims in published maps and institutional affiliations.

1. Introduction

In recent years, many high-end pieces of equipment have been working under high operating temperature in industrial applications such as hot metal forming and aerospace [1,2]. The beneficial lubrication of self-lubricating materials may be lost due to high temperature friction heating during sliding contact in addition to high environmental temperature. How to achieve good lubrication performances of the friction system at high temperature has become a very urgent tribological problem [3–5]. The lubricating oils and greases have good lubricating performances at ambient temperature, but the lubrication performance is reduced or even completely fails due to the decomposition or deterioration of the lubricating oils and greases at high temperature especially in extremely harsh environments [6,7]. The operating temperature of the machine parts in the field of hot metal forming and aerospace is higher than 300 °C. There is observable oxidation, even failure, of high performance solid lubricants, such as DLC films, around 300 °C [8–10]. Therefore, other solid lubricants, such as 2D materials, are proposed to improve the lubrication performances of the machine parts above 300 °C [11,12]. The friction heating of the friction pair is frequently generated during sliding and a tribochemical reaction at the interface of the friction pair is found at the worn surface and a tribofilm may be formed, improving the tribological properties of the friction pair. This layer of tribofilm as a protective film separates the two rubbing surfaces from direct contact reducing friction and wear. However, many solid lubricants (e.g., graphite, MoS$_2$ and WS$_2$) can be degraded or oxidized in ambient air at high temperature [13–15].

Hexagonal boron nitride (h-BN) exhibits a good combination of the properties of high chemical stability, high mechanical strength, high thermal conductivity and low density of surface dangling bonds. The h-BN has a lamellar crystal structure and is served as solid lubricant with low coefficient of friction (CoF), even superlubricity [16–21]. The h-BN as solid lubricant is used in two ways. One is that h-BN powders are used as solid lubricants in the metal or ceramic matrix materials. Chen et al. studied the friction behaviors of the SiC/h-BN composites with different volume fractions of h-BN from room temperature to 900 °C [22]. The ZrO_2/h-BN/SiC composite exhibits good lubricating behaviors, as low as below 0.3 of CoF above 800 °C [23]. CoF and wear rate of Ag/h-BN-containing Ni-based composites were found to decrease with the increasing temperature from room temperature to 600 °C [24]. TiB_2 and TiN playing a role in the wear-resistance were synthesized in situ in the Ni60/h-BN coatings on Ti-6Al-4V alloys during high power laser irradiation, and the metal oxides (TiO_2, Al_2O_3, NiO and Fe_2O_3) were synthesized in situ during high temperature sliding, exhibiting good anti-friction behavior at 600 °C [25]. The h-BN powders were used to the sliding interface of Si_3N_4 against die steel H13 at 800 °C. The anti-wear behaviors of h-BN lubricating film are featured to the gradual undermining and complete damage [26]. Moreover, h-BN has been found to perform worse than graphite in terms of friction and material transfer in high temperature aluminum forming [27]. The other way is that h-BN is used as solid lubricant coating with good tribological properties for the sliding components made of the substrate steels at high temperature [28,29]. The h-BN coating prepared from a polyborazylene (PBN) polymeric precursor was deposited on titanium-based substrates and annealed via infra-red irradiation in a rapid thermal annealing (RTA) furnace. CoF was reduced from 0.72 for the Ti/stainless tribosystem to 0.35 for the Ti/h-BN/stainless tribosystem at 360 °C [19]. The tribological characteristics of single-layer h-BN were investigated to elucidate the feasibility as a protective coating layer and solid lubricant for micro- and nano-devices. The results indicate that the friction of h-BN before failure was orders of magnitude smaller than that of a SiO_2/Si substrate and the feasibility of atomically thin h-BN as a protective coating layer and solid lubricant [30]. It is well known that the tribological properties are strongly dependent on the operating parameters including temperature and counterpart materials [31–33]. However, the limited papers discussed the anti-friction mechanism of h-BN coatings at high temperature.

In the present paper, the h-BN coatings were deposited on the high-speed tool steel by radio frequency magnetron sputtering method, and the tribological properties of the friction pair of h-BN coatings on the disc against ZrO_2 ball in the range of temperatures from 500 °C to 800 °C was systemically studied. The anti-friction behaviors and mechanisms of the friction pair at high temperature are systemically discussed.

2. Experimental Details

The h-BN coatings were deposited on the high-speed tool steel discs by radio frequency magnetron sputtering system in an Ar atmosphere with an h-BN target ($\Phi76.2 \times 3$ mm with the purity of 99.9% and the back target of the copper with the thickness of 2 mm, Quanzhou Qijin New Material Technology Co. Ltd., Quanzhou, China). The chamber was pumped to 2×10^{-5} Pa prior to deposition. The heat treatment of the steel is the heating quenching and tempering. The steel discs, which is steel with high hardness, high wear resistance and high heat resistance, were polished by SiC paper with $400^{\#}$, $600^{\#}$, $800^{\#}$, $1200^{\#}$, $1500^{\#}$ and $2400^{\#}$ and chenille sandpaper to surface roughness around 0.02 μm. The discs with a radius of 30 mm and a height of 5 mm were ultrasonically cleaned around 10 min with absolute ethanol before sputtering. The substrates were mounted on the rotating sample holder. The sample holder was mounted motionless during the deposition. The distance between the target and the target was around 20 mm. The steel discs were firstly etched for 10 min by Ar ions to remove the contamination on surface. Then, Ti intermediate layer of 100 nm in thickness was prepared to enhance the bonding strength between the coatings and steel substrate by Ti target ($\Phi76.2 \times 3$ mm, purity of 99.995%, Zhongnuo Advanced Material Co. Ltd., Beijing, China). Ti intermediate layer buffers the lattice matching and stress between the BN layer and steel surface, and the intermediate layer can improve the service life of the h-BN coatings. Finally, h-BN coatings were deposited for 180 min at 300 °C and the flow rate was about 15 sccm. The power was 300 W. The thickness of h-BN coatings was about 600 nm. A ZrO_2 ball of 9.5 mm diameter was used as a rubbing pair.

The tribotests were performed by the tribometer at different temperatures. The samples were ultrasonically cleaned with absolute ethanol for 10 min and dried with dry air before tribotests. The load is 2 N and the sliding speed was 0.1 m/s for all tribotests in order to ensure the same friction environmental conditions. The Poisson's ratio and elastic modulus of ZrO_2 ball were 0.3 and 220 GPa and the Poisson's ratio and elastic modulus of the steel were 0.3 and 206 GPa, the maximum contact pressure was about 0.95 GPa according to the Hertz contact theory. All friction tests were repeated three times to obtain a high level of reproducibility of results. The testing temperatures were 500 °C, 600 °C, 700 °C and 800 °C. The samples were heated to the preset temperature and maintained at this temperature for 15 min to make sure the heating balance and eliminate the effect of the thermal deformation of the substrate and coatings. Then, load was applied to the samples, the tribotest was conducted and data was recorded. A new sample was cleaned and used for each tribotest condition to compare the experimental results.

The microstructure of h-BN coatings was studied by X-ray diffraction (XRD, D8-Advance, Bruker, Saarbrücken, Germany) with a scan rate of 0.5° per second in the range of 2θ form 20–90°. Scanning electron microscope (SEM, MALA3 LMH, TESCAN, Brno, Czech) and Laser scanning confocal microscopy (LSCM, OLS4000, Olympus, Tokyo, Japan) were used to observe the morphologies of the worn surface and the depth of the wear scar, and the wear rate of the coatings was calculated after tribotests. After the tribotesting, SEM was immediately used to observe the morphologies of the worn surface.

The mechanical properties of h-BN coatings were studied using the nanoindentation test method to measure elastic modulus and hardness by a nanoindentation tester (Ti950, Hysitron, Eden Prairie, Hennepin, MN, USA) with a Berkovich diamond indenter. The effective tip radius was 50 nm and the indenter depth was 55 nm. The loading and unloading were both carried out at a constant indentation speed of 5 nm/s, and the loading time and the unloading time were both 11 s. In order to eliminate the influence of the substrate on the hardness and elastic modulus of h-BN coating, the indentation depth was larger than 10% of the actual thickness of h-BN coating. Scratch testing was carried out on the sample using the micro scratch tester. The tester was used as a progressive loading device. The initial load was 0, the final load was 5 N and a loading rate of 50 N/mm was used for the scratch tests. The friction force was measured and an accelerometer detected the acoustic emission produced as the coating is damaged. The value of the critical load was then determined using these traces in conjunction with an optical microscope. Surface roughness is most important in influencing friction. The surface morphology and roughness of h-BN coating was observed and characterized by atomic force microscope (AFM, Innova, Bruker, Saarbrücken, Germany). Raman spectroscopy with 633 nm He-Ne laser excitation source and 1 cm^{-1} resolution (HR800, Horiba Jobin Yvon, Paris, France) was used to analyze the microstructure and surface morphology of the friction pair.

3. Results and Discussion

Figure 1 shows the XRD pattern of h-BN coatings. The characterization diffraction peaks at 28°, 41° and 43° were detected, and assigned to (002), (100) and (101) crystallographic planes of h-BN, respectively. These particles are crystalline in coatings with sizes in the range of 80–120 nm calculated by the Debye-Sherrer equation [34]. The peaks at 45°, 65° and 82° are the characteristic peaks of Fe from the steel substrate. The three peaks at 37°, 44° and 73° maybe the characteristic peaks of Fe_3O_4. Moreover, the weak peak at 46° about TiO_2 phase is also observed in the XRD pattern corresponding to TiO_2 (004) phase from Ti intermediate layer. The weak peak at 35° is due to the presence of γ-Fe_2O_3 phase matched with the standard [35]. It is clearly concluded from the measurement of XRD that h-BN coatings are prepared on the steel substrate and there is also γ-Fe_2O_3, Fe_3O_4 and TiO_2 on the steel surface. The magnetron sputtering is a clean process. However, the sputtering temperature is around 300 °C, the steel substrate maybe oxidized to Fe_2O_3, Fe_3O_4 and Ti to Ti-O in the coating. These peaks of Fe_2O_3, Fe_3O_4 and Ti-O in XRD are relatively weak, which means there are only a few in the coating. Figure 2 shows the surface topography of h-BN coatings scanned on the area of 10 μm× 10 μm sizes, which gives an indication of the height and depth dimensions of h-BN coatings in AFM images. The peak-to-valley roughness was found in the range of 5–15 nm. The surfaces of h-BN coatings were slightly smooth.

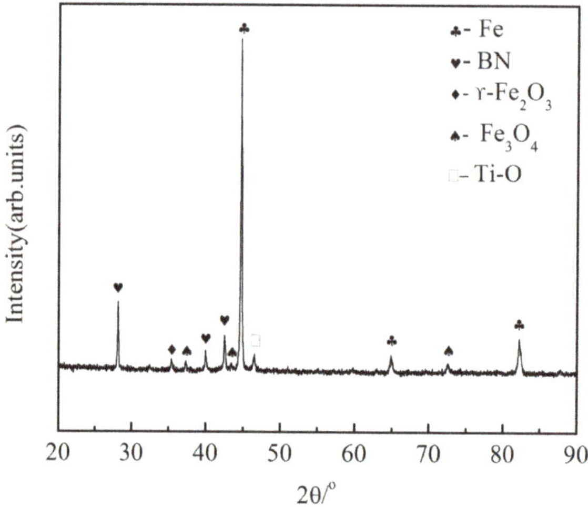

Figure 1. XRD of h-BN coatings.

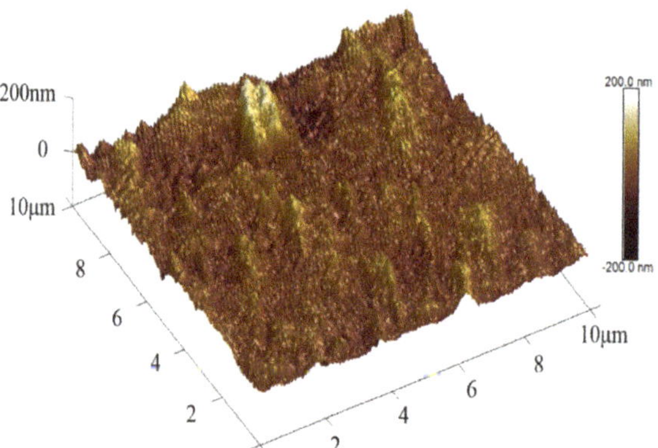

Figure 2. AFM image of h-BN coatings.

Figure 3 shows the loading-unloading curve of h-BN coatings. The h_f and h_{max} are the residual indentation depth and the maximum indentation depth, respectively. The P_{max} is the peak load at h_{max}. The results show that the hardness of the coatings is 3.42 GPa, the elastic modulus is 46.29 GPa, and the H/E value of the coating is 0.074. It is found that the maximum displacement of h-BN coatings is 55.5 nm and h_f is 8nm, and the elastic recovery rate is 85.6%. It is well known that the large recoverable strain is important for improving the tribological properties of the friction system apart from high H/E value because large recoverable strain is restored at initial state.

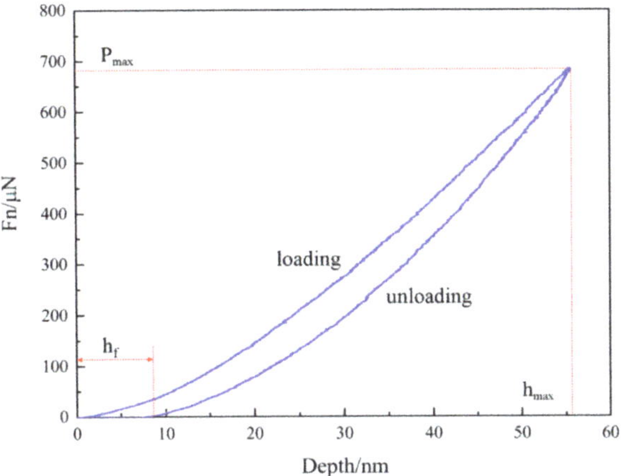

Figure 3. The representative load-depth curves of h-BN coatings.

Figure 4 shows the surface topography and the acoustic emission signal of scratches of h-BN coatings. The initial load is 0, the final load is 5 N and the loading rate is 50 mN/s. The total length of the scratch is set to 2 mm. The force L_{c1} of BN coatings was measured. It is found from the curve that the acoustic emission signals begin to fluctuate when the load is 2.3 N, meaning that this load reaches the critical load of h-BN coatings. The acoustic emission signal of the indenter prior to the critical load is very weak and h-BN coatings are peeled off slightly from the scratch morphology. The indenter was penetrated to h-BN coatings and even contacted the transition layer or the substrate with the increase in load.

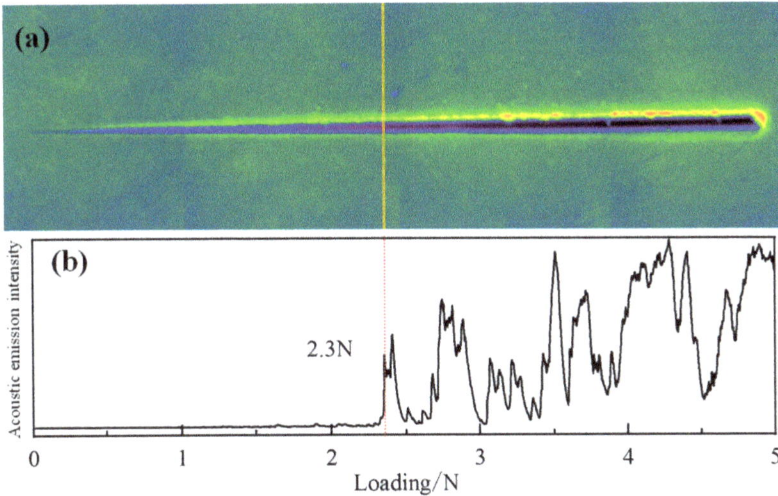

Figure 4. Detailed analysis of h-BN coatings. (**a**) microscopic record, (**b**) the acoustic emission record showing the critical load of the coatings first penetration.

Figure 5 shows the CoF curves of h-BN coatings in the atmospheric environment from 500 to 800 °C. The tendency of the CoF curve is different in each CoF curve among all the temperatures according to Figure 5. Figure 6 shows the details of the CoF curves of h-BN coatings in the atmospheric environment from 500 to 800 °C. It shows that h-BN coatings are beneficial to improve high temperature antifriction behaviors of steel. The tendency of the CoF curve is slightly different although all initial CoFs are the maximum value in each CoF curve among all the temperatures. At 500 °C, CoF decreased first and then increased slightly with the increase in sliding time. The initial CoF was 0.39 at 500 °C, and then decreased gradually, and the average CoF was around 0.13 at the steady stage. At 600 °C, the initial CoF was around 0.48, and then decreased to a low value; however, this CoF value at the same time was little higher than that at 500 °C. At about 150 s, CoF reached around 0.45 at 610 s, and finally decreased around 0.05, as shown in Figure 6b. At 700 °C, CoF decreased from 0.3 to 0.05 for the period of 140 s, and then fluctuated with the sliding time, average CoF was about 0.07. At 800 °C, CoFwas as high as 0.72 in the initial stage, and then decreased and increased alternately fluctuations, and CoF reached to the minim value of 0.017 around 1800 s, and then CoF increased again with the increase in the remaining sliding time.

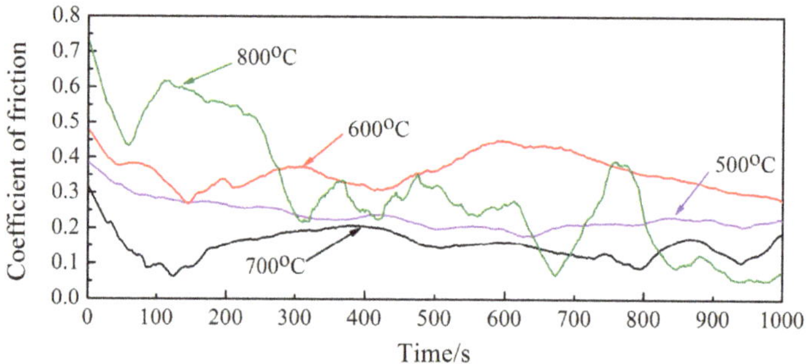

Figure 5. CoF of h-BN coatings under different temperatures.

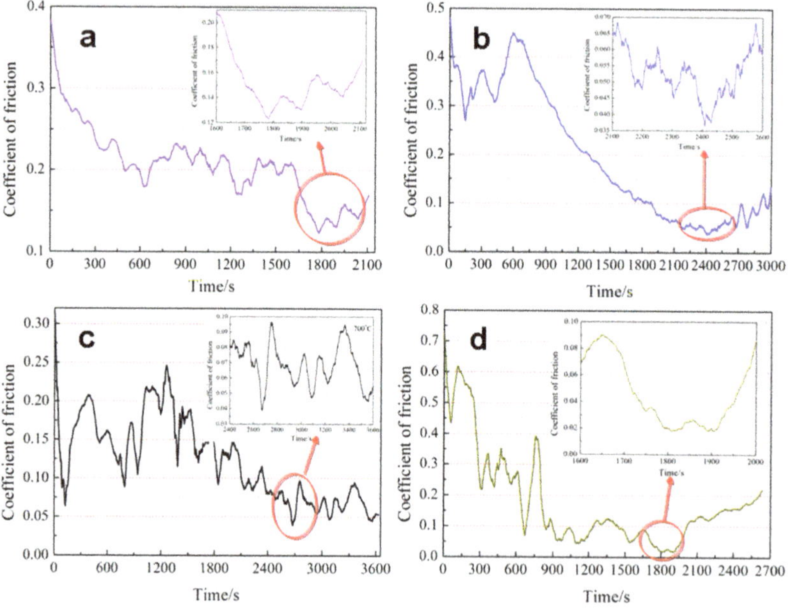

Figure 6. CoF of BN coatings under different temperatures in details: (**a**) 500 °C, (**b**) 600 °C, (**c**) 700 °C and (**d**) 800 °C.

Figure 7 shows the average CoF of the uncoated steel and h-BN coatings on the steel under different temperatures. CoFs of the uncoated steel are also listed in the reference [3].

It is found that CoFs of the steel decrease and CoFs of h-BN coatings decrease first and then fluctuate from 500 °C to 800 °C. Moreover, CoFs of h-BN coatings are much lower than those of the uncoated steel at the same temperature. It is of interest that h-BN coatings exhibit high temperature superlubricity at 800 °C.

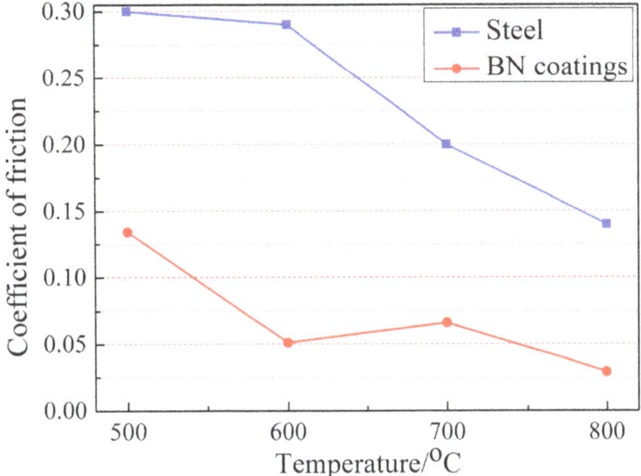

Figure 7. CoF of the uncoated steel and h-BN coatings under different temperatures.

Figures 8 and 9 show the surface topography of the worn surface of the friction pair, respectively. There was an obvious circle wear scar and wear debris on the ball surface. The wear debris were not only adhered to the worn surface and but also pushed on the edge of the worn surface along the sliding direction. The widths of disc are 818 μm, 1163 μm, 909 μm and 788 μm and the widths of ZrO_2 ball are 831 μm, 1273 μm, 918 μm, and 1202 μm from 500 °C to 800 °C, respectively.

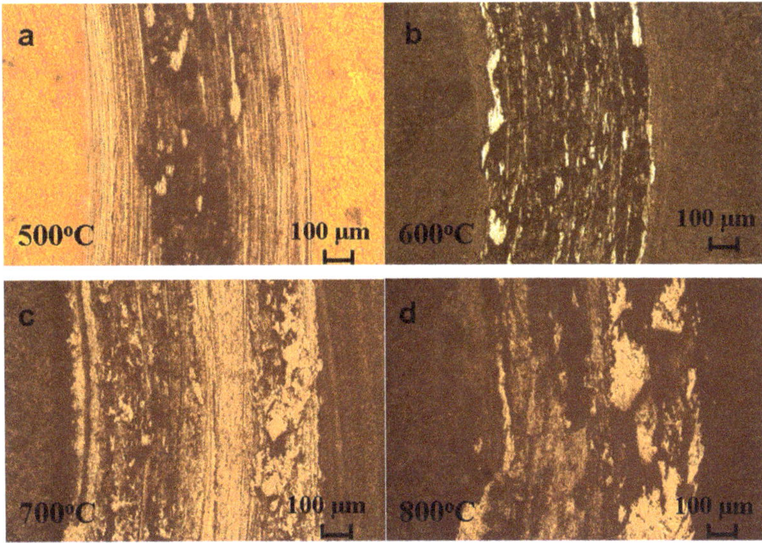

Figure 8. Optical images of flat worn surface under different temperatures: (**a**) 500 °C, (**b**) 600 °C, (**c**) 700 °C and (**d**) 800 °C.

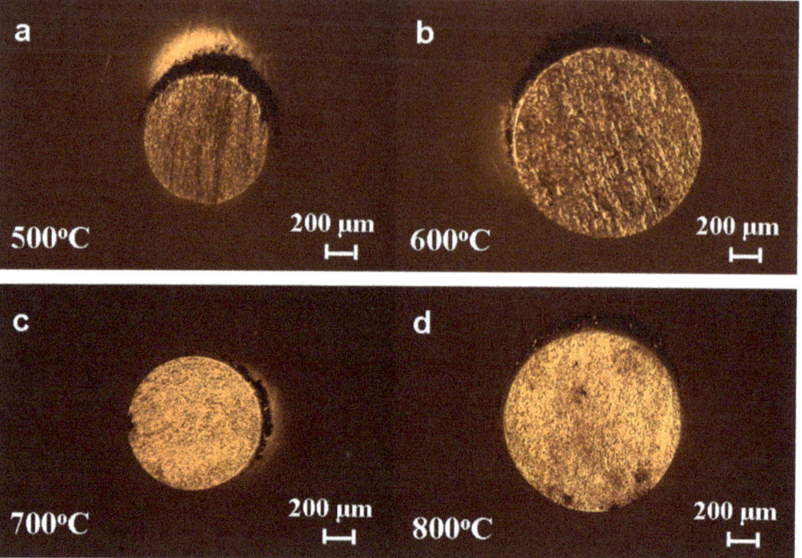

Figure 9. Optical images of ZrO_2 ball worn surface under different temperatures: (**a**) 500 °C, (**b**) 600 °C, (**c**) 700 °C and (**d**) 800 °C.

On the worn surface topography of disc, there was obvious adhesion wear on the wear scars from 500 to 800 °C, as well as abrasive wear at the temperatures of 600–800 °C. There was a wear scratch on the worn surface of disc along the sliding direction and the corresponding scratch on ball at 600 °C.

The three-dimensional morphology of the worn surface on the steel disc was observed with a laser confocal microscope, as shown in Figure 10. The wear rate of h-BN coatings was calculated according to the cross-sectional area of the worn surface at each temperature. The wear rates are 2.89×10^{-4} mm^3 (Nm)$^{-1}$, 1.5×10^{-4} mm^3 (Nm)$^{-1}$, 4.01×10^{-5} mm^3 (Nm)$^{-1}$ and 7.25×10^{-5} mm^3 (Nm)$^{-1}$ from 500 to 800 °C. It is found that the depth is large, although the width of the wear scar of ball and disc is narrower at 500 °C, and the wear rate is high.

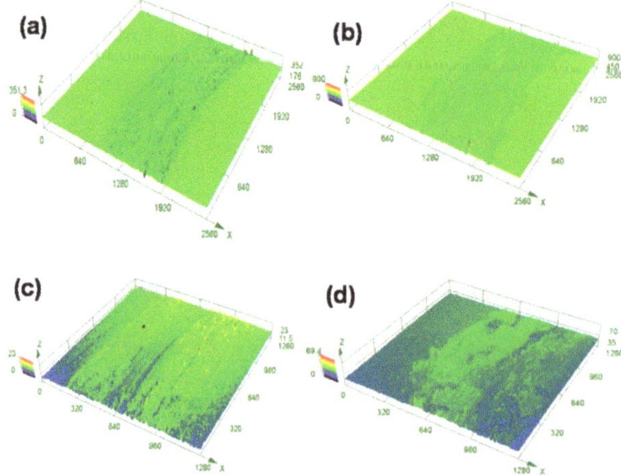

Figure 10. Three-dimensional images of the worn surface of BN coatings under different temperatures: (**a**) 500 °C, (**b**) 600 °C, (**c**) 700 °C and (**d**) 800 °C.

The wear scar of h-BN coatings was measured by XRD to discuss the friction mechanism of the friction pair at high temperature. Figure 11 shows the XRD pattern of the wear scar of h-BN coatings under different temperatures. It was found that there wash-BN and α-Fe$_2$O$_3$ in the worn surface of disc under different temperatures. The steel substrates were easily oxidized in ambient environment once h-BN coatings were worn out or destroyed during sliding. There was a small amount of boron oxide in the wear scar, which is helpful to reduce friction at the temperatures of 500 °C, 700 °C and 800 °C. However, there was no boron oxide measured in the worn surface of disc at 600 °C, although CoF is low at the steady stage. Moreover, it was found that there was γ-Fe$_2$O$_3$ in the worn surface at 800 °C.

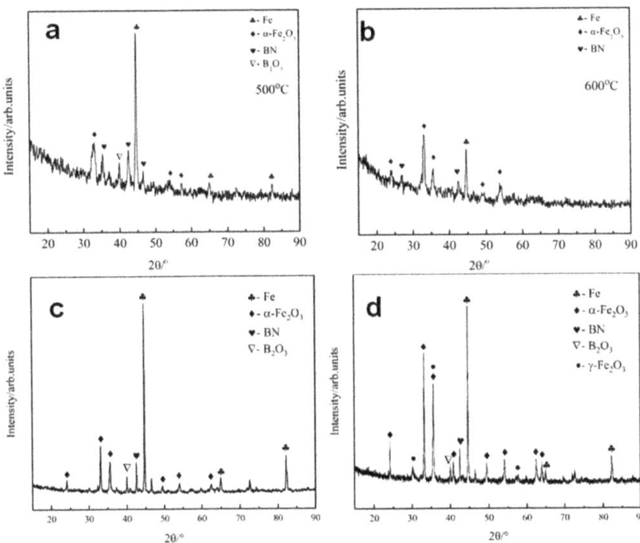

Figure 11. XRD of the worn surface of h-BN coatings under different temperatures: (**a**) 500 °C, (**b**) 600 °C, (**c**) 700 °C and (**d**) 800 °C.

The wear scar of h-BN coatings was observed by SEM, as shown in Figure 12. At 500 °C, there was a compacted layered material around the center of the wear scar, and there were also block products. At 600 °C, there was a lot of granular debris at the wear scar, while needle-like wear debris appeared on the surface of the ink mark at 700 °C. It shows that more α-Fe$_2$O$_3$ was produced at this temperature, the width of the wear scar at 800 °C was narrowed and the surfaces also had a more obvious compaction layer and wear debris particles.

Raman spectroscopy was used to analyze the microstructure of material. Figure 13 and Table 1 show Raman spectra of the friction pair at 700 °C and 800 °C, respectively. All obvious peaks were listed. Raman spectra of the disc worn surface exhibited bands at 217 cm^{-1}, 281 cm^{-1}, 287 cm^{-1}, 397 cm^{-1}, 590 cm^{-1}, 644 cm^{-1}, 1304 cm^{-1} and 1551 cm^{-1} at 700 °C. Raman spectra of the ball worn surface exhibited bands at 219 cm^{-1}, 264 cm^{-1}, 319 cm^{-1}, 463 cm^{-1}, 642 cm^{-1}, 848 cm^{-1}, 1308 cm^{-1} and 1555 cm^{-1} at 700 °C. Raman spectra of the disc worn surface exhibited bands at 207 cm^{-1}, 225 cm^{-1}, 264 cm^{-1}, 291 cm^{-1}, 406 cm^{-1}, 613 cm^{-1}, 658 cm^{-1}, 1307 cm^{-1} and 1554 cm^{-1} at 800 °C. Raman spectra of the ball worn surface exhibited bands at 207 cm^{-1}, 264 cm^{-1}, 383 cm^{-1}, 574 cm^{-1}, 748 cm^{-1}, 852 cm^{-1}, 950 cm^{-1}, 1067 cm^{-1}, 1304 cm^{-1}, 1416 cm^{-1}, 1554 cm^{-1} and 1764 cm^{-1} at 800 °C.

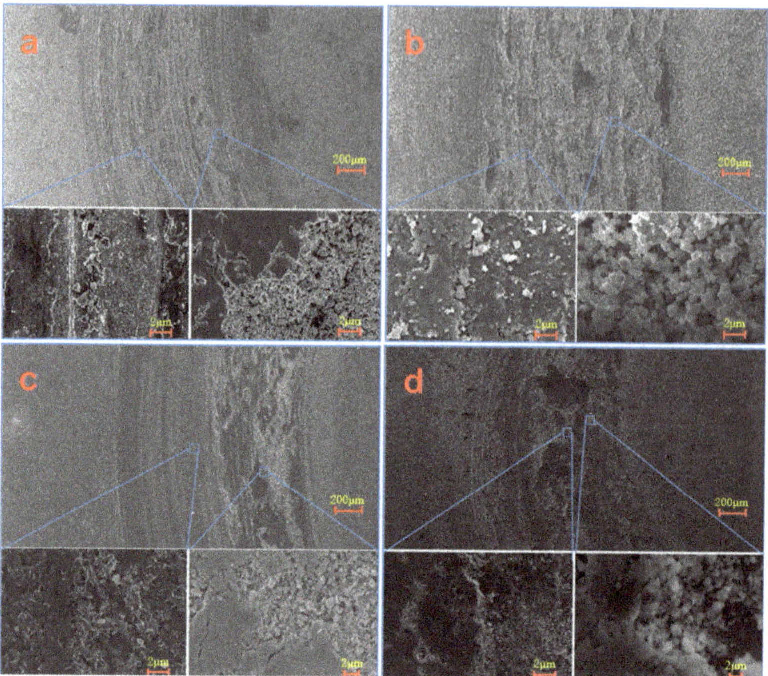

Figure 12. SEM images of the worn surface of BN coatings under different temperatures: (**a**) 500 °C, (**b**) 600 °C, (**c**) 700 °C and (**d**) 800 °C.

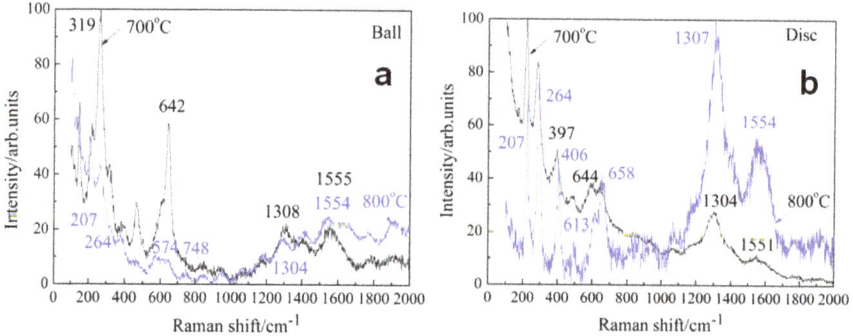

Figure 13. Raman spectra of the worn surface at 700 °C and 800 °C: (**a**) ball and (**b**) disc.

For the ball worn surface at 700 °C, the characteristic band was visible at position 642 cm^{-1}, which may be caused by TiO_2 [35]. The Raman spectra of 219 cm^{-1} and 319 cm^{-1} are characteristic of the hematite phase. The sharp and strong intensity band at 463 cm^{-1} indicates the metal oxygen (Fe-O) vibrations in hematite. Seen from the Raman spectra, α-Fe_2O_3 exhibited 848 cm^{-1} [36]. The band at 1308 cm^{-1} may be assigned to γ-FeOOH [37]. For the disc worn surface at 700 °C, TiO_2 was also characterized by bands at 397 and 644 cm^{-1}. The band at 1304 cm^{-1} is assigned to second harmonic vibrations of α-Fe_2O_3. The hematite bands were clearly visible at 281 and 287 cm^{-1} derived from symmetric Fe-O bending mode. All these Raman peaks are attributed to the modes arising due to the presence of α-Fe_2O_3.

Table 1. Raman spectra of friction pair at 700 °C and 800 °C.

FrictionPair	Ball		Disc	
Temperature (°C)	700 °C	800 °C	700 °C	800 °C
	219	207	217	207
				225
	264	264	281	264
	319	383	287	291
	463	574	397	406
	642	748	590	613
Wave shift (cm^{-1})	848	852	644	658
		950		
		1067		
	1308	1304	1304	1307
		1416		
	1555	1554	1551	1554
		1764		

For the worn surface at 800 °C, we assign the peaks at 207 and 264 cm^{-1} to the Fe-O mode. For the disc worn surface at 800 °C, the peak at 658 cm^{-1} is characteristic of Fe$_3$O$_4$. It can be clearly seen that the peaks appearing at 225, 291, 406 and 613 cm^{-1} correspond to the characteristic peaks of α-Fe$_2$O$_3$. The wave number was measured at 1307 cm^{-1}, which represents the metal-oxygen stretching vibrations (Fe-O) and can be assigned to a cubic-phase of α-Fe$_2$O$_3$. For the ball worn surface at 800 °C, the 383 cm^{-1} and 1304 cm^{-1} bands are possibly assignable to α-Fe$_2$O$_3$, in agreement with the results of XRD. The peak at 1304 cm^{-1} may be characteristic of γ-FeOOH, indicating that some of Fe$_3$O$_4$ are transformed into γ-FeOOH due to tribochemical reactions. The peak at 1554 cm^{-1} is characteristic of γ-Fe$_2$O$_3$ [38]. This material is generated by the thermal dehydration of γ-FeOOH. The γ-Fe$_2$O$_3$ is partially transformed into α-Fe$_2$O$_3$, which is the most stable form of iron oxide. The transformation from γ-Fe$_2$O$_3$ to α-Fe$_2$O$_3$ occurred usually at 800 °C. Therefore, this phase transformation during the friction process could be due to an instantaneous high temperature. The Fe-N-O bending vibrations are located at 574 cm^{-1} [39]. Other bands at 748 cm^{-1} and 852 cm^{-1} represent the Raman spectra of the iron complex. The peak at 1067 cm^{-1} corresponds to Ti-O stretching modes. The band occurs a peak of 950 cm^{-1}, which is also the main band of iron titanium oxides [40]. In fact, when Ti-O bonds are present, they are the most prominent features at 1416 cm^{-1}.

There was α-Fe$_2$O$_3$ on the disc and ball worn surface at 700 °C and 800 °C, which is in agreement with XRD measurements. The γ-Fe$_2$O$_3$ peak was also detected from the disc worn surface at 800 °C. There was α-Fe$_2$O$_3$ and TiO$_2$ on the disc worn surface, and γ-FeOOH, α-Fe$_2$O$_3$ and TiO$_2$ on the ball worn surface at 700 °C. There was α-Fe$_2$O$_3$, Fe$_3$O$_4$ and γ-Fe$_2$O$_3$ on the disc worn surface, and α-Fe$_2$O$_3$, γ-FeOOH, γ-Fe$_2$O$_3$, Fe-N-O and TiO$_2$ on the ball worn surface at 800 °C. Therefore, it is considered that tribooxidation was the main reason leading to antifriction behavior of h-BN coatings.

The h-BN coatings were prepared on the steel substrate to improve high temperature tribological properties of steel. The h-BN coatings were partially worn out during sliding, or there were some cracks due to the thermal expansion between h-BN coatings and substrate under high temperature. The steel substrate was oxidized to iron oxidessuch as magnetite (Fe$_3$O$_4$), hematite (α-Fe$_2$O$_3$) and magnetic maghemite (γ-Fe$_2$O$_3$) according to thermodynamic calculation, as shown in Table 2. Fe$_3$O$_4$was not stable in the ambient atmosphere, and further oxidized to α-Fe$_2$O$_3$ and γ-Fe$_2$O$_3$ at high temperature. According to XRD and Raman measurements, there was h-BN and α-Fe$_2$O$_3$ on the worn surface and Fe-N-O species were also observed. It is inferred that the composite of Fe$_2$O$_3$/h-BN is formed during sliding although there is also few B$_2$O$_3$. This composite is helpful to improve the high temperature antifriction behavior even high temperature superlubricity of the friction pair. Moreover, γ-Fe$_2$O$_3$crystalline has FCC crystal microstructure and 12 kinds of slip plane, which leads to the good antifriction behaviors comparing with α-Fe$_2$O$_3$, which is beneficial to achieve superlubricity at 800 °C. However, as shown in Figures 8 and 9, the wear debris were accumulated on the wear scar of disc and these wear debris were partially transferred to the worn surface of the ball, which may result in the fluctuation of CoF.

Table 2. The Gibbs free energy of the friction pair ($\Delta_r G_m^\theta$).

Reaction	500 °C	600 °C	700 °C	800 °C
$3Fe + 2O_2 = Fe_3O_4$	-857.39	-827.29	-797.67	-767.94
$4Fe + 3O_2 = 2Fe_2O_3$	-615.91	-590.77	-565.94	-541.16
$2Fe + O_2 = 2FeO$	-427.55	-414.64	-401.68	-388.52
$6FeO + O_2 = 2Fe_3O_4$	-432.13	-410.65	-390.31	-370.32
$4Fe_3O_4 + O_2 = 6Fe_2O_3$	-265.91	-235.47	-204.99	-175.23
$2BN + 3H_2O = B_2O_3 + 2NH_3$	-123.82	-130.50	-134.90	-137.25
$B_2O_3 + 3H_2O = 2H_3BO_3$	-2.67	7.27	18.62	31.23

High temperature friction results prove that h-BN is very suitable to be served as high temperature solid lubricant. The h-BN coatings show excellent high temperature tribological properties due to the intrinsic interlayer slipping reducing the sliding resistance. The superb chemical inertness and high temperature resistance of h-BN can be applied in many harsh conditions [41]. The h-BN has its distinctive advantages at 500 °C and above. The friction mechanism of h-BN coatings is attributed to the lamellar structures of h-BN at relatively low temperature. The lamellar structures have strong covalent bonds in the plane and weak van der Waalsforces between the planes. During sliding, h-BN coatings were partially worn out and the steel substrate was unavoidably oxidized by oxygen in ambient environment at high temperature. There was α-Fe_2O_3 on the worn surface of disc and ball according to XRD and Raman measurements. Theα-Fe_2O_3weretransferred and adhered to form a dense lubrication layer on the worn surface on ball, as shown in Figure 9, which makes low CoF of the friction system. Therefore, it is considered that the tribooxidation was the main reason leading to low CoF of the friction system [42–45]. Oxygen atoms tended to be embedded in B and N vacancies and reacted with B atoms to form B_2O_3 at high temperature. B_2O_3 was detected in the wear scar of disc at the temperatures of 500 °C, 700 °C and 800 °C. It is well known that h-BN can be functionalized with iron oxide [46]. At high temperature, there is much more friction heating, which improves the surface activity of h-BN. Therefore, the composite of α-$Fe_2O_3$and BN was generated in situ during sliding as h-BN is functionalized with α-Fe_2O_3. The XRD measurement of the disc worn surface shows the characterization peaks of h-BN andα-Fe_2O_3, which signifies α-$Fe_2O_3$was successfully prepared on the surface of h-BN. The h-BN is expected to be used as a supporter of α-Fe_2O_3. Theα-Fe_2O_3/h-BN composite achieves an excellent high temperature antifriction performance. Therefore, this composite combines the advantageous mechanical properties of α-Fe_2O_3 and h-BN, forming the synergistic effect in improving high temperature tribological performances of h-BN coatings. The results verify that the hybrid materials exhibit the superior tribological performance than h-BN andα-Fe_2O_3 alone.

At 500 °C, friction occurred between h-BN coatings and ZrO_2 at the initial stage. The initial CoF is high, about 0.39, because the friction occurred between h-BN and ZrO_2 ceramics. There maybe at hermalm is match expansion and stress behavior of h-BN coatings and steel; therefore, the cracks were formed and oxygen was penetrated into coatings at high temperature. The oxides were generated on the disc due to high friction heating and oxides were partially transferred to the worn surface of ball. The steel substrate was oxidized to iron oxide, such as Fe_3O_4, since the Gibbs free energy of the chemistry reaction of Fe_3O_4 is low at initial stage. However, Fe_3O_4 is easily oxidized into Fe_2O_3 in open air. The contact friction pair was iron oxide on ZrO_2 ball and the composite of α-Fe_2O_3/h-BN; therefore, CoF decreased, although CoF fluctuated slightly and was relatively stable due to the good sliding between plans of h-BN with the increase in the sliding time. At the final stage, h-BN was oxidized to B_2O_3 according to XRD measurement under high friction heating and high environment temperature. The contact friction pair became α-Fe_2O_3 on

ZrO_2 ball and the composite of α-Fe_2O_3/h-BN and B_2O_3; therefore, CoF decreased further to the minim value around 1800s and increased slowly again along the sliding time. At 600 °C, there was Fe_3O_4 and α-Fe_2O_3 on the steel because the steel was sufficiently oxidized at higher temperature. The friction occurred between ZrO_2 on ball and mostly iron oxides with h-BN on steel at the initial stage, which causes a high initial CoF of 0.48 compared with that at 500 °C. It is interesting that there is more wear debris and the transferred films on ball at 600 °C. The transferred films are probably α-Fe_2O_3 [3] and covered evenly all the worn surfaces of ball, as shown in Figure 9b. CoF decreased to a low value with the adding of α-Fe_2O_3 on ball. The contact friction pair became α-Fe_2O_3 and ZrO_2 on ball and the composite of Fe_3O_4, α-Fe_2O_3 and h-BN, which is probably wrapped in iron oxides and as hardening phase due to high hardness, as shown in Figure 3. CoF increased slowly and fluctuated to a high value around 600 s with the full coverage of α-Fe_2O_3 on ball. At this point, the temperature at the friction interface was higher than 600 °C because there was high friction heating and the environmental temperature was 600 °C. It is well known that Fe_3O_4 is stable at high temperature and the transition from Fe_3O_4 to γ-Fe_2O_3 was performed at 650 °C [46]. The contact friction pair was α-Fe_2O_3 on ball and the composite of Fe_3O_4, γ-Fe_2O_3, α-Fe_2O_3 and h-BN on steel, which resulted in high CoF at this point. The γ-Fe_2O_3 acts as a soft solid lubricant between the contact surfaces at high temperature [47]. CoF decreased with increase in the transition from Fe_3O_4 to γ-Fe_2O_3 until all Fe_3O_4 was worn out. CoF decreased to the minimum value around 0.03 due to the phase transformation of iron oxides and increased slowly because the soft γ-Fe_2O_3 was possibly worn out and there was no more transition from Fe_3O_4 to γ-Fe_2O_3. It is surprising that there is no boron oxide according to XRD measurement, because h-BN was probably wrapped in iron oxides and the steel was much easily oxidized than h-BN according to Table 2. At the final stage, CoF was low because the contact pair was α-Fe_2O_3 on the ball and the composite of α-Fe_2O_3/h-BN according to XRD. With temperature reaching 700 °C, Ti was oxidized to TiO_2 according to Raman spectra and XRD measurement. Fe_3O_4 was oxidized to α-Fe_2O_3 and friction occurred between ZrO_2 on ball and the mixtures of iron oxides, TiO_2 and h-BN on steel, which results in a low initial CoF of 0.31. There were few oxides on the worn surface of the ball, as shown in Figures 9 and 13, and then α-Fe_2O_3 reacted with ZrO_2 to form the composite of α-Fe_2O_3/ZrO_2 under the synergistic effect of high friction heating and high environment temperature [3], which results in low CoF. CoF fluctuated with the sliding time and was relatively stable according to tribochemistry reaction of Fe, Ti and h-BN under high friction heating and high environment temperature. The average CoF was about 0.07 at final and stable stage. The contact pair was α-Fe_2O_3, TiO_2 and γ-FeOOH on ball and the composite of α-Fe_2O_3/h-BN, B_2O_3 and TiO_2 according to XRD and Raman spectra. The phase transition from metastable γ-Fe_2O_3 to stable α-Fe_2O_3 with rhombohedral crystal structure was approved at 800 °C. At 800 °C, there was mostly α-Fe_2O_3 on the disc surface. The contact pair was ZO_2 on ball and the mixtures of iron oxides, h-BN and TiO_2, which leads to high initial CoF of 0.73. The iron oxide was transferred to ball and there is the tribochemistry reaction of h-BN, the formation of Fe-N-O and the phase transformation of iron oxide due to high friction heating and high environment temperature. CoF fluctuated and decreased to the minima value of 0.02 although CoF increased slightly. The contact pair was α-Fe_2O_3, TiO_2, γ-FeOOH, γ-Fe_2O_3 and Fe-N-O species on the ball and the mixtures of α-Fe_2O_3/h-BN, γ-Fe_2O_3, B_2O_3 and Fe_3O_4 on steel according to XRD and Raman spectra. There was complex tribooxidation and the phase transformation during sliding, resulting in high temperature superlubricity at 800 °C.

According to the analysis, the possible high temperature superlubricity mechanisms can be concluded as shown in Figure 14. Firstly, h-BN exhibits excellent antifriction behaviors inherently due to its unique lamellar structure. Secondly, Fe_2O_3, especially in γ-Fe_2O_3, has good high temperature plasticity and lubricity. Thirdly, α-Fe_2O_3 supported by h-BN plays an important synergistic effect in enhancing high temperature antifriction behavior. Finally, the friction occurred between the composite of α-Fe_2O_3/ZrO_2 on ball and the composite of γ-Fe$_2$O$_3$/h-BN on the disc, which results in high temperature superlubricity at 800 °C, although there maybe h-BN, oxides of Fe and Ti and boron oxide. Actually, these oxides of Fe and Ti, boron oxide and BN are beneficial to improve the high temperature tribological properties of steel, or not superlubricity in the friction system. Therefore, the composite of α-Fe$_2$O$_3$/h-BN against the composite of α-Fe$_2$O$_3$/ZrO$_2$ can significantly improve high temperature antifriction behaviors of steel.

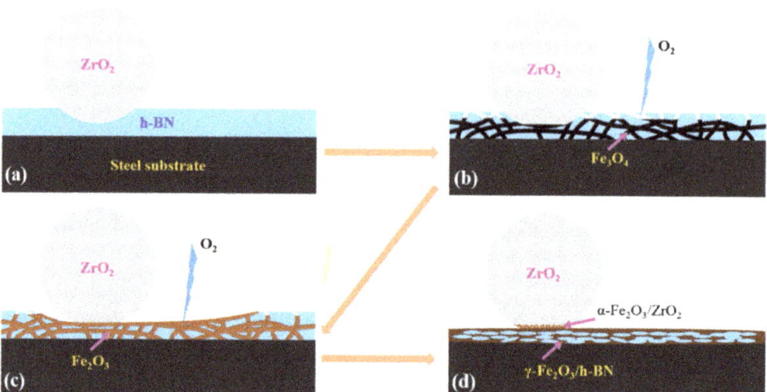

Figure 14. Schematic diagram of the friction process at 800 °C: (**a**) at initial stage, (**b**) the oxidation stage, (**c**) the phase change stage and (**d**) superlubricity stage.

4. Conclusions

The hexagonal boron nitride coatings were prepared on the steel substrate by using radio frequency magnetron sputtering technology. The tribological properties of hexagonal boron nitride coatings and the influence mechanism of temperature on the tribological properties were investigated. Microstructure, morphology, mechanics and high temperature low friction mechanism of h-BN coatings on steel were discussed in depth. These results provide significant theoretical and practical implications of h-BN coatings on steel and of great practical value in industries help to understand the tribological properties. The main findings in this work are summarized as follows:

(1) The h-BN coatings are beneficial to improve the high temperature tribological properties of steel. CoFs of h-BN coatings are much lower than those of the uncoated steel at the same temperature;

(2) The wear mechanism of h-BN coatings is tribooxidation, resulting in super low friction of the friction pair at high temperature. There are α-Fe$_2$O$_3$, Fe$_3$O$_4$ and γ-Fe$_2$O$_3$ on the disc worn surface, and α-Fe$_2$O$_3$, γ-FeOOH,γ-Fe$_2$O$_3$ and TiO$_2$ on the ball worn surface at 800 °C. The oxidation is the main factor for the antifriction and anti wear behavior of the friction pair at high temperature;

(3) The h-BN coatings on steel exhibit high temperature superlubricity at 800 °C. CoFs of the friction pair are as super low as about 0.02 at 800 °C at the stable stage. The superlubricity mechanism is attributed to the formation of the composite of γ-Fe$_2$O$_3$/h-BN due to tribochemistry.

Funding: This research was funded by National Natural Science Foundation of China (51675409 and 51805409), National Science and Technology Major Project (j2019-IV-0004-0071) and the Natural Science Foundation of Chongqing, China (cstc2019jcyj-msxmX0577) and Natural Science Basic Research Plan in Shaanxi Province of China (2022JM-251).

Institutional Review Board Statement: Not applicable.

Informed Consent Statement: Not applicable.

Data Availability Statement: Not applicable.

Acknowledgments: Not applicable.

Conflicts of Interest: The authors declare no conflict of interest.

References

1. Trzepiecinski, T.; Lemu, H. Recent developments and trends in the friction testing for conventional sheet metal forming and incremental sheet forming. *Metals* **2019**, *10*, 47. [CrossRef]
2. Voevodin, A.; Zabinski, J. Nanocomposite and nanostructured tribological materials for space applications. *Compos. Sci. Technol.* **2005**, *65*, 741–748. [CrossRef]
3. Zeng, Q.; Zhu, J.; Long, Y.; Bouchet, M.I.d.; Martin, J.M. Transformation-induced high temperature low friction behaviors of ZrO$_2$-steel system at temperatures up to 900 °C. *Mater. Res. Express* **2019**, *6*, 0865f5. [CrossRef]
4. DellaCorte, C. The effect of counterface on the tribological performance of a high temperature solid lubricant composite from 25 to 650 °C. *Surf. Coat. Technol.* **1996**, *86*, 486–492. [CrossRef]
5. Zeng, Q. High-temperature superlubricity behaviors of γ-Fe$_2$O$_3$@SiO$_2$ nanocomposite coatings. In *Composite Materials*; Elsevier: Amsterdam, The Netherlands, 2021; pp. 489–501.
6. Nowak, P.; Kucharska, K.; Kamiński, M. Ecological and health effects of lubricant oils emitted into the environment. *Int. J. Environ. Res. Public Health* **2019**, *16*, 3002. [CrossRef]
7. Wu, Y.Y.; Tsui, W.C.; Liu, T.C. Experimental analysis of tribological properties of lubricating oils with nanoparticle additives. *Wear* **2007**, *262*, 819–825. [CrossRef]
8. Fontaine, J. Towards the use of diamond-like carbon solid lubricant coatings in vacuum and space environments. *Proc. Inst. Mech. Eng. Part J J. Eng. Tribol.* **2008**, *222*, 1015–1029. [CrossRef]
9. Zeng, Q.; Erdemir, A.; Eryilmaz, O. Ultralow friction of ZrO$_2$ ball sliding against DLC films under various environments. *Appl. Sci.* **2017**, *7*, 938. [CrossRef]
10. Liu, Y.; Yu, B.; Cao, Z.; Shi, P.; Zhou, N.; Zhang, B.; Zhang, J.; Qian, L. Probing superlubricity stability of hydrogenated diamond-like carbon film by varying sliding velocity. *Appl. Surf. Sci.* **2018**, *439*, 976–982. [CrossRef]
11. Vazirisereshk, M.; Martini, A.; Strubbe, D.; Baykara, M.Z. Solid lubrication with MoS$_2$: A review. *Lubricants* **2019**, *7*, 57. [CrossRef]
12. Manu, B.; Gupta, A.; Jayatissa, A. Tribological properties of 2D materials and composites—A review of recent advances. *Materials* **2021**, *14*, 1630. [CrossRef]
13. Zhu, L.; Wang, C.; Wang, H.; Xu, B.; Zhuang, D.; Liu, J.; Li, G. Microstructure and tribological properties of WS$_2$/MoS$_2$ multilayer films. *Appl. Surf. Sci.* **2012**, *258*, 1944–1948.
14. Uzoma, P.C.; Hu, H.; Khadem, M.; Penkov, O.V. Tribology of 2D nanomaterials: A review. *Coatings* **2020**, *10*, 897. [CrossRef]
15. Zhu, J.; Zeng, Q.; He, W.; Zhang, B.; Yan, C. Elevated-temperature super-lubrication performance analysis of dispersion-strengthened WSN coatings: Experimental research and first-principles calculation. *Surf. Coat. Technol.* **2021**, *406*, 126651. [CrossRef]
16. Zeng, Q.; Yu, F.; Dong, G. Superlubricity behaviors of Si$_3$N$_4$/DLC Films under PAO oil with nano boron nitride additive lubrication. *Surf. Interface Anal.* **2013**, *45*, 1283–1290. [CrossRef]
17. Podgornik, B.; Kosec, T.; Kocijan, A.; Donik, C. Tribological behaviour and lubrication performance of hexagonal boron nitride (h-BN) as a replacement for graphite in aluminium forming. *Tribol. Int.* **2015**, *81*, 267–275. [CrossRef]
18. Podgornik, B.; Kafexhiu, F.; Kosec, T.; Jerina, J.; Kalin, M. Friction and anti-galling properties of hexagonal boron nitride (h-BN) in aluminium forming. *Wear* **2017**, *388*, 2–8. [CrossRef]
19. Yuan, S.; Toury, B.; Benayoun, S. Novel chemical process for preparing h-BN solid lubricant coatings on titanium-based substrates for high temperature tribological applications. *Surf. Coat. Technol.* **2015**, *272*, 366–372. [CrossRef]
20. Li, J.; Luo, J. Advancements in superlubricity. *Sci. China Technol. Sci.* **2013**, *56*, 2877–2887. [CrossRef]
21. Niu, Z.; Chen, F.; Xiao, P.; Zhuan, L.; Lang, P.; Yang, L. Effect of h-BN addition on friction and wear properties of C/C-SiC composites fabricated by LSI. *Int. J. Appl. Ceram. Technol.* **2022**, *19*, 108–118. [CrossRef]
22. Chen, J.; Chen, J.; Wang, S.; Sun, Q.; Cheng, J.; Yu, Y.; Yang, J. Tribological properties of h-BN matrix solid-lubricating composites under elevated temperatures. *Tribol. Int.* **2020**, *148*, 106333. [CrossRef]
23. Chen, J.; Sun, Q.; Chen, W.; Zhu, S.; Li, W.; Cheng, J.; Yang, J. High-temperature tribological behaviors of ZrO$_2$/h-BN/SiC composite under air and vacuum environments. *Tribol. Int.* **2021**, *154*, 106748. [CrossRef]
24. Tyagi, R.; Xiong, D.; Li, J.; Dai, J. High-temperature friction and wear of Ag/h-BN-containing Ni-based composites against steel. *Tribol. Lett.* **2010**, *40*, 181–186. [CrossRef]
25. Zhang, D.; Cui, X.; Jin, G.; Song, Q.; Yuan, C.; Fang, Y. Microstructure and tribological performance of laser-cladded Ni60/h-BN coatings on Ti-6Al-4V alloy at high temperature. *Tribol. Trans.* **2019**, *62*, 779–788. [CrossRef]
26. Zhang, Y.; Wang, W.; Hu, Z.; Liu, K.; Chang, J. Investigation of hBN powder lubricating characteristics of die steel H13-ceramic Si$_3$N$_4$ tribopairat 800 °C. *Proc. Inst. Mech. Eng. Part J J. Eng. Tribol.* **2020**, *234*, 622–631. [CrossRef]
27. Torres, H.; Podgornik, B.; Jovičević-Klug, M.; Ripoll, M.R. Compatibility of graphite, hBN and graphene with self-lubricating coatings and tool steel for high temperature aluminium forming. *Wear* **2022**, *490*, 204187. [CrossRef]

28. Wang, Y.; He, W.; Zeng, Q.; Zhu, J. Effects of multiple post weld heat treatments on microstructure and precipitate of fine grained heat affected zone of P91 weld. *Steel Res. Int.* **2019**, *90*, 1800607. [CrossRef]
29. Mukhtiar, S.; Hitesh, V.; Ravinder, K. Micro structural characterization of BN thin films using RF magnetron sputtering method. *Mater. Today Proc.* **2020**, *26*, 2277–2282.
30. Tran, T.; Chung, K. Tribological characteristics of single-layer h-BN measured by colloidal probe atomic force microscopy. *Coatings* **2020**, *10*, 530. [CrossRef]
31. Yang, Y.; Chen, J.; Guo, L.; Tan, D.; Zeng, Q.; Liu, Q.; Liu, K.; Chen, Z.; Zou, J.; Lu, D. Effects of initial texture on rolling texture and property of electrodeposited nickel plate. *J. Comput. Theor. Nanosci.* **2015**, *12*, 2643–2647. [CrossRef]
32. Zhao, Y.; Zhang, G.; Liang, M.; Zeng, Q. Study on the friction and wear performance of lightly loaded reciprocating carbon/aramid-based composites. *Adv. Mater. Sci. Eng.* **2021**, *2021*, 9924690. [CrossRef]
33. Nasr, M.; Soussan, L.; Viter, R.; Eid, C.; Habchi, R.; Miele, P.; Bechelany, M. High photo degradation and antibacterial activity of BN-Ag/TiO_2 composite nanofibers under visible light. *New J. Chem.* **2018**, *42*, 1250–1259. [CrossRef]
34. Yang, X.; Xu, L.; Choon, N.; Chen, S. Magnetic and electrical properties of poly pyrrole-coated γ-Fe_2O_3 nanocomposite particles. *Nanotechnology* **2003**, *14*, 624.
35. Khaldi, O.; Majouri, A.; Larbi, T. Theoretical and experimental investigation of the electronic, optical, electric, and elastic properties of Zn-doped anatase TiO_2 for photocatalytic applications. *Appl. Phys. A* **2021**, *127*, 1–8. [CrossRef]
36. Harraz, F.; Faisal, M.; Jalalah, M.; Almadiy, A.A.; Sayari, S.A. Conducting polythiophene/α-Fe_2O_3 nanocomposite for efficient methanol electrochemical sensor. *Appl. Surf. Sci.* **2020**, *508*, 145226. [CrossRef]
37. Dong, H.; Zhao, F.; Zeng, G.; Tang, L.; Fan, C.; Zhang, L.; Zeng, Y.; He, Q.; Xie, Y.; Wu, Y. Aging study on carboxymethyl cellulose-coated zero-valent iron nanoparticles in water: Chemical transformation and structural evolution. *J. Hazard. Mater.* **2016**, *312*, 234–242. [CrossRef]
38. Tokubuchi, T.; Arbi, R.; Zhenhua, P.; Katayama, K.; Turak, A.; Sohn, W.Y. Enhanced photoelectrochemical water splitting efficiency of hematite (α-Fe_2O_3)-Based photoelectrode by the introduction of maghemite (γ-Fe_2O_3) nanoparticles. *J. Photochem. Photobiol. A Chem.* **2021**, *410*, 113179. [CrossRef]
39. Benko, B.; Yu, N. Resonance Raman studies of nitric oxide binding to ferric and ferrous hemoproteins: Detection of Fe (III)-NO stretching, Fe (III)-N-O bending, and Fe (II)-N-O bending vibrations. *Proc. Natl. Acad. Sci. USA* **1983**, *80*, 7042–7046. [CrossRef]
40. Leon, Y.; Lofrumento, C.; Zoppi, A.; Carles, R.; Castellucci, E.M.; Sciau, P. Micro-Raman investigation of terra sigillata slips: A comparative study of central Italian and southern Gaul productions. *J. Raman Spectrosc.* **2010**, *41*, 1550–1555. [CrossRef]
41. Zhang, R.; Ding, Q.; Yang, L.; Zhang, S.; Niu, Q.; Ye, J. A novel sonogel based on h-BN nanosheets for the tribological application under extreme conditions. *Tribol. Int.* **2019**, *138*, 271–278. [CrossRef]
42. Lavrenko, V.; Alexeev, A. High-temperature oxidation of boron nitride. *Ceram. Int.* **1986**, *12*, 25–31. [CrossRef]
43. Zeng, Q.; Qin, L. High temperature anti-friction behaviors of a-Si: H films and counterface material selection. *Coatings* **2019**, *9*, 450–460. [CrossRef]
44. Zeng, Q.; Chen, T. Superlow friction and oxidation analysis of hydrogenated amorphous silicon films under high temperature. *J. Non-Cryst. Solids* **2018**, *493*, 73–81. [CrossRef]
45. Zeng, Q.; Erdemir, A.; Eryilmaz, O. Superlubricity of the DLC films-related friction system at elevated temperature. *RSC Adv.* **2015**, *5*, 93147–93154. [CrossRef]
46. Thangasamy, P.; Sathish, M. Dwindling the re-stacking by simultaneous exfoliation of boron nitride and decoration of α-Fe_2O_3 nanoparticles using a solvo thermal route. *New J. Chem.* **2018**, *42*, 5090–5095. [CrossRef]
47. Kazeminezhad, I.; Mosivand, S. Phase transition of electro oxidized Fe_3O_4 to γ and α-Fe_2O_3 nano particles using sintering treatment. *Acta Phys. Pol. A* **2014**, *125*, 1210–1214 [CrossRef]

coatings

Article

Study of Tribotechnical Properties of Multilayer Nanostructured Coatings and Contact Processes during Milling of Titanium Alloys

Mars Sharifullovich Migranov *, Semen Romanovich Shehtman, Nadezhda Aleksandrovna Sukhova, Artem Petrovich Mitrofanov, Andrey Sergeevich Gusev, Arthur Marsovich Migranov and Denis Sergeyevich Repin

Department of High-Efficiency Processing Technology, Institute of Production Technology and Engineering, Moscow State Technological University STANKIN, 127055 Moscow, Russia
* Correspondence: migmars@mail.ru

Abstract: The paper presents the results of theoretical and experimental research on tribotechnical characteristics: tool wear on the back surface, tool durability period, critical length of the cutting path before blunting, adhesion component of the friction coefficient, contact processes, temperature, and force dependences for the application of innovative nanostructured multilayer composite coatings on a tool for milling of titanium alloys. The proposed thermodynamic model of cutting tool wear allows us to determine the ways by which cutting tool wear intensity decreases and the conditions of increase in cutting tool wear resistance with wear-resistant coatings. A substantial increase in wear resistance of end mills when processing titanium alloys with the use of innovative multilayer nanostructured coatings is established, in particular an improvement of an average of 1.5–2 times. These positive results are related to a significant decrease in temperature–force loading in the cutting zone, a decrease in the friction coefficient (adhesion component), and the phenomenon of adaptation (self-organization) of friction surfaces during cutting by tools with wear-resistant coatings, contributing to the formation of films of various compounds with shielding, protective, and lubricating properties.

Keywords: counter milling; nanostructured composite multilayer coatings; titanium alloys; wear resistance; tool durability period; cutting path length; adhesion component of friction coefficient; temperature; cutting forces

Citation: Migranov, M.S.; Shehtman, S.R.; Sukhova, N.A.; Mitrofanov, A.P.; Gusev, A.S.; Migranov, A.M.; Repin, D.S. Study of Tribotechnical Properties of Multilayer Nanostructured Coatings and Contact Processes during Milling of Titanium Alloys. *Coatings* **2023**, *13*, 171. https://doi.org/10.3390/coatings13010171

Academic Editor: Yanxin Qiao

Received: 28 December 2022
Revised: 9 January 2023
Accepted: 10 January 2023
Published: 12 January 2023

1. Introduction

The rapid growth of modern machine-building production places increased demands on machining. High-productivity mechatronic systems require increasing the reliability and serviceability of cutting tools by, first, using various methods to improve the cutting properties of tools (hardening technologies and the use of coolant and wear-resistant coatings and their combinations); second, developing modern methods to evaluate the tribotechnical properties of cutting tools; and, third, making scientifically justified choices of cutting modes, taking into account the characteristics of contact processes (temperature and force of cutting).

Regarding the problem of improving the serviceability of cutting tools and technological reliability of multitool and multitransition machining on machine tools with control systems, the issues of wear, fracture, strength, and stress–strain state of cutting tools and the study of thermal and contact phenomena are topics of the fundamental research of many scientists [1–5]. The results of these works allowed the study of the physical mechanisms of cutting tool wear in a wide range of changing elements of the cutting regime and solutions for optimization and control problems of the cutting process and played a major role in the formation of modern concepts in the theory of material cutting.

At the same time, the use of mechatronic systems equipped with high-speed numerical control (NC) and adaptive control (AdC) [6–8] machines solves extremely important problems of increasing machining productivity and automation of individual and small-scale production widespread in modern mechanical engineering, but metal-cutting equipment of turning and milling groups has a considerable cost, which has led to high requirements for the scientific validity of cutting modes.

In the manufacture of parts of heavily loaded high-temperature tribocouplings, more and more are made of innovative materials: with unique physical and mechanical properties and, therefore, with low machinability by cutting; in terms of their geometric shape, they mainly represent large-sized complex shaped bodies: bodies, disks, shafts, rotors, flanges, etc. Their work takes place in harsh conditions of simultaneous and long exposure to high temperatures and specific force loads, with high requirements in terms of both accuracy and quality of surface processing. The small dimensional tolerances are due not only to the requirements of interchangeability of the mating surfaces but also to the specific conditions of "high accuracy", due to which even free dimensions must have small tolerances. The machining of the parts listed above constitutes a significant share of milling operations. The technological problems encountered in the production of each of the above varieties of parts are somewhat different. For example, end milling of a long-sized workpiece with a relatively small thickness is characterized by the large suppleness of the part, so special devices must be developed to create constant stiffness across the entire blade. When machining hollow shafts and rotors made of titanium alloys, it can be difficult to mill slotted and toothed parts due to their long length. However, there is also a general problem of machining such parts associated with the low wear resistance of metal-cutting tools [7–13].

Nanostructured coatings are one of the typical small-sized objects, which are intensively studied due to the interest in identifying the features of the nanocrystalline state, characterized usually by sizes less than 100 nm [14–16]. As was already noted, metal-cutting tools, in most cases and for a few reasons, work in conditions of intermittent cutting and are exposed to alternating thermomechanical influence, leading to their rather intensive destruction in the form of chipping and active cracking. Changes in thermal–physical, physical–mechanical, and crystallochemical characteristics influence tool serviceability. It is known [17,18] that the modification of cutting tool properties by applying multifunctional nanostructured coatings on its working surfaces is one of the most effective ways to improve its reliability. The currently available equipment and technological processes allow the synthesis of multilayer coatings with multicomponent composite architecture, with a nanoscale thickness of nanolayers on modern and highly efficient facilities by various methods, including PVD with magnetic arc filtration, magnetron, and multifunctional high-energy, to modify the surfaces of metal-cutting tools. At the same time, many studies [18–20] have established unique properties that are unique only to modern multifunctional nanostructured coatings. Nano-objects are interesting, on the one hand, as metal-like compounds and, on the other hand, as typical brittle phases, not to mention the numerous applications of materials based on embedding phases. In this regard, data on the structure and properties of these compounds in the nanocrystalline state appear to be important for both theoretical materials science and applications. The specific properties of nanostructured coatings are largely due to the peculiarities of their structure: the high-volume fraction of interfaces, the strong binding energy of adjacent phases, the absence of dislocations within nanocrystallites, the implementation of deformation by grain boundary slip, the presence of intergranular amorphous interlayers, and the change in the mutual solubility of components in the embedding phases. All these features allow the achievement of record values of physical, chemical, mechanical, and tribological properties of the material during the transition to the nanostructured state [17,18].

According to research [21,22], control of the technological process of the formation of nanostructured coatings allows the reception of compositions with variable properties depending on the percentage ratio of components. The hardness, friction coefficient, surface roughness, and coating color change. All this makes it possible to optimize the coating properties for a specific task. Applications include cutting, shaping, stamping, and other tools. For example, for TiZrN, high hardness, thermodynamic stability, and the strength of connection are caused by the high similarity of structures and the close sizes of atoms providing the presence of significant areas of mutual solubility of atoms Ti and alloying component Zr in corresponding nitrides. The maximum microhardness of coating TiZrN of 32.2 GPa is observed at the content of 6%–10% Zr, which exceeds the value of Hμ for coating TiN of 22%. The covering (Ti-Zr)CN on the basis of carbonitrides Ti and Zr, the microhardness of which corresponds to 64 GPa, is relative. In research [13–15], the extreme character of change in the intensity of wear of the cutting tool depending on the structure of the nanostructured coating is established. During the processing of workpieces from steels $12 \times 18H10T$, the minimum intensity of wear was observed at plates from alloy BK6 with TiZrN coating at a content of 8%–16% Zr. It is established that the minimum wear rate of the plates with carbonitride coating (Ti-Zr)CN corresponds to the same content of the alloying component as that of the plates with nitride-based coating. The minimum wear is observed at the content of acetylene of 25%–35% and 20%–25% when processing workpieces made of 12X18H10T steel. Studies of the machining process and the thermal and stress state of the cutting tool wedge have established that the coatings TiZrN and (Ti-Zr)CN increase the number of cycles before all longitudinal cracks escape to the back surface compared to the cutting tool with conventional coating by five times, and for nanostructured coatings TiZrN and (Ti-Zr)CN, by 8 and 6.25 times, respectively [13]. Thus, it can be concluded that nanostructured coatings allow an increase in the serviceability of cutting tools due to management of the parameters of coating structure, directionally controlled, which can significantly change coating properties; provide high crack resistance, strength properties, and a high level of compressive residual stresses restraining processes of the formation and development of cracks; and provide a low level of contact and thermal loads, minimizing processes of coating destruction.

Proceeding from the above statements, the purpose of this work is to increase the efficiency of milling of titanium alloys by investigating tribotechnical properties and developing modern innovative wear-resistant coatings based on thermodynamic analysis of contact processes.

The set goal of the work is achieved by describing the thermodynamic model of cutting tool wear, considering the influence of temperature, force parameters, and contact processes during blade cutting by cutting tools with nanostructured multilayer wear-resistant coatings; developing experimental research methods (wear resistance, temperature and force, adhesion, metal science, and others) and conducting them; and analyzing the obtained results.

2. Theoretical and Experimental Background

Metal cutting, by its physical nature, is a complex process [12–15]; high pressures (up to 2000 MPa and more), high strain rates (up to 10^6 s^{-1}), increased temperature (up to 1200–1500 K), etc., occur on the contact surfaces of the cutting tool with the machined material. This creates favorable conditions for the development of adhesion, mutual diffusion, oxidation, and hydrogen saturation of surfaces, changes in their structural and phase composition, and generation of electromotive force (EMF). These phenomena have a significant impact on the state and properties of the contact surfaces of the tool and the machined material.

One of the main sources of heat and factors of the formation of near-surface layers during cutting is friction. Since it occurs at high temperatures under conditions of contact juvenility and with the presence of plastic deformations, the adhesion (molecular) component dominates in the adhesion processes [16–18]. The relative sliding of the contact surfaces of the tool and the workpiece is accompanied by a continuous process of adhesion spot formation and shearing. The tool surface is under the action of shear stresses, causing the material particles to shear away from the surface in places. Usually, such shearing is much greater on the softer material's side, but studies using [19–22] electron microscopy show that there is always some transfer of particles from the harder material (tooling) to the softer material (machinable) at the same time.

At the present time, when describing contact processes and the wear of a cutting tool, a complex power representation proposed in the works of Professors Y.G. Kabaldin, A.D. Makarov, V.N. Poduraev, S.S. Silin, N.V. Talantov, and other scientists is used [1–7].

To describe the contact processes, friction, and wear of cutting tools with wear-resistant coatings, a dynamic process such as cutting is considered from the position of nonequilibrium thermodynamics, taking into account and specifying the components of energy balance of the process of interaction of the machining material and surfaces of the cutting tool.

It is known [10,20] that the first law of thermodynamics for friction and wear processes can be written down by the equation:

$$Wf = Q + \Delta U \tag{1}$$

where Wf is the work of friction forces; Q is the amount of heat released under friction; and ΔU is the change in the internal energy of the friction contact zone.

The change in internal energy can be represented as

$$\Delta U = \Delta W_{mv} + \Delta W_{ph} + \Delta W_{\Delta d} + \Delta W_{\Delta f} \tag{2}$$

where ΔW_{mv} is energy spent on the separation of the wear particle, ΔW_{ph} is energy spent on structural phase transformations, $\Delta W_{\Delta d}$ is energy spent on plastic deformation, $\Delta W_{\Delta f}$ is energy spent on the formation of rubbing surfaces, etc.

Since, at present, there are no analytical dependences by which all components of expression (2) can be estimated, let us assume, at first approximation, that during friction, there will be dissipation of a part of internal energy at the expense of

- Plastic deformation of a single microuneven (microcutting) as a result of shear by the average diameter of the contact spot;
- The form modification of the surface layer of the wear material as a result of the formation of a wear fragment.

Using the principles of nonequilibrium thermodynamics, an irreversible dynamic process such as wear can be most effectively described by means of dissipative functions (DF), $\overline{\Psi}_i$ which are the rate of change $\Delta W_i/d\tau$ of energy spent on a process (for example, plastic deformation of the machined material), related to a unit of actual contact area A_r.

$$\overline{\Psi}_i = \frac{dW_i}{d\tau} \cdot \frac{1}{A_r} \tag{3}$$

According to the laws of nonequilibrium thermodynamics, to ensure the irreversibility of a process, it is necessary to have a generalized thermodynamic flux and force, i.e., supported by gradients of the state values of the thermodynamic system, preventing the inverse process and taken with the opposite sign. The DF of such a process is equal to the product of the generalized flux and force

$$\overline{\Psi}_i = J_i(-\Delta F_i) \tag{4}$$

where J_i is generalized flux and ΔF_i is generalized force.

In the first approximation, the change in internal energy of the frictional thermodynamic system ΔU can be represented as the sum of energy spent on dispersion (wear) of the tool and energy of elastoplastic deformation of the processed material. Unfortunately, it is currently impossible to make a more precise quantitative assessment of the energies expended on structural phase transformations, etc., but an analysis of literature data allows us to qualitatively combine these moments and simplify and account for the other components. In this case, the DF balance equation can be represented as

$$\overline{\Psi}_T = \overline{\Psi}_d + \overline{\Psi}_s \tag{5}$$

where is the DF of the friction process, ψ_d is the DF of the shear plastic deformation process of the machined material by the mean diameter of the contact spot, and ψ_s is the DF of dispersion and shape change of the surface layer of the wearable tool material.

Using the methods of the theory of adaptation (self-organization) of nonequilibrium systems, physical kinetics, and force laws of the theory of strength and plasticity, the mechanochemistry of metals, thermal physics and cutting mechanics, and the components of the balance equation for the working conditions of tools with wear-resistant coatings are considered and specified.

Based on the analysis of contact processes, the friction process DF is presented as a change in the specific work of external forces, referred to as the unit of contact area of the tool with chips and workpieces, considering the influence of the shape and radius of the cutting edge and the current value of cutting force:

$$\overline{\Psi}_T = \frac{P_z(\tau) \cdot V_i \cdot f}{(l_1 + c) \cdot b} \tag{6}$$

where $P_z(\tau)$ is the current value of cutting force; V_i is the current value of cutting speed; l_1 and c are the contact lengths of the cutter with the chip and with the workpiece, respectively, on the front and back surfaces; b is the width of the cut layer; and f is the friction factor.

The plastic deformation of the processed material is determined by the current $\theta(\tau)$ and initial θ_0 values of the cutting temperature, the degree of plastic deformation $\Delta\gamma$, the short-term yield strength $\sigma_{T.\text{д}.}$ and the shear modulus $G_{\text{д}}$ of the processed material, and the resonant frequency of self-oscillations in the contact zone f_r:

$$\overline{\Psi}_d = 10^{-4}\theta(\tau) \cdot \Delta\gamma \cdot \left[f_r \cdot \frac{\sigma_{T.\text{д}.}}{G_{\text{д}}} \cdot \frac{\theta(\tau)}{\theta_0} \cdot \exp\left(\frac{\theta_0}{\theta(\tau)}\right) \right]^{\frac{1}{n}} \tag{7}$$

The DF of the form change of the tool material is determined based on the analysis of the form stability of the cutting edge, the model of damage accumulation in the contact layers of the tool, and the probabilistic nature of the wear particle separation set out in the works of Professor T.N. Loladze, as well as the specification of the wear scheme:

$$\overline{\Psi}_s = J_h \cdot V \cdot \left(\frac{HV_t}{HV_m}\right)^{\alpha} \cdot \text{erf}(P(\tau)) \cdot \left[P_r + \frac{12(1+\mu)\sigma_T^2}{E} \right] \tag{8}$$

where J_h is wear intensity, V is cutting speed, HV_t/HV_m is the ratio of microhardness of the tool and machined materials in the conditional plane of shear [4], α is the index, taking into account the influence of cutting temperature on microhardness, $\text{erf}(P(\tau))$ is the probability of wear particle separation, p_r is normal specific load in the contact zone, E is the modulus of elasticity of the machined material, σ_T is the short-term yield strength of the tool material, and μ is Poisson's coefficient.

From the solution of the system of Equations (6)–(8) concerning the cutting tool wear rate, the expression is obtained:

$$J_h = \frac{P_z(\tau) \cdot V \cdot f/(l_1 + c) \cdot b - 0.186\theta(\tau)J_d J_\theta}{V \cdot \left(\frac{HV_t}{HV_m}\right)^{\alpha} \cdot \text{erf}P(\tau) \cdot \left[P_r + \frac{12(1+\mu)}{E}\sigma_T^2\right]} \tag{9}$$

where $J_d J_\theta$ are dimensionless criteria, respectively, considering physical and mechanical properties of the machined material and the character of cutting temperature change.

From Equation (9), it follows that the main ways of reducing the intensity of blade cutting tool wear are

-   Reducing the temperature and cutting force.
-   Reducing the friction coefficient f (adhesive component) of the frictional contact with the machined material (due to the formation of secondary structures and phases on the working surfaces as products of adaptation (self-organization) of the tribosystem).
-   Increasing the hardness ratio of the tool's contact surfaces HV_m and the workpiece HV_t (by reducing the dependence of the physical and mechanical properties of the contact surfaces on the temperature in the working zone, considering the phenomena of adaptation (self-organization) during friction) [12,13,23–25].

At the present time, the above can be achieved by developing, implementing, and researching innovative multilayer composite coatings on the cutting tool for blade cutting processing, which have not only unique strength properties but also the ability to adapt to external influences with subsequent improvement of operational properties.

It is known that the temperature and cutting forces can have a two-fold influence on the intensity of cutting tool wear: on the one hand, increasing temperature reduces the direct effect of the force factor, and on the other hand, increasing temperature leads to the formation of secondary structures and phases on frictional contact with a corresponding change in physical and mechanical properties. Consequently, when creating tool materials and wear-resistant coatings with a friction adaptation effect, it is necessary to consider the specified region of cutting modes and the cutting temperature interval corresponding to this region [26–31].

3. Materials and Methods

Experimental research during milling was carried out on certified equipment to ensure the reliability of the obtained test results, specifically on vertical milling machines of normal accuracy and rigidity "VM 127M" (Votkinskiy Machine Building Plant, Votkinsk, Russia) and "Knuth WF 4.1" (KNUTH, Wasbek, Germany) (Figure 1). Figure 1 of vertical milling machine "VM 127M" shows the complexity of the equipment for measuring and recording the components of forces and cutting temperature, from left to right: dynamometer amplifier; PC with software "KISTLER" (version 2825A-03-2); PC for recording cutting temperature with TEMS converter for the temperature; millivoltmeter on the machine table; and three-component dynamometer "KISTLER" with a workpiece mounted on it.

Figure 1. Vertical milling machine VM 127M.

To investigate wear laws of cutting tools at face milling with carbide end mills (diameter d = 12 mm; number of teeth z = 4; with geometry of cutting part of helix angle $\omega = 30°$; forward angle $\gamma = 10°$; and back angle $\alpha = 4°$) of H10F grade with various coatings, titanium alloys—Ti64 and Ti811—were used as a machined material. Chemical composition, physical and mechanical properties, and heat treatment of the alloys under study are presented in Tables 1 and 2.

Table 1. Chemical composition of the materials to be machined.

	Main Components, % by Weight				
Alloy	Ti	Al	V	Mo	Others
Ti64	Basis	5.5–6.75	3.5–4.5	-	-
Ti811	Basis	7.5–8.5	0.75–1.25	0.75–1.25	-

Table 2. Physical and mechanical properties of machined materials.

Parameters	Alloys	
	Ti64	Ti811
σ, MPa	910	1141–1175
δ, %	7	7.2–9
ψ, %	15	14–16
KCU, J/m^2	2.5	0.19–0.24
KCT, J/m^2	7.85	2.5–4.1
σ_{100}/σ_T		
300 °C	793	-
400 °C	725	-
500 °C	509/705	-
600 °C	372/676	690/725
σ_{-1}, MPa	284	-
t (N = 2 × 10^7) °C	550	-
σ_{-1}^H, MPa	147	-
t (N = 2 × 10^7, K$_t$ = 3.35) °C	550	-
σ_0, MPa	539	-
t (N = 2 × 10^4) °C	550	-
σ_{-1}^H, MPa	284	-
t (N = 2 × 10^4, K$_t$ = 3.35)	550	-
K$_{1C}$, MPa·m$^{1/2}$	59–68.7	-

The study of the machinability of titanium alloys in full-scale tests was carried out as applied to the conditions of semifinishing and finishing machining: at n = 2000 rpm; S = 200 mm/min; a_e = 5 mm; and a_p = 1.5 mm. In this case, as shown by preliminary experiments and analysis of literature data [32–36], in the conditions of finishing and semifinishing cutting, the determining element of tool wear is the wear chamfer on the rear surface of the cutting wedge. Analysis of the rear surface wear profile showed that the average wear of the rear surface along the main cutting edge is characterized by the smallest variability of the wear measurement results. This parameter at constant values of the front and rear angles of the cutting wedge reflects the dimensional wear resistance of the whole tool. Based on the above, the average width of the rear surface wear chamfer (excluding notches) was used as the investigated tool wear parameter. In the process of milling to ensure the identity of the results and to exclude errors of measurement, the width h_r of tool wear chamfer on the back surface was measured using the reading microscope MIR-2M with nozzle MOB-I5 accurate to 0.002 mm at the workplace, and when it reached h_r = 0.3–0.4 mm, for additional control and photofixation of the wear pattern, a universal motorized stereo microscope with the possibility of telecommunication "Carl Zeiss Stereo Discovery V12" (ZEISS Microscopy, Oberkochen, Germany) with a visualization system based on the video camera "Zeiss Axiocam 503 Color" (ZEISS Microscopy, Oberkochen,

Germany) was used (Figure 2). In this case, the wear of the cutter was measured at certain number of passes, and the cutting path length was set to obtain a picture of all stages of the wear curve (section of the running-in, normal, and catastrophic wear).

Figure 2. Working area with motorized stereo microscope "Carl Zeiss Stereo Discovery V12".

The wear resistance of a cutting tool is characterized by the period of its durability T and is defined as [3,4]:

$$T = 1/S_m \tag{10}$$

where T is the durability period (min), l is the cutting path length (m), S_m is the minute feed rate (m/min), h_{rl} is the relative linear, and h_{sw} is the surface wear:

$$h_{rl} = \frac{h_{r.f} - h_{r.in}}{l_f - l_{in}}$$
$$\text{or}$$
$$h_{sw} = \frac{(h_{r.f} - h_{r.in}) \cdot 100}{(l_f - l_{in}) \cdot S} \; . \tag{11}$$

The criterion for the blunting of cutting tools was taken to be $h_r^{cr} = 0.3$ mm.

The schematic diagram of temperature and force studies during milling is shown in Figure 3.

To obtain information about the average contact temperature during cutting with sufficiently high accuracy, we used the simple and reliable natural thermocouple method. The part (2) and cutter (3) were isolated from each other by an insulator (1) to exclude errors from the so-called "parasitic" thermo-EMF. Measurement of thermo-EMF was carried out in intervals of 10–15 s after the beginning of cutting. The mercury current collector (5), digital voltmeter "Endim" (8), and PC (9) were used for registration and recording of the thermo-EMF value [10].

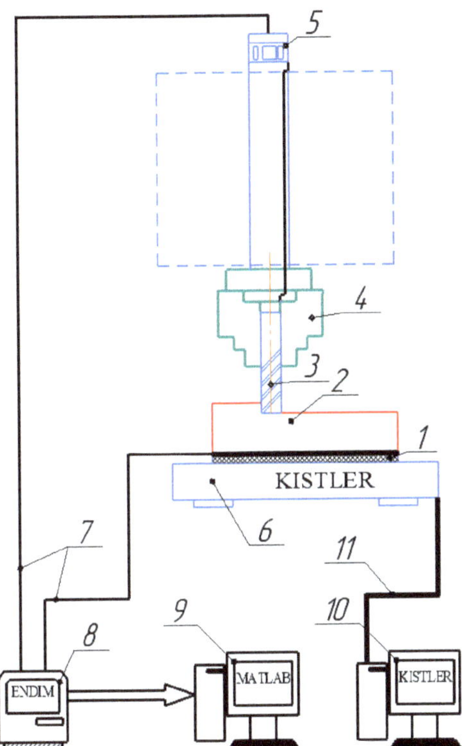

Figure 3. Schematic diagram of measuring temperature and cutting force components by natural thermocouple method. 1—insulation; 2—workpiece; 3—insulated cutter; 4—chuck; 5—mercury current collector; 6—dynamometer "Kistler"; 7—connecting wire; 8—amplifier–converter; 9—computer for registration of thermal emfs; 10—computer for registration of cutting force components; and 11—connecting bus.

Currently, for research purposes, a set of equipment "Kistler" (Figure 3) is used to measure the components of the cutting force, which provides a high degree of accuracy. The main element of the measuring system is a three-component dynamometer by "Kistler" in the form of a bed for fixing the workpiece from the machined material, model 9253B23 (Kistler Group, Winterthur, Switzerland), the appearance of which is shown in Figure 4.

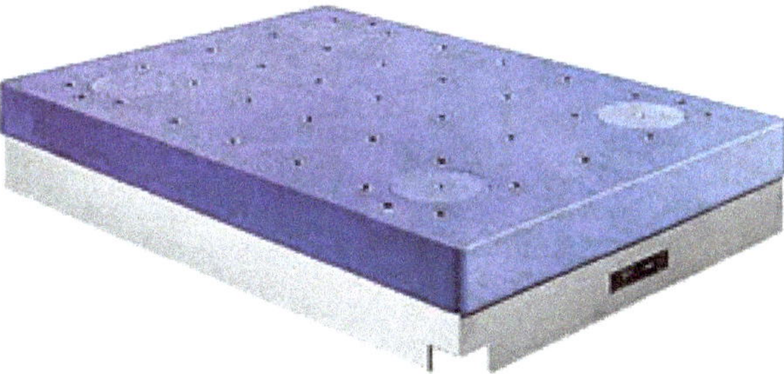

Figure 4. External view of the 9253B23 dynamometer.

The connection diagram of the main elements of the dynamometer is shown in Figure 5.

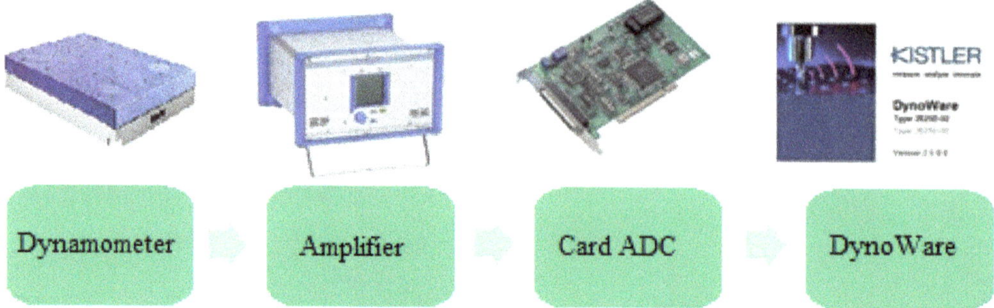

Figure 5. Wiring diagram.

The resulting force acting on the dynamometer is proportional to the algebraic sum of the corresponding components of the individual forces, which are formed because of the parallel arrangement, as shown in Figure 6. Thus, the dynamometer is a multicomponent force sensor, independent of the point of its application.

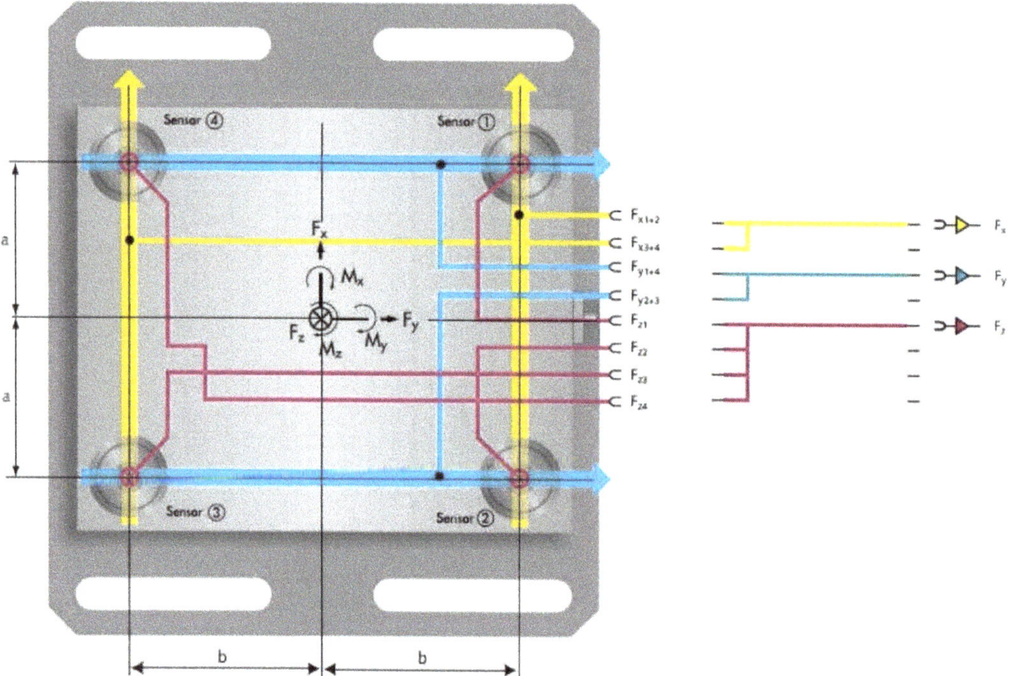

Figure 6. Structure of a three-component dynamometer. (a, b is the distance from the machining point to the center of the sensor; 1–4 is force sensors; M is moment of force).

To amplify and convert the signal, we used an 8-channel amplifier and converter type 5070A01110 from "Kistler" (Kistler group, Schweiz), the appearance of which is shown in Figure 7.

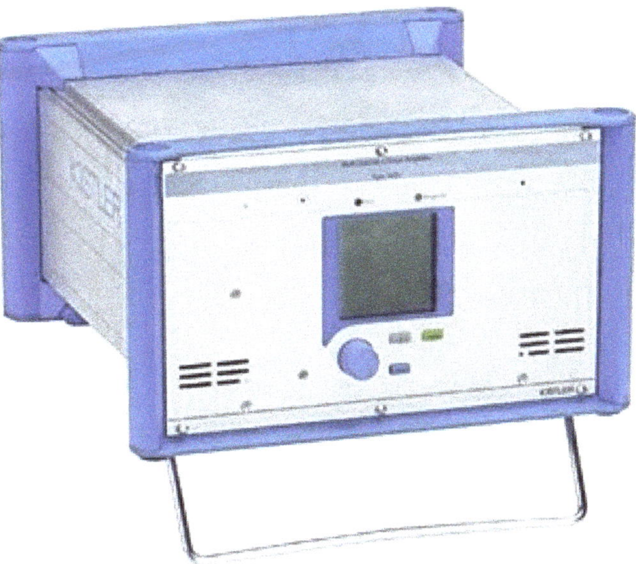

Figure 7. Exterior view of the amplifier and converter type 5070A01110.

Kistler "DynoWare" software (version 2825A-03-2) is used for data collection and analysis. The "DynoWare" software is versatile, easy to use, and compatible with dynamometers or single- and multicomponent force transducers. "DynoWare" provides continuous visualization of the measured curves during signal analysis and has all necessary mathematical and graphical functions. In addition to the simple configuration of the most important measurement tools, "DynoWare" supports the documentation of measurement processes and the storage of configuration and measurement data.

Easy operation, setup, and monitoring of "Kistler" measuring instruments via RS-232C or IEEE 488 interfaces; powerful graphic capabilities; useful functions for evaluation and calculations; simultaneous data logging from up to eight measuring channels; and recording and evaluation of any physical quantities are possible.

It is known that adhesion wear is most prevalent in machining, and at the same time, to reduce the time to determine the effective coating composition and application modes, as well as reduce the cost of both the tool and cathode material, several adhesion tests were carried out. To estimate the tribotechnical parameters (τ_{nn}, p_{rn}, and τ_{nn}/p_{rn}), an experimental method [25] was used. This method is based on the physical model (Figure 8) and the developed setup (Figure 9) [12,26], which in the first approximation reflects real friction and wear conditions at local contact. According to this model, a spherical indentor (1) made of tool material H10F with different coatings (simulating a single rough spot of rubbing solids), squeezed by two plane-parallel samples (2) made of titanium alloys Ti64 and Ti811 (with high accuracy and cleanness of contact surfaces), rotates under the load N around its own axis. The force F_{ex} expended on the indentor rotation and applied to the tether (3) placed in the disk slot (4) is mainly related to the shear strength τ_{nn} of the adhesion bonds. The other end of the cable is connected to the elastic elements (9, 10) and the recording device by means of the wire (11) on which the drive forces for the rotation of the indentor (1) are recorded. Ensuring the temperature regime in the contact zone of the indentor with the samples of the processed material in a wide range of variation (0–1300 °C) uses the electrical contact method, and a high voltage is supplied from the power-controlled transformer to the terminals (5) isolated from the housing.

nACo3–(TiAlSi)N + TiN or AlSiTiN(up) + TiAlSiN + TiN and nACRo–TiAlN or AlCrN is embedded in the amorphous matrix SiN.

4. Experimental Results and Discussion

The analysis of the works of researchers allows us to conclude that the solution to the problem of increasing the wear resistance of cutting tools with different coatings is developing in different directions. At the same time, there is practically no data on the results of experimental research with the indication of cutting modes and specific relationships between "tool-machined material", application modes, and percentage content of wear-resistant coating elements, as well as their operational characteristics (adhesion coefficient, service life, critical cutting path length, and wear on the rear surface of the cutting tool); this does not allow a comparative analysis of the results obtained in this case. In the works [36–41], in search of solutions to the problem, we followed a way of modeling the alloying components (Ti, Cr, Al, and Si) in the composition of nanocomposite coatings (nc-TiAlN/a-SiN, nc-AlCrN/a-SiN, and nc-AlTiCrN/a-SiN), and the increase in the tool service life by 10%–15% on average does not allow substantiation of the improvement of tribotechnical characteristics of wear-resistant coatings, with regard for possible measurement errors.

At the same time, it is known [36,42–48] that most modern nanostructured multi-layer composite coatings used on the cutting tool for blade cutting, under certain contact processes and temperature and force conditions, can improve their tribotechnical characteristics. In particular,

- The nanocomposite coating based on aluminum nitride and titanium nACo3 is a high-tech solid coating applied by physical vacuum deposition and consisting of nanoparticles in a binding amorphous matrix. The uniqueness of the coating lies in the successful combination of physically mutually exclusive parameters: while increasing the hardness, its elasticity increases at the same time;
- The nanocomposite coating based on chrome and aluminum carbonitrides nACRo has very high hardness at elevated cutting temperatures;
- Titanium diboride TiB_2 nanocomposite coating is a synthetic, particularly hard, heat-resistant, refractory, and wear-resistant material. These coatings are advantageous due to their high hardness and resilience as well as good abrasion resistance.
- Diamond-like coating (DLC) is a diamond-like carbon material with a tough and friction-reducing coating on top of other conventional wear-resistant coatings in most cases.

The results of experimental studies of the wear resistance of cutting tools with different coatings during milling (Figures 11–16) of titanium alloys are presented in the form of graphs of the dependence of tool wear on the rear surface (h_r, mm) (Figures 11 and 14) on the critical length of cutting path (l, m) (Figures 13 and 16) and also graphs of the dependence of the durability period on the coating applied to the cutting tool (Figures 12 and 15). The best indices of cutting tool wear resistance (wear on the back surface, critical cutting path length, and tool durability period) were measured as follows:

- Milling of titanium alloy Ti64 was ensured when using wear-resistant coatings: improvement by 2.4 times with "nACRo+TiB_2" coating, by 2.2 times with "nACo3+TiB2", and by 2 times with "nACRo" coating compared to milling without coating;
- Milling of titanium alloy Ti811 was provided with wear-resistant coating: improvement by 4 times with "nACRo" coating and improvement by 2 times with "nACRo+TiB_2" coating in comparison with milling without coating.

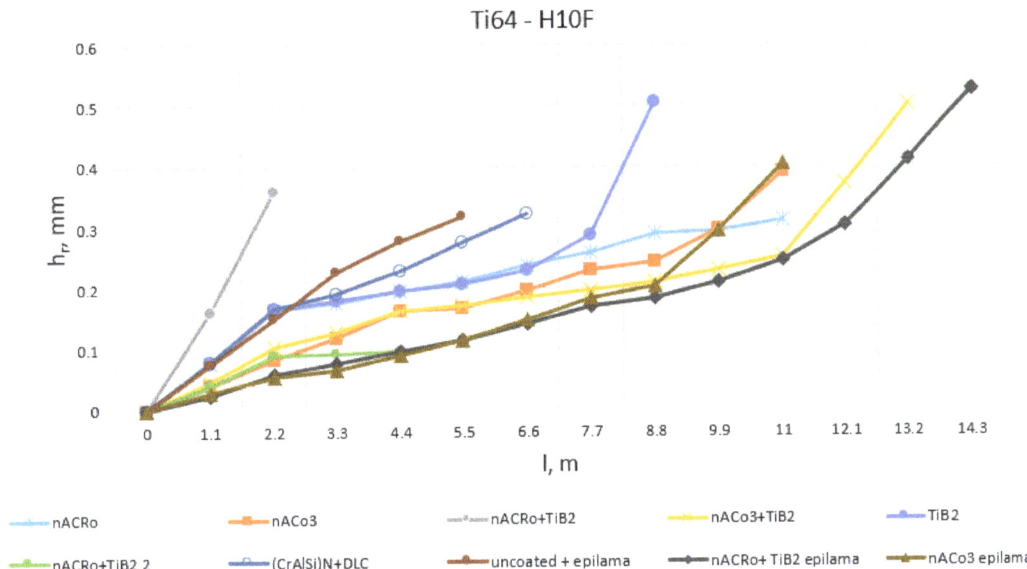

Figure 11. Effect of the cutting path length (l, m) on the value of wear on the rear surface (h_r, mm) during milling of titanium alloy Ti64 by carbide cutters of grade H10F with different coatings (n = 2000 rpm, S_{min} = 200 mm/min, a_e = 5 mm, a_p = 1.5 mm).

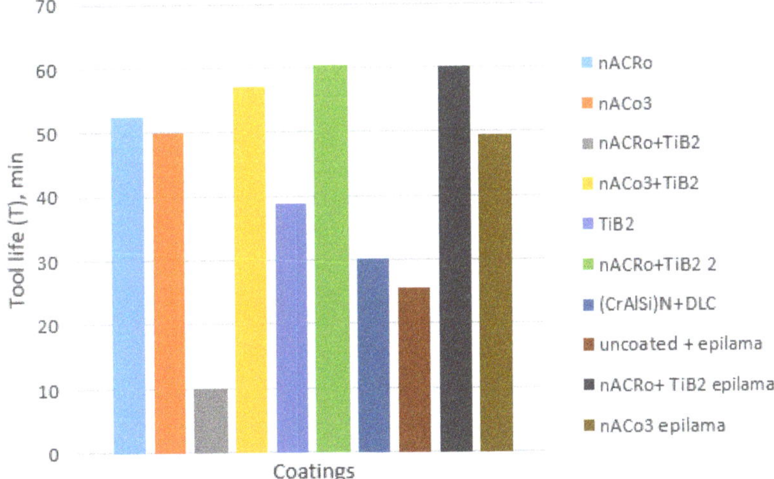

Figure 12. Durability period when milling titanium alloy Ti64 with tungsten carbide cutters of grade H10F with different coatings.

According to the results of experimental studies, the best indicators of cutting tool wear resistance with different coatings were found as follows:

- Milling of titanium alloy Ti64 was provided when using wear-resistant coatings: improvement by 18% with "nACRo + TiB2" coating and by 17% with "nACRo + TiB2 + epilama" coating in comparison with "nACo3" coating;
- Milling of titanium alloy Ti811 was ensured when using wear-resistant coatings: improvement by 1.8 times with "nACRo" coating and by 1.5 times with "nACo3 + TiB2 + epilama" coating in comparison with "nACo3" coating.

According to the results of field experiments during milling of titanium Ti64 by milling cutters with different durations of titanium diboride application, it was found that

- For milling of Ti64 titanium alloy, an increase in TiB_2 coating time of more than 30 min does not contribute to the increase in wear resistance of the milling cutter.

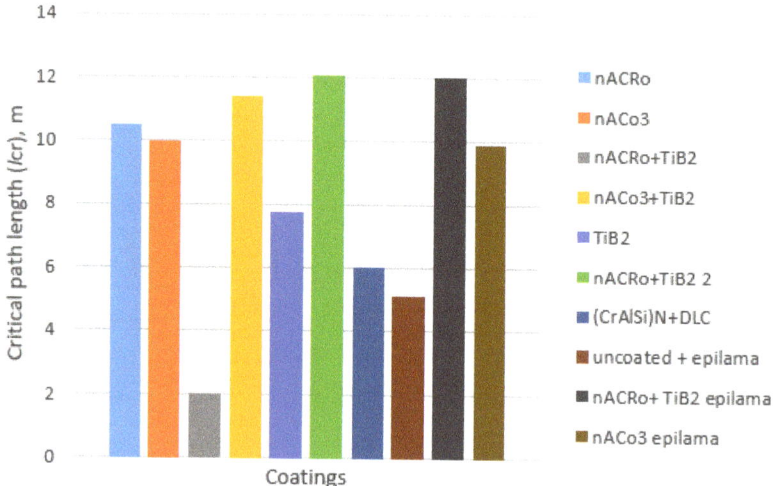

Figure 13. Critical path length when milling titanium alloy Ti64 with H10F carbide cutters with different coatings.

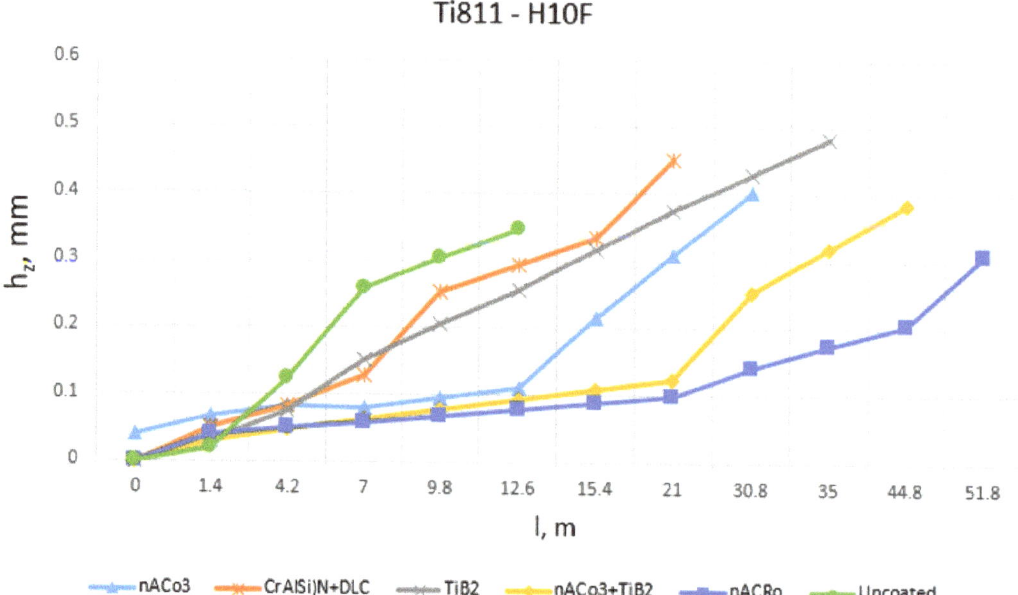

Figure 14. Effect of the cutting path length (l, m) on the value of wear on the rear surface (h_z, mm) during milling of titanium alloy Ti 811 by carbide cutters of grade H10F with different coatings (n = 2000 rpm, S_{min} = 200 mm/min, a_e = 5 mm, a_p = 1.5 mm).

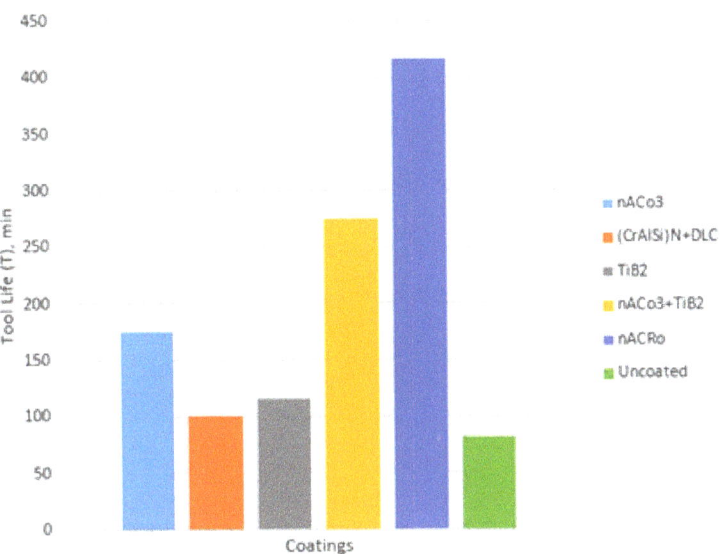

Figure 15. Durability period when milling titanium alloy Ti811 with tungsten carbide cutters of grade H10F with different coatings.

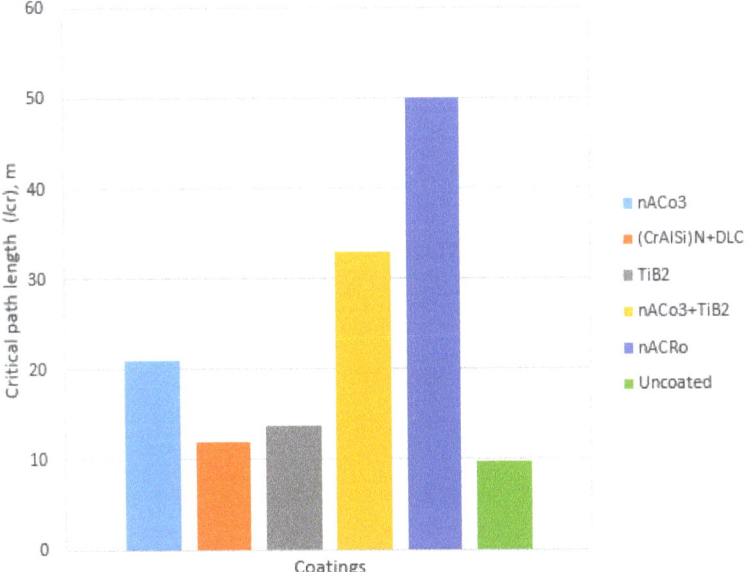

Figure 16. Critical path length when milling titanium alloy Ti811 with H10F carbide cutters with different coatings.

The results of the temperature experiments are shown in Figures 17 and 18.

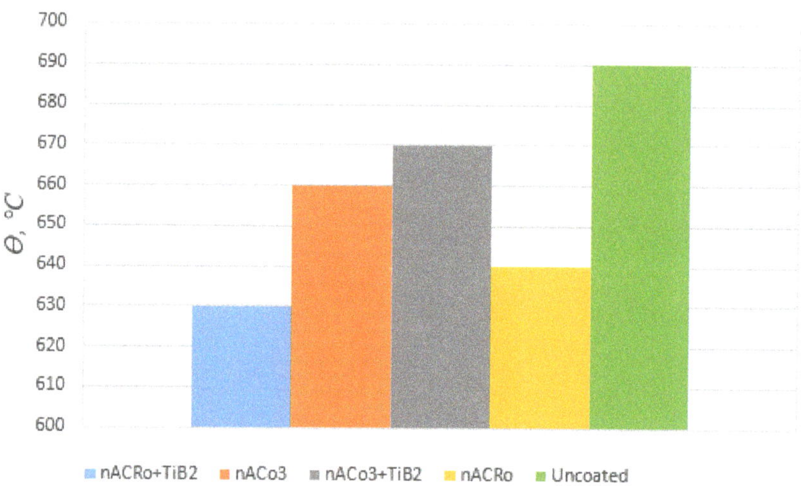

Figure 17. Cutting temperature (thermal emf) when milling titanium alloy Ti64 at different spindle speeds (n = 2000 rpm; S_{min} = 200 mm/min, a_e = 5 mm, a_p = 1.5 mm).

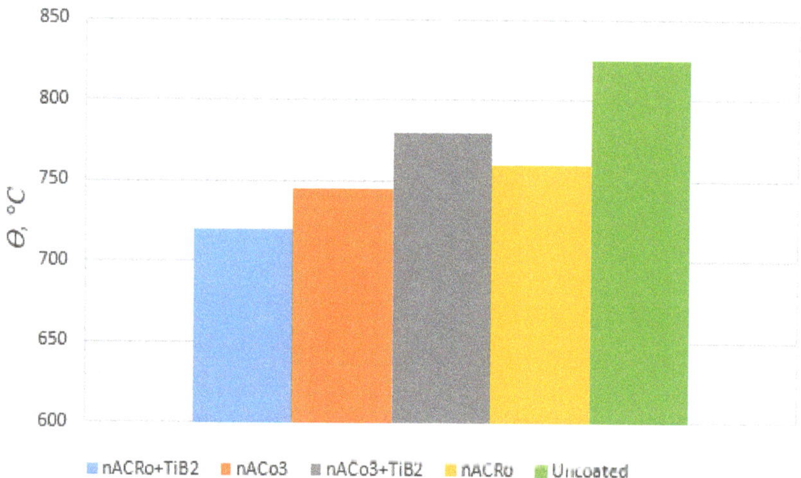

Figure 18. Cutting temperature when milling titanium alloy Ti811 at different spindle speeds (n = 2000 rpm; S_{min} = 200 mm/min, a_e = 5 mm, a_p = 1.5 mm).

According to the results of temperature studies, it was found that

-   The lowest cutting temperature value is provided when using "nACRo + TiB_2" coating when milling Ti64 and Ti811 titanium alloys, and they are lower by 9% and 12%, respectively, relative to uncoated machining;
-   Milling of Ti64 titanium alloy is less heat-intensive compared to Ti811.

The results of the performed force experiments are shown in Figures 19 and 20.

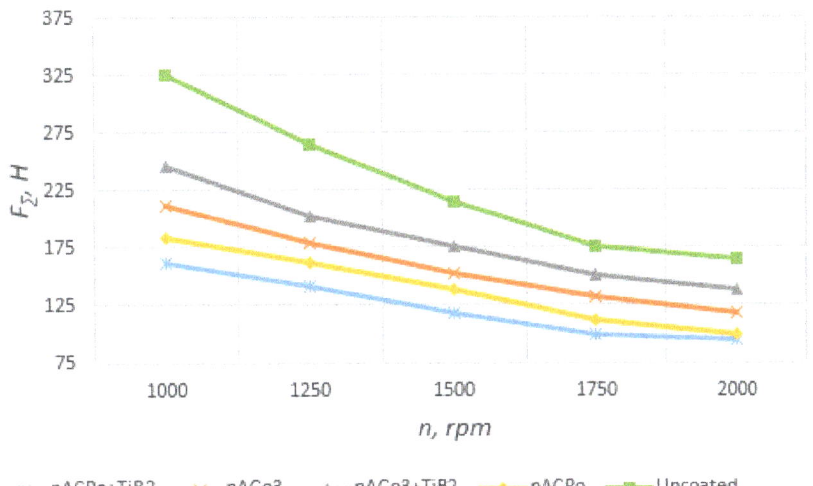

Figure 19. Total component of cutting force when milling titanium alloy Ti64 at different spindle speeds (S_{min} = 200 mm/min, a_e = 5 mm, a_p = 1.5 mm).

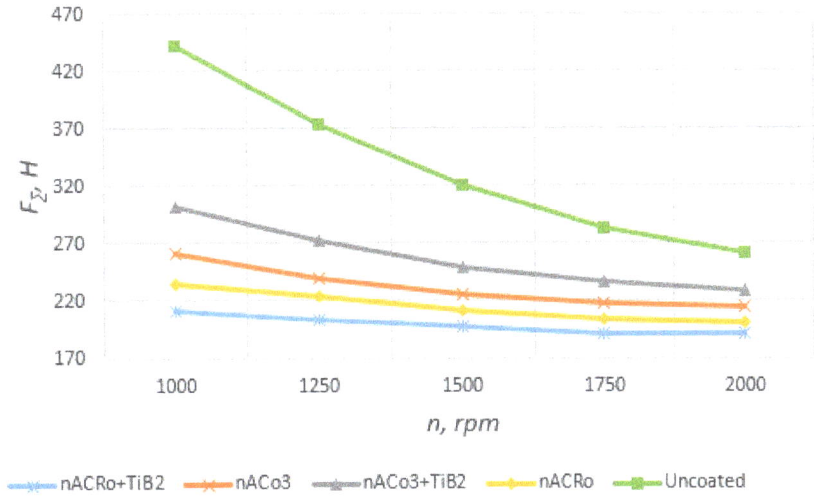

Figure 20. Total component of cutting force when milling titanium alloy Ti811 at different spindle speeds (S_{min} = 200 mm/min, a_e = 5 mm, a_p = 1.5 mm).

According to the results of the conducted force tests, it was found that

- The lowest value of cutting force is provided when using "nACRo3 + TiB2" coating for both titanium alloy Ti64 and alloy Ti811 with 1.5–2 times less force in comparison with milling without coating;
- Milling of titanium alloy Ti64 has less force loading in comparison with Ti811 by 1.5–2 times.

The results of adhesion studies are shown in Figures 21 and 22.

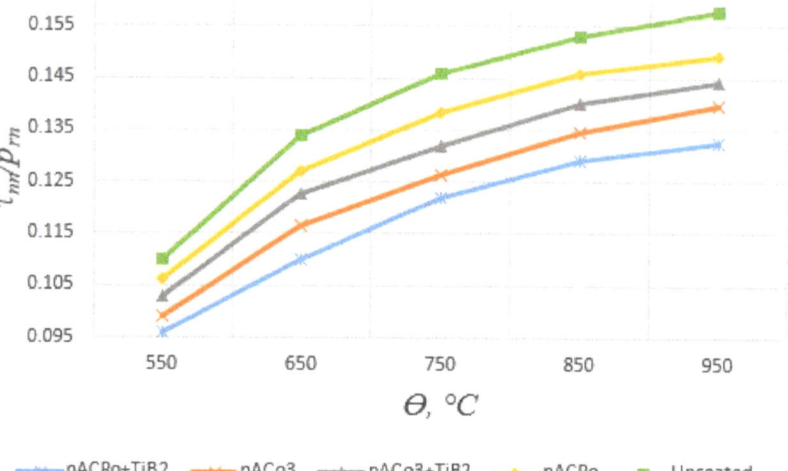

Figure 21. Temperature dependence of frictional characteristics of plastic contact "Ti64-H10F with coatings".

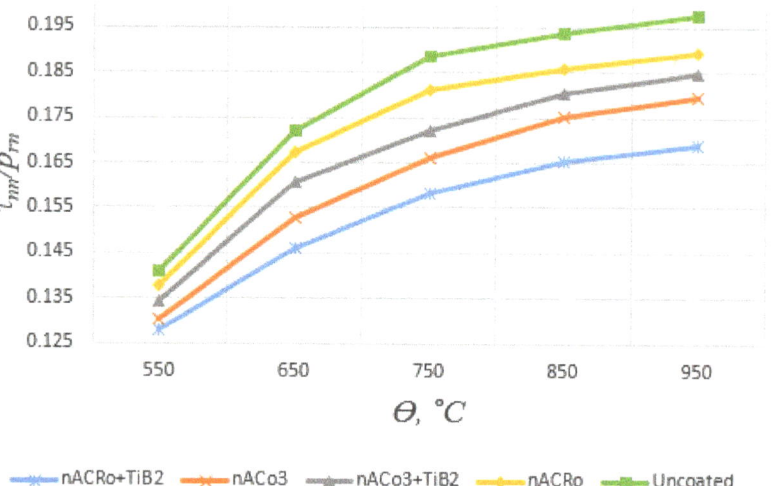

Figure 22. Temperature dependence of frictional characteristics of plastic contact "Ti811-H10F with coatings".

Analysis of experimental data has shown that, with increasing contact temperature, the adhesion component of the friction coefficient increases monotonically for all the coatings under study at temperatures corresponding to optimal cutting speeds (according to wear intensity), and the smallest value corresponds to the most favorable composition and technology of their applications. It was found that

- The most effective coatings in terms of contact processes on a single contact spot and adhesion coefficient are the coatings applied on "Platit π411": $nACo_3$, nACRo, and $nACRo + TiB_2$;
- The lowest coefficient of adhesion interaction between the tool material and the machined material in the whole investigated temperature range is provided by coating $nACRo + TiB_2$ on titanium alloys Ti64 and Ti811, which confirms the best coating adhesion with the substrate and allows using them for high-speed milling.

To explain the mechanism of increasing tribotechnical characteristics of multilayer nanostructured coatings, a series of material science studies of the wells of samples of machined and spherical surfaces of tool materials was carried out (Figures 23–25).

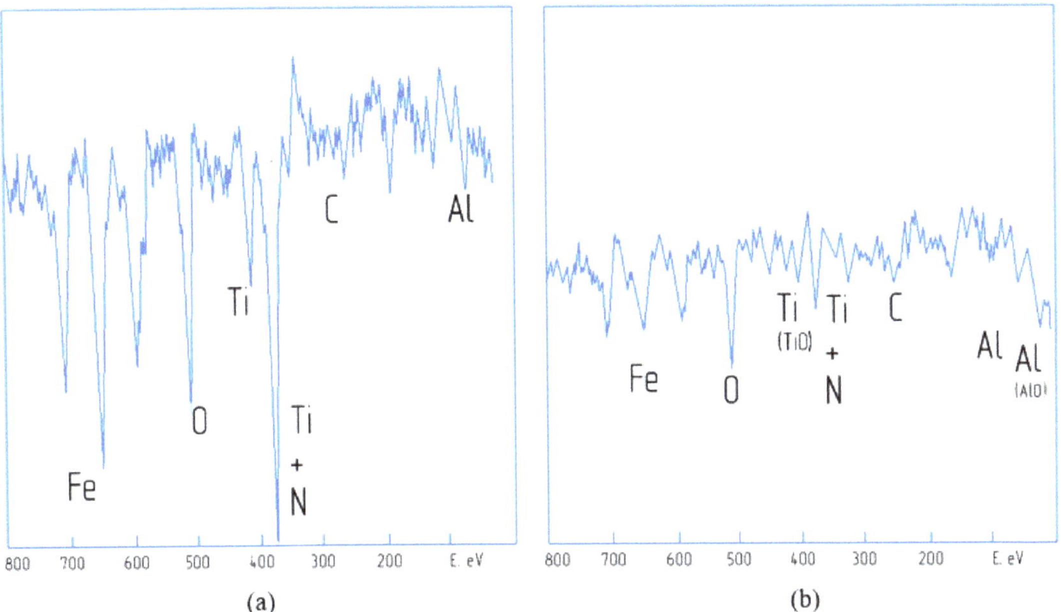

Figure 23. Auger electron spectra of surfaces of worn nACRo-coated inserts: (**a**)—at the running-in stage after 3 min from the beginning of cutting; (**b**)—at the steady-state stage after 20 min from the beginning of cutting.

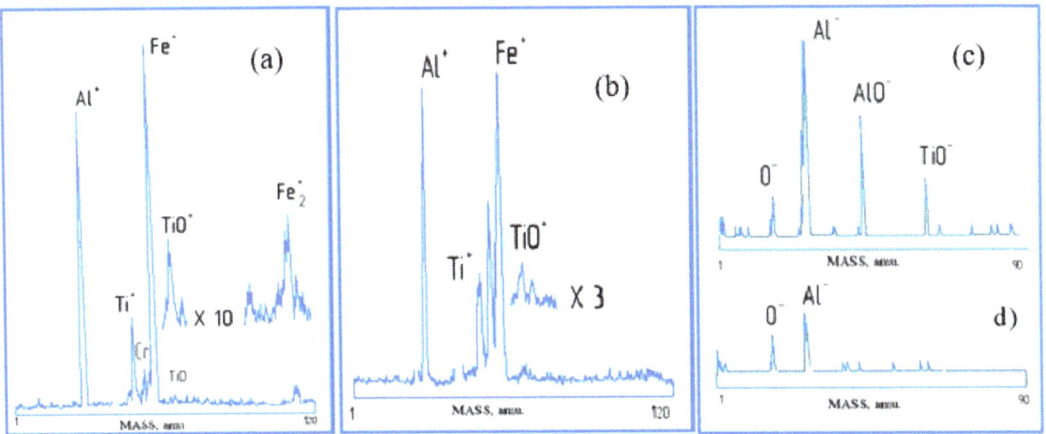

Figure 24. Spectra of positive (**a**,**b**) and negative (**c**,**d**) secondary ions of the surface of worn nACRo-coated plates: (**a**,**d**)—after 3 min of cutting; (**b**,**c**)—after 20 min of cutting.

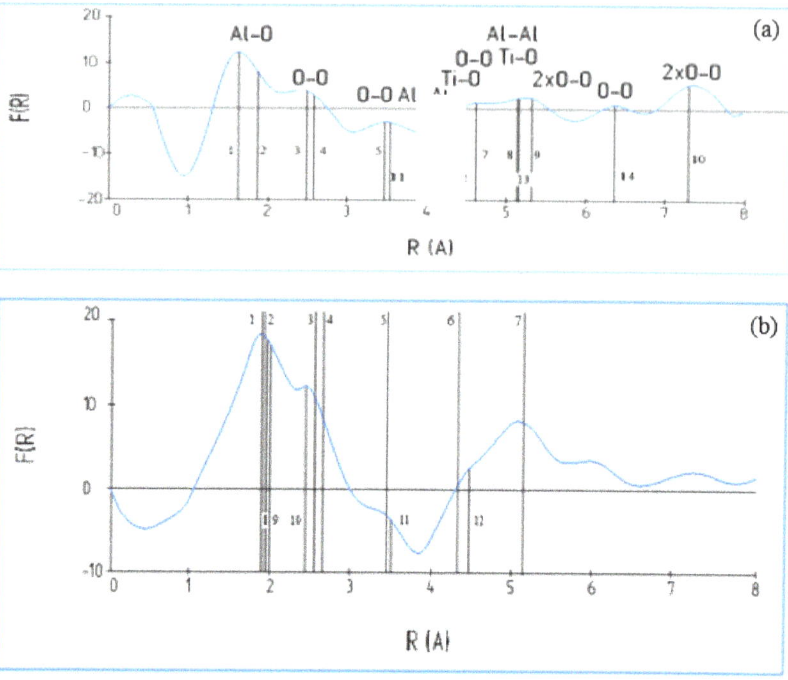

Figure 25. Fourier transforms on the EELFS close to the elastic scattering electrons for films formed on the surface: (**a**)—nACRo-coated cutting tool; (**b**)—binary TiAl alloy after isothermal oxidation in air.

The surface morphology and microstructure of the coatings were investigated by transmission electron microscopy on a JEOL JEM-201 OE unit with an accelerating voltage of 200 kV. At the same time, the coatings were deposited on the cemented carbide bottom layer with a thickness of 2.5 to 3 μm for observations. Samples were prepared using the FIB ion beam focusing method on the JEOL-JFIM-2100 system and were thinned to 0.1 μm using Ga ions with an accelerating voltage of 30 kV and a current of 2.0 mA from a Ga ion source. The chemical composition of the secondary structures appearing on the cutter surface during cutting was investigated by the SIMS method. The ion etching rate was on the order of 0.2 monolayers per minute, and the analysis was performed in static mode. The atomic structure of the films formed on the tool surface during cutting was investigated using EELFS and an ESCALAB MK 2 (VG) electron spectrometer. The friction surface was investigated in areas free from adhesion of the workpiece material. A high magnification ratio (2000×) was used. The primary electron energy was $E_p = 1000$ eV. A fine spectral energy loss pattern was recorded close to the elastically scattering electron line in the 250 eV range. The conditions for the analysis were chosen to provide the best energy resolution with a good signal intensity ratio [18,48–51].

Intensive tribo-oxidation of the cutting tool surface occurs during high-speed machining. Figure 24 shows Auger electron spectra for tools with nACRo wear-resistant coating. The oxidation of the contact surfaces is obvious, as evidenced by the presence of large amounts of oxygen in both spectra. The intense ionic peaks correspond to the adhesion zones of the part material. The Ti line is significantly stretched (Figure 24b) in this zone; this is the result of the oxidation process. An increased amount of aluminum oxide is observed in the spectrum of the filtered coatings, which is shown as a shift in the aluminum line to a lower energy zone (60 eV). At the same time, the intensity of the metallic Al LMM line level around 68 eV decreases.

Figure 24a,b show a series of spectra of positive secondary ions for both 3 min of cutting and after 20 min of coating cutting, and Figure 24c,d show spectra of negative secondary ions for both coatings. In both positive secondary spectra, the TiO line intensity is high, and this is due to intense tribo-oxidation, which forms rutile-like films. However, some aluminum oxide is formed only on the surface of the coatings after 20 min, and this effect can be observed on the spectrum of negative secondary ions (Figure 24c). The formation of aluminum oxide films on the cutter surface significantly changes the heat fluxes and heat dissipation into the chip. This is confirmed by images of chip cross-sections after scanning on an electron microscope, and three different zones can be seen in the chip cross-sections.

The atomic structure of the films appearing on the wear zone surface during tribo-oxidation of the nACRo coating was compared with the oxide layer obtained during oxidation of the binary TiAl alloy under equilibrium conditions. Figure 26 shows the Fourier transforms because of mathematical consideration of the fine structure of electron spectra close to the line of elastically scattering electrons. The positions of the peaks on the Fourier transforms correspond to the interatomic distances for the closest coordination spheres. The decoding of the data was based on the analysis of known crystal characteristics of oxides formed on the surface. The positions of the main peaks of the Fourier transforms correspond to the interatomic distances associated with the Al_2O_3 and TiO_2 lattices (Figure 25) [51–54].

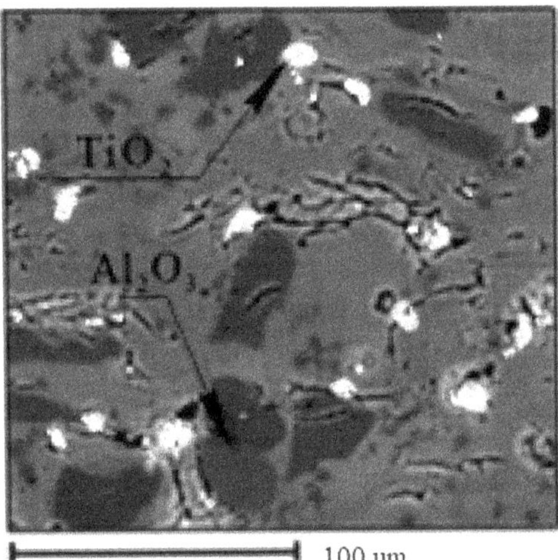

Figure 26. Results of the micro-X-ray spectral analysis.

In general, the results of metallurgical studies state that the use of such coatings contributes to the reduction in friction forces and cutting tool wear due to the formation of titanium and aluminum oxides. Studies of protective film formation at low and moderate cutting speeds show that there is only one type of protective film formed on the surface because of self-organization phenomena [17,28–30]. These films have an amorphous-like structure with high ductility and improved lubricity. More complex phenomena occur during high-speed processing. These are, first, the low-intensity peaks that are found at far atomic distances in the Fourier transforms shown in Figure 25. From this figure, it can be assumed that the films that form during tribo-oxidation under high-speed processing conditions are amorphous-like. These films of aluminum oxide contribute to the reduction in wear because, due to the low thermal conductivity, they prevent the intensive removal of heat generated during cutting into the body of the cutting tool [55–59].

Figure 26 shows the results of the micro-X-ray spectral analysis of the worn surfaces of the cutting tools with nanostructured coatings. Worn and oxidized areas of the cutter surface revealed aluminum and titanium oxides formed during milling with nACRo nanostructured coatings.

5. Conclusions

1. The thermodynamic model of cutting tool wear is proposed, allowing us to determine ways to reduce the intensity of cutting tool wear and the dissipative function of the tool material shape change during tool wear, and the conditions for improving the wear resistance of the cutting tool using the phenomenon of adaptation (self-organization) under friction are formulated.
2. Experimental studies on the wear resistance of cutting tools with different coatings have shown that, during milling of titanium alloys, a significant increase in wear resistance is provided when using innovative multilayer nanostructured coatings, with an improvement on average of 1.5–2 times.

These positive results are associated with a complex set of phenomena occurring in the contact processes:

Firstly, there is a significant decrease in temperature–force loading in the cutting zone by an average of 15%–25% both in terms of cutting temperature and cutting force components, and this phenomenon, according to numerous studies, can be explained by the formation of secondary structures on the friction surfaces in the form of aluminum and titanium oxides, which have heat-reflecting (shielding) and lubricating properties;

Secondly, it is because of the above reduction in the friction coefficient (adhesive component) in a wide range of temperature changes from 5500 C to 9500 C by 13%–17%.

3. Important in the analysis are the results of metallurgical research, as the obtained data allow one to assert that, in contact processes, there is a phenomenon of adaptation (self-organization) of friction surfaces at cutting by a tool with wear-resistant coatings, promoting the formation of films of various compounds with shielding, protective, and lubricating properties, in particular Al_2O_3, TiO_2, etc. These films have an amorphous-like structure with high plasticity and improved lubricity. This is confirmed by spectroscopic studies and above all by the low-intensity peaks found at long atomic distances on Fourier transforms.

Author Contributions: Conceptualization, M.S.M.; Methodology, S.R.S. and A.M.M.; Software, A.S.G.; Formal analysis, N.A.S.; Writing—review & editing, A.P.M.; Visualization, D.S.R. All authors have read and agreed to the published version of the manuscript.

Funding: The research was funded by Grant No. 22-19-00670 of the Russian Science Foundation, https://rscf.ru/project/22-19-00670/ (accessed on 27 December 2022).

Institutional Review Board Statement: Not applicable.

Informed Consent Statement: Not applicable.

Data Availability Statement: Not applicable.

Conflicts of Interest: The authors declare no conflict of interest.

References

1. Vereshchaka, A.S.; Kushner, V.S. *Cutting Materials*; Higher School: Moscow, Russia, 2009; 336p.
2. Kabaldin, Y.G.; Semibratova, M.V.; Kirichenko, V.V. Self-organization in friction processes during cutting. In Proceedings of the Tomsk Polytechnic University Proceedings of TPU, Tomsk, Russia, 25–26 April 2002; Volume 305, pp. 95–100.
3. Makarov, A.D. *Optimization of Cutting Processes*; Mechanical Engineering: Moscow, Russia, 1976; 150p.
4. Loladze, T.N. *Durability and Wear Resistance of Cutting Tools*; Mechanical Engineering: Moscow, Russia, 1982; 320p.
5. Matalin, A.A. *Engineering Technology: University Textbook*; Mechanical Engineering: Moscow, Russia, 1985; 512p.
6. Zoritkuev, V.C. *Identification and Automatic Control of Technological Processes in Machine-Tool Systems: A Training Manual*; USATU: Ufa, Russia, 1993; 118p.
7. Andrievsky, R.A.; Ragulia, A.V. *Nanostructured Materials*; Academia Publishing Center: Moscow, Russia, 2005; 192p.
8. Gusev, A.I.; Rempel, A.A. *Nanocrystalline Materials*; Fizmatlit: Moscow, Russia, 2004; 327p.

9. Panov, D.O.; Simonov, Y.N.; Leont'ev, P.A.; Smirnov, A.I.; Zayats, L.T. A study of phase and structural transformations of hardened low-carbon steel under conditionsof multiple intense heat effect. *Met. Sci. Heat Treat.* **2013**, *54*, 582–586. [CrossRef]

10. Beresnev, V.M.; Pogrebniak, A.D.; Azarenkov, N.A.; Farenik, V.I.; Kirik, G.V. Nanocrystalline and nanocomposite coatings, structure, properties. *FIP* **2007**, *5*, 4–27.

11. Levashov, E.A.; Shtanskii, D.V. Multifunctional nanostructured films. *Uspekhi Chem.* **2007**, *76*, 501–509. [CrossRef]

12. Shuster, L.S. *Adhesion Interaction of Solid Metal Bodies*; Publishing House GILEM: Ufa, Russia, 1999; 199p. Available online: http://www.prometeus.nsc.ru/exhibit/28-11-00/index2.ssi (accessed on 27 December 2022).

13. Tabakov, V.P.; Smirnov, M.Y.; Tsirkin, A.V. *Serviceability of End Mills with Multilayer Wear-Resistant Coatings*; Ulyanovsk State Technical University: Ulyanovsk, Russia, 2005; 151p.

14. Cantero, J.L.; Diaz-Alvarez, J.; Migueles, M.J.; Marin, N.K. Analysis of tool wear patterns in Inconel 718 turning. *Wear* **2013**, *297*, 885–894. [CrossRef]

15. Migranov, M.S. Enhancement of Tool Wear Resistance Based on the Prediction of Processes of Adaptation of Friction Surfaces When Cutting Metal. Ufimsky iz-"Gilem". Ph.D. Thesis, 2011; p. 229. Available online: https://www.dissercat.com/content/povyshenie-iznosostoikosti-instrumentov-na-osnove-prognozirovaniya-protsessov-adaptatsii-pov (accessed on 27 December 2022).

16. Vereshchaka, A.; Aleksandrov, I.; Muranov, A.; Mikhailov, M.; Tatarkanov, A.; Milovich, F.; Andreev, N.; Migranov, M. Study of tribological and performance properties of (mex, moy, al1-(x+y))n (me -ti, zr or cr) coatings. *Tribol. Int.* **2022**, *165*, 107305. [CrossRef]

17. Fox-Rabinovich, G.S.; Weatherly, G.C.; Dodonov, A.I.; Kovalev, A.I.; Shuster, L.S.; Veldhuis, S.C.; Dosbaeva, G.K.; Wainstein, D.L.; Migranov, M.S. Nanocrystalline filtered arc TiAlN PVD coatings for high-speed machining. *Surf. Coat. Technol.* **2004**, *177*, 800–811. [CrossRef]

18. Fox-Rabinovich, G.S.; Kovalev, A.I.; Shuster, L.S.; Bokiy, Y.F.; Dosbaeva, G.K.; Vainshtein, D.L.; Mishina, V.P. Characteristics of VSS alloying—Based on deformed composite powder materials taking into account the self-organization of the tool during cutting. *Wear* **2006**, *1997*, 214. [CrossRef]

19. Prengel, H.G.; Jindal, P.C.; Wendt, K.H.; Santhanam, A.T.; Hegde, P.L.; Penich, R.M. A new class of high-performance PVD coatings for carbide cutting tools. *Surf. Coat. Technol.* **2001**, *139*, 25–34. [CrossRef]

20. Haubner, R.; Lessiak, M.; Pitonak, R.; Köpf, A.; Weissenbacher, R. Evolution of traditional hard coatings for use on cutting tools. *Int. J. Refract. Met. Hard. Mater.* **2017**, *62*, 210–218. [CrossRef]

21. Zhao, D.X.; Bui, C.T.; Zeng, X.T. Abrasion wear resistance of Ti1—XAlXN hard coatings deposited by vacuum arc system with side rotating cathodes. *Surf. Coat. Technol.* **2008**, *203*, 680–684. [CrossRef]

22. LaGrange, D.D.; LaGrange, T.; Santana, A.; Jähnig., R. Macroparticle formation during cathodic-arc deposition of nitride coatings from alloy cathodes TiNb. *J. Vac. Sci. Technol. A Vac. Surf. Film.* **2017**, *35*, 021309. [CrossRef]

23. Alami, J.; Eklund, P.; Emmerlich, J.; Wilhelmsson, O.; Jansson, U. Powerful pulsed magnetron sputtering of Ti-Si-C thin films from a composite Ti$_3$SiC$_2$ target. *Thin Solid Film.* **2006**, *515*, 1731–1736. [CrossRef]

24. Zhou, H.; Zheng, J.; Gui, B.; Geng, D.; Wang, Q. AlTiCrN deposited by HIPIMS/DC hybrid magnetron sputtering. *Vacuum* **2017**, *136*, 129–136. [CrossRef]

25. Shuster, L.S. *Adhesive Interaction of Solid Metal Bodies*; Gilem Publishing House: Ufa, Russia, 1999; 198p. Available online: https://search.rsl.ru/ru/record/01000652403 (accessed on 27 December 2022).

26. Shuster, L.S.; Migranov, M.S. Device for the Study of Adhesive Interaction. Useful Model Patent No. 34249, 24 June 2003.

27. Haršáni, M.; Ghafoor, N.; Calamba, K.; Zacková, P.; Sahul, M.; Vopát, T.; Satrapinskyy, L.; Čaplovičová, M.; Čaplovič, Ľ. Adhesion-strain relationships and mechanical properties of nc-AlCrN/a-SiNXhard coatings deposited at different displacement stresses. *Thin Solid Film.* **2018**, *650*, 11–19. [CrossRef]

28. Migranov, M.S. Study of wear of tool materials and coatings from the position of thermodynamics and self-organization. Izvestiya vysshee uchebnykh obucheniya Proceedings of higher educational institutions. *Mech. Eng.* **2006**, *11*, 65–70.

29. Migranov, M.S.; Mukhamadeev, V.R.; Migranov, A.M.; Mukhamadeev, I.R.; Khazgalieva, A.A. Structure and phase changes in nanostructured combined coatings in the collection. *IOP Conf. Ser. Mater. Sci. Eng.* **2018**, *447*, 012082. [CrossRef]

30. Migranov, M.S.; Shuster, L.S. Wear resistance of cutting tools with multilayer coatings. *Frict. Wear* **2005**, *26*, 304–307.

31. Cselle, T.; Coddet, O.; Galamand, C.; Holubar, P.; Jilek, M.; Jilek, J.; Luemkemann, A.; Morstein, M. TripleCoatings3®-New Generation of PVD-Coatings for Cutting Tools. *J. Mach. Manuf.* **2009**, *49*, 19–25.

32. Veprek, S.; Jilek, M. Superhard and functional nanocomposites formed by self-organization versus hardening of coatings by energy ion bombardment during their deposition. *Rev. Adv. Mater. Sci.* **2003**, *5*, 6–16.

33. Sobol, O.V.; Andreev, A.A.; Grigoriev, S.N.; Volosova, M.A.; Gorban, V.F. Vacuum arc multi-layer nanostructured TiN/Ti coatings: Structure, stress state, properties. *Met. Sci. Heat Treat.* **2012**, *54*, 28–33. [CrossRef]

34. Grigoryev, S.N.; Vereshchaka, A.A.; Fedorov, S.V.; Sitnikov, N.N.; Batako, A.D. Comparative analysis of cutting properties and wear character of carbide cutting tools with multilayer nanostructure and gradient coatings obtained using different spraying methods. *Int. J. Adv. Manuf. Technol.* **2017**, *90*, 3421–3435. [CrossRef]

35. Vereschaka, A.; Tabakov, V.; Grigoriev, S.; Sitnikov, N.; Milovich, F.; Andreev, N.; Bublikov, J. Investigation of wear mechanisms of a cutting tool surface with multilayer composite nanostructured Cr-CrN-(Ti,Cr,Al,Si)N coating during high-speed steel turning. *Wear* **2019**, *438–439*, 203069. [CrossRef]

36. Vereshchaka, A.; Tabakov, V.; Grigoryev, S.; Sitnikov, N.; Milovich, F.; Andreev, N.; Sotova, K.; Kutina, N. Study of the in-fluence of nanolayer thickness in Ti-TiN-(Ti,Cr,Al)N coating wear layers on cutting failure and wear of carbide cutting tools. *Surf. Coat. Technol.* **2020**, *385*, 125402. [CrossRef]
37. Grigoriev, S.N.; Volosova, M.A.; Vereschaka, A.A.; Sitnikov, N.N.; Milovich, F.; Bublikov, J.I.; Fyodorov, S.V.; Seleznev, A.E. Properties of composite coatings (Cr,Al,Si)N-(DLC-Si) deposited on a cutting ceramic substrate. *Ceram. Int.* **2020**, *46*, 18241–18255. [CrossRef]
38. Grigoryev, S.N.; Vereshchaka, A.A.; Milovich, F.; Bublikov, Y. Investigation of multicomponent nanolayer coatings based on Cr, Mo, Zr, Nb and Al nitrides. *Surf. Coat. Technol.* **2020**, *401*, 126258. [CrossRef]
39. Grigoryev, S.N.; Vereshchaka, A.A.; Milovich, F.; Sitnikov, N.; Ohanyan, G.V. Study of tribological properties of Ti-TiN-(Ti,Al,Nb,Zr)N composite coating and its effectiveness in increasing the wear resistance of metal cutting tools. *Tribol. Int.* **2021**, *164*, 107236. [CrossRef]
40. Migranov, M.; Migranova, R. Tool coatings with the effect of adaptation to cutting conditions. *Key Eng. Mater.* **2012**, *496*, 75–79. [CrossRef]
41. Grigoriev, S.; Vereschaka, A.; Zelenkov, V.; Sitnikov, N.; Bublikov, J.; Milovich, F.; Andreev, N.; Mustafaev, E. Specific features of the structure and properties of arc-PVD coatings depending on the spatial arrangement of the sample in the chamber. *Vacuum* **2022**, *200*, 111047. [CrossRef]
42. Grigoriev, S.; Vereschaka, A.; Zelenkov, V.; Sitnikov, N.; Bublikov, J.; Milovich, F.; Andreev, N.; Sotova, C. Investigation of the influence of the features of the deposition process on the structural features of microparticles in PVD coatings. *Vacuum* **2022**, *202*, 111144. [CrossRef]
43. Fox-Rabinovich, G.S.; Yamamoto, K.; Veldhuis, S.K.; Kovalev, A.I.; Dosbaeva, G.K. Tribological adaptability of TiAlCrN PVD coatings under high-performance dry processing. *Surf. Coating. Technol.* **2005**, *200*, 1804–1813. [CrossRef]
44. Vereshchaka, A.; Grigoriev, S.; Milovich, F.; Sitnikov, N.; Migranov, M.; Andreev, N.; Bublikov, Y.; Sotova, K. Study of tribological and functional properties of Cr, Mo-(Cr,Mo)N-(Cr,Mo,Al)N multilayer composite coating. *Tribol. Int.* **2021**, *155*, 106804. [CrossRef]
45. Yamamoto, K.; Kujime, S.; Takahara, K. Structural and mechanical properties of Si (Ti, Cr, Al)N inclusion coatings deposited by arc ion plating. *Surf. Coating. Technol.* **2005**, *200*, 1383–1390. [CrossRef]
46. Zhang, J.; Lv, H.; Cui, G.; Jing, Z.; Wang, C. Effect of displacement stress on microstructure and mechanical properties of (Ti,Al,Cr)N solid films with Ngrandent distribution. *Thin Solid Films* **2011**, *519*, 4818–4823. [CrossRef]
47. Forsén, R.; Johansson, M.P.; Odén, M.; Ghafoor, N. Effect of Ti doping of AlCrN coatings on thermal stability and oxidation resistance. *Thin Solid Films* **2013**, *534*, 394–402. [CrossRef]
48. Grigoryev, S.N.; Abadias, G.; Saladukhin, I.A.; Uglov, V.V.; Zlotski, S.V.; Eyidi, D. Thermal stability and oxidation behavior of quaternary TiZrAlN magnetron sputtered thin films: Influence of the pristine microstructure. *Surf. Coat. Technol.* **2013**, *237*, 187–195.
49. Nguyen, T.D.; Kim, Y.J.; Han, J.G.; Lee, D.B. Oxidation of TiZrAlN nanocomposite thin films in air at temperatures between 500 and 700 °C. *Thin Solid Films* **2009**, *517*, 5216–5218. [CrossRef]
50. Sergeyevnin, V.S.; Blinkov, I.V.; Volkhonskii, A.O.; Belov, D.S.; Kuznetsov, D.V.; Gorshenkov, M.V.; Skryleva, E.A. Wear-resistant adaptive nanomultilayer Ti-Al-Mo-N coatings. *Appl. Surf. Sci.* **2016**, *388*, 13–23. [CrossRef]
51. Bobzin, K.; Brogelmann, T.; Kalscheuer, K.; Stah, L.K.; Lochner, T.; Yilmaz, M. Effect of (Cr,Al)N and (Cr,Al,Mo)N coatings on friction under minimum lubrication. *Surf. Coating. Technol.* **2020**, *402*, 126154. [CrossRef]
52. Ju, H.; Yu, D.; Xu, J.; Yu, L.; Bin, Z.; Geng, Y.; Huang, T.; Shao, L.; Ren, L.; Du, C.; et al. Crystal structure and tribological properties of ZrAlMoN composite films deposited by magnetron sputtering. *Mater. Chem. Phys.* **2019**, *230*, 347–354. [CrossRef]
53. Ilo, S.; Tomala, A.; Badish, E. Oxidative wear kinetics in sliding contact of unlubricated steel. *Tribol. Int.* **2011**, *44*, 1208–1215. [CrossRef]
54. Chen, Z.; Lou, M.; Geng, D.; Xu, Y.X.; Wang, Q.; Zheng, J.; Zhu, R.; Chen, Y.; Kim, K.H. Effect of the modulation geometry on mechanical and tribological properties of TiSiN/TiAlN nano-multilayer coatings. *Surf. Coating. Technol.* **2021**, *423*, 127586. [CrossRef]
55. Povstugar, I.; Choi, P.-P.; Tytko, D.; Ahn, J.-P.; Raabe, D. Interface-directed spinodal decomposition in TiAlN/CrN multilayer hard coatings studied by atom probe tomography. *Acta Mater.* **2013**, *61*, 7534–7542. [CrossRef]
56. Han, J.; Cao, K.; Xiao, L.; Tan, X.; Li, T.; Xu, L.; Tang, Z.; Liao, G.; Shi, T. In situ measurement of cutting-edge temperature during turning using near infrared fiber optic two-color pyrometer. *Meas. J. Int. Confed. Meas-Urem.* **2020**, *156*, 107595.
57. Baltatu, M.S.; Vizureanu, P.; Sandu, A.V.; Munteanu, C.; Istrate, B. Microstructural Analysis and Tribological Behavior of Ti-Based Alloys with a Ceramic Layer Using the Thermal Spray Method. *Coatings* **2020**, *10*, 1216. [CrossRef]
58. Baltatu, M.S.; Sandu, A.V.; Nabialek, M.; Vizureanu, P.; Ciobanu, G. Biomimetic Deposition of Hydroxyapatite Layer on Titanium Alloys. *Micromachines* **2021**, *12*, 1447. [CrossRef] [PubMed]
59. Baltatu, I.; Sandu, A.V.; Vlad, M.D.; Spataru, M.C.; Vizureanu, P.; Baltatu, M.S. Mechanical Characterization and In Vitro Assay of Biocompatible Titanium Alloys. *Micromachines* **2022**, *13*, 430. [CrossRef]

Article

Improving the Efficiency of Metalworking by the Cutting Tool Rake Surface Texturing and Using the Wear Predictive Evaluation Method on the Case of Turning an Iron–Nickel Alloy

Mikhail Stebulyanin, Evgeny Ostrikov, Mars Migranov and Sergey Fedorov *

Department of High-Efficiency Machining Technologies, Moscow State University of Technology STANKIN, Vadkovskiy per. 3A, 127055 Moscow, Russia
* Correspondence: sv.fedorov@icloud.com; Tel.: +7-916-290-26-07

Abstract: The article proposes and substantiates the function of predictive evaluation using the criterion of the relative efficiency of using a cutting tool with a microtextured rake surface based on tangential force and cutting temperature. Comprehensive durability tests carried out under various processing modes with the measurement of heat power parameters made it possible to create an experimental base for mathematical modeling. An empirical model of cutting parameters based on modified multiplicative functions with non-constant indicators in the form of linear dependencies on processing factors was used based on planning an experiment for processing a heat-resistant alloy for predictive wear assessment in order to determine rational cutting modes. Predicting the rational use of a cutting tool with a microtextured work surface made it possible to obtain a 1.3-fold increase in durability.

Keywords: microtexturing; carbide tool; thermal force parameters; mathematical modeling

Citation: Stebulyanin, M.; Ostrikov, E.; Migranov, M.; Fedorov, S. Improving the Efficiency of Metalworking by the Cutting Tool Rake Surface Texturing and Using the Wear Predictive Evaluation Method on the Case of Turning an Iron–Nickel Alloy. *Coatings* **2022**, *12*, 1906. https://doi.org/10.3390/coatings12121906

Academic Editor: Jinyang Xu

Received: 9 November 2022
Accepted: 2 December 2022
Published: 6 December 2022

Publisher's Note: MDPI stays neutral with regard to jurisdictional claims in published maps and institutional affiliations.

1. Introduction

Approaches to improving the efficiency of blade processing are quite diverse. They can associate with the use of new cutting strategies, the development of new tool materials and systems, progressive lubricating and cooling technological means (LCTM), and methods of modifying the working surfaces of the tool, which are one of the foremost modern directions for improving efficiency. At the same time, approaches are used for surface alloying [1,2], application of protective wear-resistant coatings [3–7], and various kinds of energy effects designed to provide the required structural and chemical changes or to affect the macro- and micro-geometry of the instrument [8–10].

The influence of microtexturing of the tool surface on the nature of the cutting process has also been the object of research by several scientists [11]. In these works, various methods of obtaining microtextures and their influence on the cutting process during the processing of steels, aluminum, and titanium alloys with and without lubricants are studied, and practical recommendations on micro geometry of textures are proposed, taking into account the thermal and mechanical factors. All these approaches achieve the goal of improving the machinability of materials and tool durability by reducing cutting forces, residual stresses, cutting temperature, friction forces, chemical activity of the material, etc. However, all these effects, being in close interaction, have different effects on the final wear of the tool and its durability period [12].

Many authors emphasize the fact that various LCTMs are better held in the microtexture on the tool surface. It is a kind of lubricating microchannel directly in the contact zone with the processed material, which influences the microtextures on the force and coefficient of friction between the materials of the workpieces and the tool.

In [13], Lei S. with co-authors, laser processing is used to obtain micro-holes on the hard alloy surface. Deng et al. [14] used a similar approach to create lubrication micro-reserves in the cutting zone. The work of Z. Wu et al. [15] investigated a way to increase the efficiency of titanium alloy processing by applying microtextures on the rake surface to retain lubrication and remove excess heat from the tool. Additionally, in other similar works [16–20], there is a decrease in the chip contact length by up to 30% compared to the original tools, which is associated with a reduction in friction forces as the effect of surface modification. These studies showed a reduction in cutting forces and a decrease in the adhesion to the processed material. Moreover, the effect was more pronounced for high cutting speeds and perpendicular arrangement of texture lines concerning the trajectory of the chip.

In the works of Deng J et al. [21], it suggested that the positive effects caused by the presence of microtextures, in the form of a decrease in cutting temperatures, can shift the zones of rational processing modes and lead to an increase in the effects of build-up due to reduced temperatures. It found [17] that tools with a combed microtexture filled with a solid lubricant reduce the cutting temperature by 5%–15% at a high cutting speed (>120 m/min). Wu Z. et al. [15] reduced the cutting temperature by 15%–20% using a specially designed cutting tool. Fatima A. et al. [22] showed that the groove textures reduce the tool temperature by 12% at lower cutting speeds compared to non-textured tools.

However, despite a significant number of publications, the systematic justification of various physical–chemical, thermal, tribological, and other phenomena explaining the effects of microtexturing of the working surfaces of the tool from the standpoint of the theory of cutting materials have not developed sufficiently. The issues of creating microtextures with parametric relief on the surface of the tool providing a change in the conditions of contact interaction with the processed material need additional studies. In addition, microtextured tool processing processes were experimentally studied for a limited range of processed materials. For example, the methods of cutting difficult-to-process iron–nickel-based alloys with such a tool, which widely used manufacturing of critical products of aviation and other high-tech industries, remain practically unexplored. There are almost no methods of the microtextured tool area's practical application predictive assessment.

It will be necessary to obtain analytical dependences of microtexturing parameters on turning modes for a predictive assessment. It is quite difficult to say exactly which effects will be prioritized for making certain assumptions and constructing an adequate theoretical dependence. As a basis, it was decided to base on empirical models for the cutting force and temperature depending on the cutting parameters. It is required to determine the relationship of the selected parameters of the tangential cutting force (F_t) and temperature (Θ) with the wear resistance of the cutting tool and select an evaluation criterion to create any predictive assessment of the operation of a microtextured tool and its wear resistance.

In this paper, it is proposed to use a complex indicator for considering the combined influence of temperature and tangential cutting force in the form of a product of these values to study the nature of the cutting process. It is proposed to evaluate the influence of these factors precisely by their product and not by their values individually since the larger the product, the greater the destructive effect of the impact of cutting conditions on the tool.

2. Materials and Methods

2.1. Tool and Workpiece

An increase in the processing efficiency of the hard-to-process iron–nickel alloy $C_{0.3}Cr_{15}Ni_{35}Mo_7Mn_7Fe$ is investigated in this work. It has a low thermal conductivity, provoking increased cutting temperatures, high strength, and mechanical hardenability during processing, complicating the cutting process and causing increased wear—also, this material contributes to increased build-up during the cutting process.

When conducting field studies on the effect of processing modes of speed, feed, and cutting depth on the cutting force and temperature for the initial tool and tool with a microtextured front surface, as well as using wear-resistant coatings, standard experimental methods were used. The tests were carried out on a universal turning machine model CU

500 M (ZMM, Sliven, Bulgaria). The machine has a spindle speed that smoothly changes the thyristor converter and sets the required cutting speed.

A blank with a diameter of 150 mm and a length of 600 mm is used when turning. Before taking measurements, the workpiece is pressed by a rotating center and pre-ground to eliminate runouts. The TK20 Sandvik alloy was selected after preliminary wear resistance tests. Square cutting plates with a flank angle of 11° and a radius at the apex of 1.2 mm of the SPGN 120, 312 shapes suitable for processing materials from the group of heat-resistant and iron–nickel alloys selected as a tool for turning. A holder for replaceable cutting plates with a rake angle of 0° and a plan angle of 45° was selected.

2.2. Microtexture Application

In this work, a laser method for producing microtextures was tested. In all tested modes, at the accepted band boundaries, there are deposits left over from the liquid phase displacement from the treatment zone as a result of the explosive evaporation of the hard alloy. In Figure 1, the appearance of growths at the borders of the grooves is noticeable due to the displaced liquid phase during the treatment with the U-15 nanosecond laser (RMI Laser LLC, Lafayette, LA, USA). Such zones may indicate the presence of undesirable temperature effects on the material and possible changes in the structure and composition of the material in the processing zone, as well as have a retarding effect on the chip coming off, causing the appearance of secondary cutting edges in the path of its coming off.

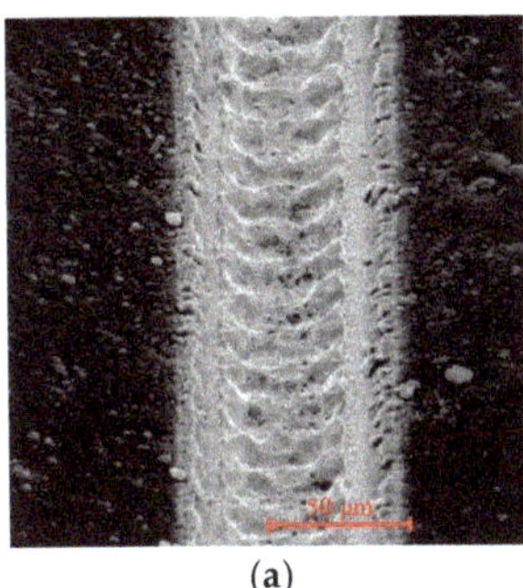

(a)

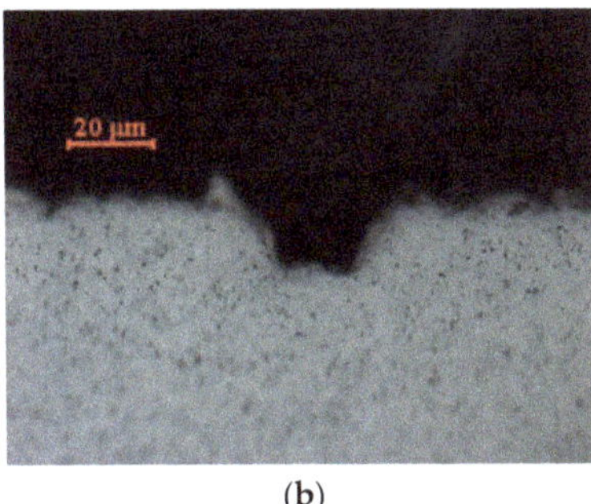

(b)

Figure 1. View of grooves on the hard alloy surface when treated with a nanosecond laser: (**a**) the top view; (**b**) the groove cross-section.

Each of the methods of applying microtexture has its drawbacks. The complexity of the indentation method lies in the fact that it is necessary to take into account factors related to the strength and flexibility of the material of the textured tool, its resistance to deformation and the effect of force on the cutting edge when moving to the actual drawing on the surface.

The shape and depth of the print will primarily depend on the indenter used. The most common and available options are Vickers and Berkovich indenters. Accordingly, in the first case, patterns of a square section are obtained, and in the second, in the form of a regular triangle. At the same time, the indentation process assumes that the adjacent prints do not intersect, and the surface is flat and perpendicular to the indenter for uniform immersion in the material and exclusion of non-axial loads on the indenter, which can lead to chipping of the pointed tip. As a result, many individual prints standing at some distance from each other can be obtained. If the depth of indentation is set constant for all prints, then the defining requirement for the texture geometry is the requirement of print structure parametric ordering, which allows for easy automation from a software (Qpix control version 2) point of view of its creation on the tool surface.

It is possible to increase or decrease the surface treatment density by changing the parameters of the fingerprint structure. A one-parameter task is preferable, in which it proposed to arrange rows of prints in two, creating pseudo-bands of dense relief areas with mutually overlapping zones on the chip escape path at the maximum possible packing density. The size of the overlap of the lines is determined by the indentation and a fixed constant step.

The Qness 10A microhardness meter (Qness Gmbh, Golling, Austria) with a Vickers indenter used (Figure 2) as equipment for applying microtextures to the surface of the cutting plates, this equipment allows performing micro indentation on a selected site in automatic mode with a load value chosen. Since the selected plates have a rear angle, and the indentation took place perpendicular to the front surface in the vicinity of the cutting edge above the overhanging part, the plates were glued to the device holder for the time of indentation.

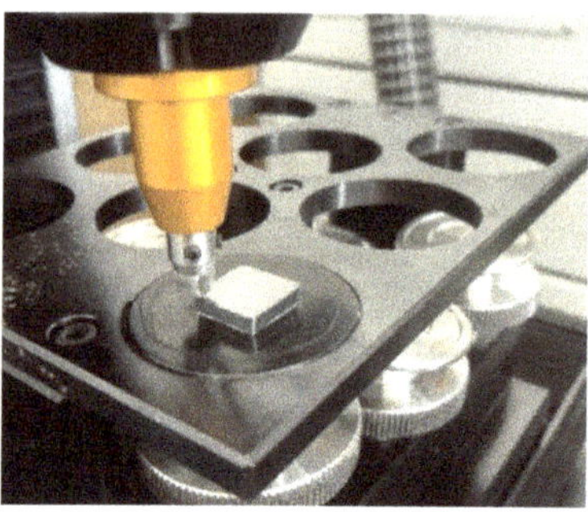

Figure 2. The Qness 10A microhardness tester used for applying the microtextures.

A force of 20 N selected. This load made it possible to obtain square-shaped prints with a width of 30 μm, a diagonal of 40 μm, and a maximum depth of 7 μm. The maximum density of tracks, which was took into account the limitations on positioning accuracy, was 50–60 μm between the centers (Figure 3).

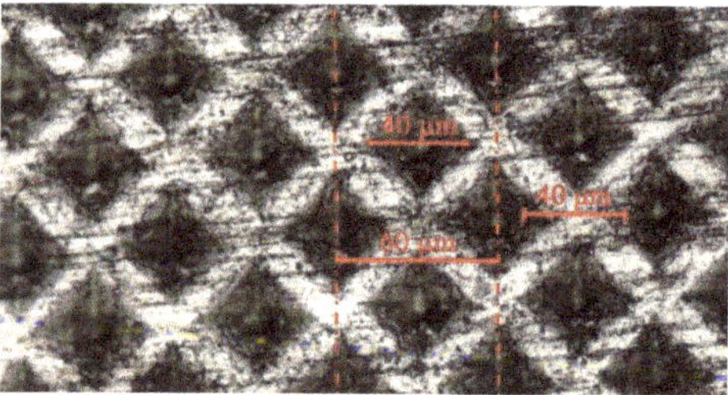

Figure 3. Type of prints on the surface of the plates after indentation.

Another limiting factor is the allowable distance to the cutting edge. It noted that the edge chipping during indentation could occur after injection at a distance of 50–70 μm. Therefore, the minimum space to the cutting edge of 100 μm is limited. The following parameter that was tried to account for during the application of microtexture is the microtexture pattern orientation concerning the cutting edge and the chip trajectory.

One of the expected effects is the function of micro-reserves for lubrication, which helps the lubricants to be held directly in the cutting area. Based on tests on various turning modes, conclusions drawn about the active contact zone of the chip and the tool with a length of 0.5–0.6 mm. It decided to take this value into account when applying microtextures to the tool surface when determining the size of the processing zone to ensure the lubricant operation over the entire contact surface in the cutting zone.

Additionally, the option of using different step densities of prints was tested when choosing the parameters of microtextures, both in terms of the step size between individual bands and when varying the step between unique prints as part of bands. During the trial tests, it was decided to focus on applying textures in the form of strips with an interval between them, approximately equal to the width of the strip, and a step between the prints of 60 μm. A lower density of microtextures ceased to have a noticeable effect on the cutting process, and a denser one sometimes enhanced the formation on the rake surface (Figure 4).

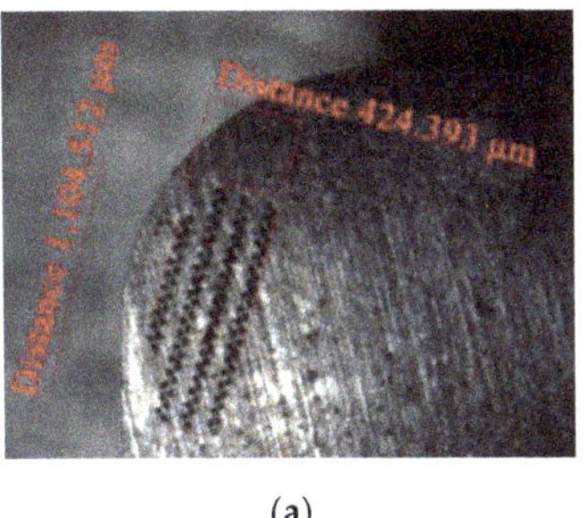

(a)

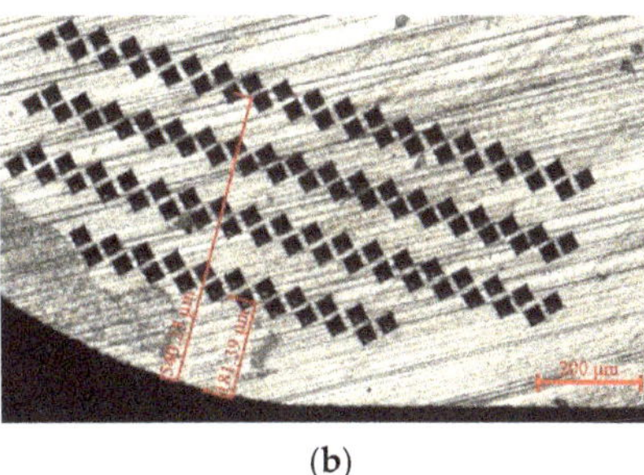

(b)

Figure 4. Final view of microtextures: (**a**,**b**) the microtexture location on the rake surface.

The possibility of combining various methods of tool modification in the form of microtexturing of the surface with a wear-resistant coating was studied. Wear-resistant coatings were applied on the Platit p311 installation (PLATIT AG, Selzach, Switzerland).

The main problem when applying prints to the surface of plates with wear-resistant coatings may be insufficient adhesion of coatings to the substrate and their cracking and peeling. The multilayer coating nATCRo³ based on (TiCrAlSi)N proved the best. The nATCRo³ coating combined (CrTi)N and (AlTi)N layer sequence and an (AlTiCr)N/SiN nanocomposite coating. A two-phase structure of the last layer is constituted by (AlTiCr)N crystals with a grain size of up to 5 nm and is distributed equally to Si_3N_4 in the amorphous phase. The phase interfaces are the area of intensive dissipation of energy, diverging the cracks in the coating from their paths, and inhibiting their propagation rate, which leads to the hardening of the material. When indenting, it turned out to choose a variant of the microtexture density that did not cause the formation of noticeable micro-cracks on the surface of the plates. Plates with this kind of coating were tested in experiments.

2.3. Selection of Modes and Measurement of Processing Parameters

The processing modes were selected from the conditions of finishing turning with small feed (0.1, 0.125, and 0.15 mm/rev) and cutting depth values (0.3, 0.4, and 0.5 mm). The cutting speed (17, and 27 m/min) was chosen with the recommended values from the manufacturer of the cutting plates and subsequently changed for turning conditions of iron–nickel alloy blanks with a durability period of the initial plates of 30 min, for which preliminary tests were carried out.

In addition, during the tests, a universal motorized stereo microscope with the possibility of telecommunications Stereo Discovery V12 with a visualization system based on the Axiocam 503 Color video camera (Carl Zeiss AG, Jena, Germany) was used to monitor and photograph the final wear of the cutting plates after the tests.

The wear resistance of the cutting tool is characterized by its durability period T

$$T = l/V \tag{1}$$

where l is the cutting path with the appropriate blunting criterion h_f^{cr} and relative linear h_{oL} and surface relative wear h_{oR}:

$$h_{oL} = \frac{h_{ff} - h_{fi}}{l_f - l_i}, \tag{2}$$

$$h_{oR} = \frac{\left(h_{ff} - h_{fi}\right) \cdot 100}{\left(l_f - l_i\right) \cdot S} \tag{3}$$

where h_{fi}, l_i, and h_{ff}, l_f are the initial and final wear values along the back surface and the cutting path, respectively.

The cutting plate failure criterion during resistance tests was $h_f^f = 0.4$ mm.

Temperature measuring during the cutting process is an ambiguous task with many features and approaches to solve it. In this regard, many methods of varying degrees of complexity for measuring this parameter. A natural thermocouple method is used based on the principle of the occurrence of a potential difference at the point of contact of different materials of the tool and the workpiece under the influence of the cutting temperature. However, this method of temperature measurement is characterized by complex calibration tests in the case of surface modification that changes their thermophysical parameters.

At the same time, to reduce measurement errors, the workpiece and the cutting tool isolated from other mechanical devices to reduce spurious signals. The measurement and fixation of the cutting temperature were carried out under a well-established cutting process at the interval of 10 s after cutting the tool into the workpiece. A current mercury collector and a digital voltmeter INSTEK GDM-8246 (Good Will Instrument Co., Ltd, Kaoshiung, China), which allows recording voltage up to 0.1 mV, are used to measure the thermo-EMF signal. The scheme for measuring the values of the average cutting temperature during turning tests shown in Figure 5.

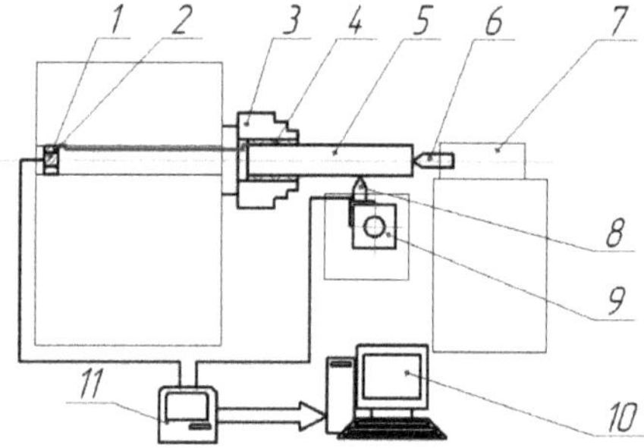

Figure 5. Scheme of temperature measurement by natural thermocouple method: 1—mercury current collector; 2, 4—insulation; 3—cartridge; 5—blank; 6—isolated rear center; 7—tailstock; 8—tool holder; 9—a cutter with insulation; 10—a computer; 11—a digital voltmeter.

The experiments were conducted to determine the components of the cutting force on a three-component dynamometer of the company Kistler model 9253B23 (Winterthur, Switzerland) to study the nature of the dependencies of cutting forces on the investigated method of modifying the surface of the cutting tool and cutting modes.

As noted above, the effect of microtextures on the cutting process is best manifested in the presence of LCTM due to better penetration and retention of these lubricants in the processing zone. Therefore, in the course of experimental studies on turning an iron–nickel alloy with carbide plates with a textured and initial surface, processing options in the absence of lubricants were considered.

As a result of these studies, it was impossible to achieve a noticeable improvement in the cutting conditions of the tool with microtexture. The considered rake surfaces of the plates showed a more significant build-up for a tool with microtextures than for the original instrument in the same processing modes. Additionally, the durability period of the modified plates under dry friction conditions turned out to be lower than for the original version of the tool.

It was decided to use dry molybdenum disulfide lubricant, which was previously applied to the rake and flank surfaces of the plates using Molykote D-321 when conducting experiments to study the properties of microtextured plates. The results of field experiments during turning have shown that such a lubricant can indeed linger in the crater of the microtexture on the cutting plate rake surface in the area of contact with the chips of the processed material. Thus, the microtextures can act as reservoirs for lubricants, feeding a dosed grease into the processing zone, and improving friction conditions (Figure 6). These modes tested for plates in the initial state, with microtexture on the rake surface, with coating, and with microtexture on the coating.

Figure 6. Grease residues inside the microtexture craters at the border of the growth zone; the build-up at V = 27 m/min, t = 0.5 mm, S = 0.15 mm/rev.

3. Results and Discussion

3.1. Determination of Cutting Process Parameters

The active formation of growths on the front surface of the tool and, as a consequence, adhesive wear is characteristic when processing heat-resistant alloys with a carbide tool. The adhesive wear prevails over other types of wear at a specific range of cutting temperatures [23,24]. The growth zone has characteristic boundaries that repeat the geometry of the rake surface microtexture, unlike a tool without it (Figure 7).

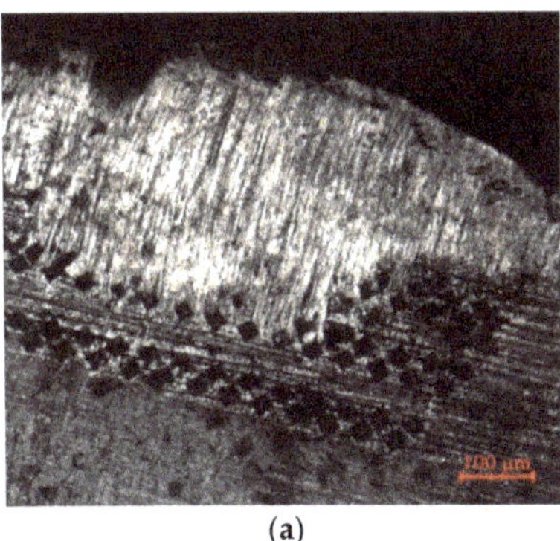
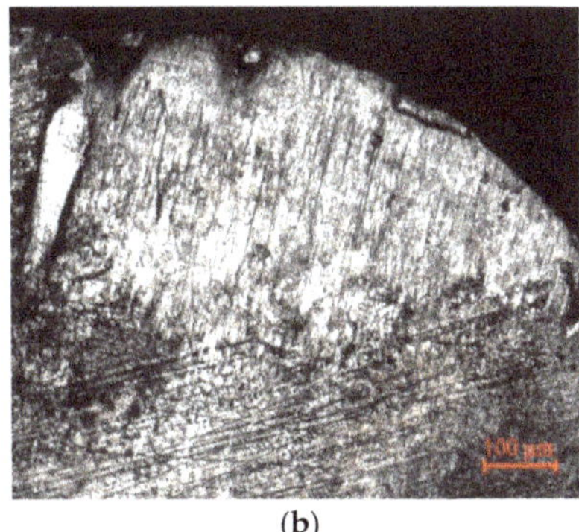

(a)   (b)

Figure 7. View of the front surface of plates with (**a**) and without (**b**) microtexture with traces of build-up after processing of heat-resistant material at V = 27 m/min, t = 0.5 mm, S = 0.1 mm/rev.

This phenomenon limits the growth zone. The cutting process becomes more controlled. Moreover, the tool surface after the growth zone in the presence of microtexture looks cleaner, without traces of adhesive setting of the material, than without microtexture on the rake surface. It is also possible to notice a closer location of scraps of solid grease residue in the presence of microtexture, which may indicate a shorter chip contact zone with the front surface of the tool.

All these effects, to one degree or another, had reflected in the same way on the cutting process parameters as cutting force and temperature. In contrast to the cutting forces, the temperature values in the cutting zone are pretty problematic to measure directly. In the applied natural thermocouple method, it is possible to talk only about the overall temperature of the cutting process. However, it is possible to compare the efficiency of the process between different tool modifications under the same processing modes even based on such an indicator.

In the results of the experiments on recording the cutting temperature presented in Figures 8 and 9, in addition to the characteristic increase in temperature, one can notice the difference between tools with different surface properties. In this case, modifications separately in the microtexture on the rake surface and coatings without microtexture show very similar results and overlap in places. The combination of these modifications gives a more pronounced drop in cutting temperature.

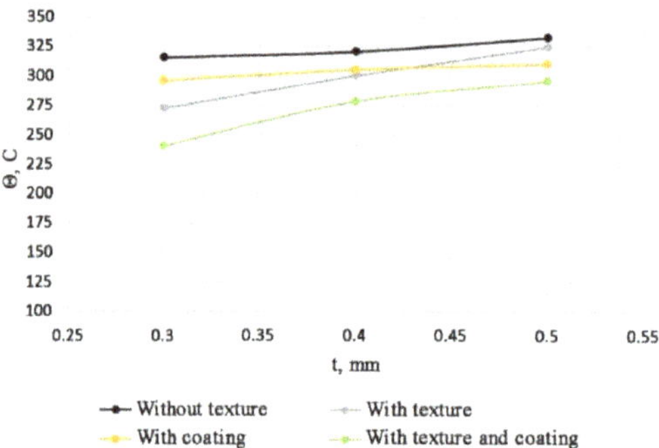

Figure 8. Dependence of the cutting temperature on the cutting depth for various tools at a cutting speed of $V = 17$ m/min and $S = 0.15$ mm/rev.

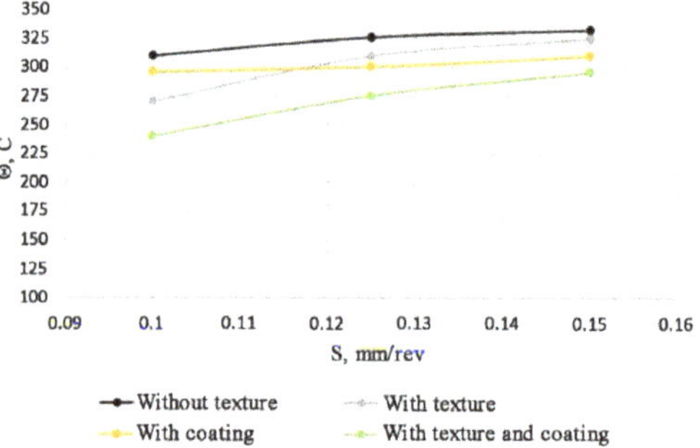

Figure 9. Dependence of the cutting temperature on the feed per revolution for various tools at a cutting speed of $V = 17$ m/min and $t = 0.5$ mm/rev.

On the one hand, the microtexture keeps the lubricant in the cutting zone and thereby affects the friction force and the interaction of the tool and workpiece materials. However, as a rule, wear-resistant coatings have the same effect on the cutting process [25,26]. Namely, they reduce the coefficient of friction and are chemically resistant. On the other hand, it can assume that the company of microtextures contributes to a reduced setting between the material of the workpiece and the tool.

It was decided to focus on studying the changes in the tangential component of the cutting force, as a force that indicates the amount of energy transferred from the machine drives to the processing zone to complete the cutting process and chip formation, and this component used to judge the nature of the cutting process when measuring the cutting forces.

In the conducted experiments, lower values of the tangential component of the cutting force are shown by a tool with a coating and the microtexture on the rake surface. Individually, these modifications showed themselves differently with a change in feed and depth of cut. With increasing cutting depth, the tool with microtexture showed values of the tangential component of the cutting force closer to the tool in the initial state, while the coated tool showed lower values (Figure 10).

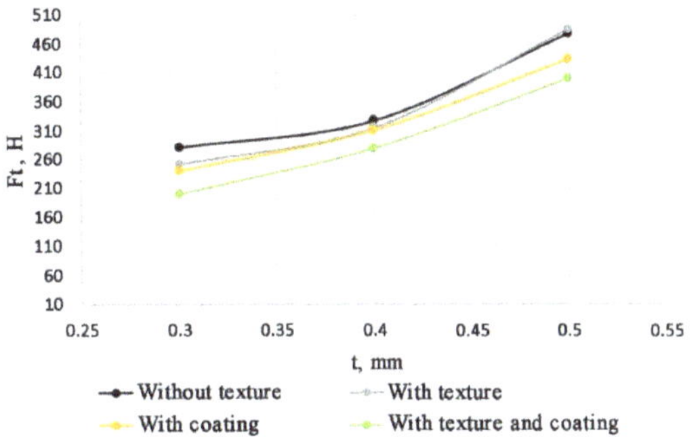

Figure 10. The tangential component of the cutting force on the cutting depth dependence for various tools at a cutting speed of $V = 17$ m/min and $S = 0.125$ mm/rev.

On the other hand, the change in feed per revolution caused similar changes in the tangential component of the cutting force between a tool with a microtexture on the rake surface and a tool with a coating (Figure 11).

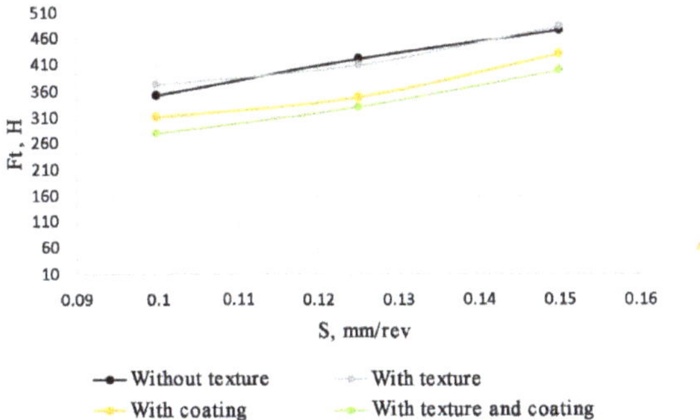

Figure 11. The tangential component of the cutting force on the feed per revolution dependence for various tools at a cutting speed of $V = 17$ m/min and $t = 0.4$ mm.

It is problematic to unambiguously judge the results of measuring the tangential component of the cutting force. The change in the cutting depth, in this case, may have a different effect on the component of the cutting force due to the influence of microtexture. However, with the same processing depth, the chip escape trajectory was constant, and when the feed changed, only the number of microtexture strips involved altered. In contrast, the effect of the microtexture effect was more stable.

There are similar patterns in a coated tool and a tool with a microtexture on the rake surface if we judge the changes in the component of the cutting force by the change in feed. It is also possible to point out the mutual addition of the effects affecting the tangential part of the cutting force when combining options for modifying tools by coating and then applying microtexture to the coated cutting surface.

An important wear parameter is worn on the flank surface due to small cross-sections of the cut when turning finishes. The results of measuring the wear of various tools are shown in Figures 12 and 13. On these dependencies, we can note the increased durability of the tool, which has both surface modifications in the form of microtexture and coating.

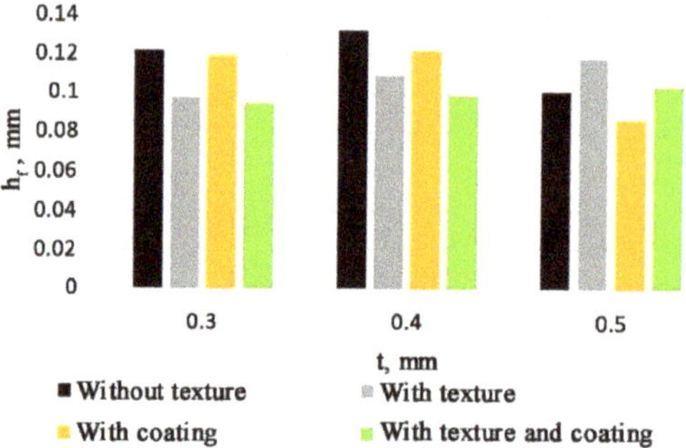

Figure 12. The wear on the flank surface on the different cutting depths dependent on various tools at a cutting speed $V = 17$ m/min and $S = 0.125$ mm/rev.

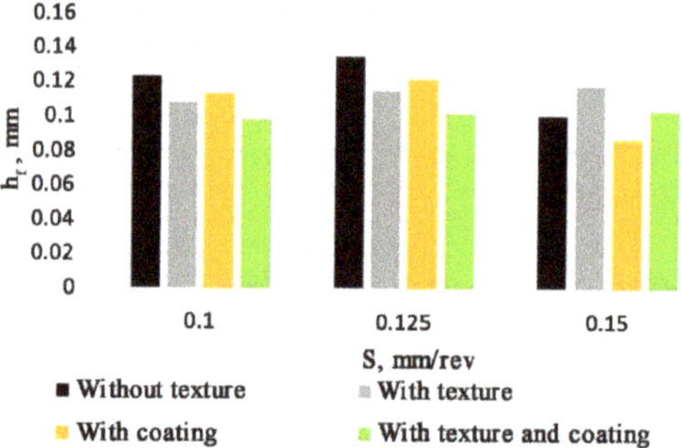

Figure 13. The wear on the flank surface on different feed per revolution values dependent on various tools at a cutting speed $V = 17$ m/min and $t = 0.4$ mm.

The study of the plate flank surface after turning shows that the wear chamfer of the original tool turned out to be broader than that of the textured one. Moreover, the slope of this chamfer also turned out to be ample for the original tool, which indicates more intensive dimensional wear of such cutting plates compared to plates having a textured surface. Thus, after the conducted research and experimental results, it concluded that the microtexture on the rake surface of the cutting plates allows influencing the friction and the adhesion zone, which has a positive effect on the temperature, and cutting force reduces tool wear.

There are summarized data obtained as a result of consistent experiments in Table 1, having considered the features and patterns of the influence of various types of modification of the tool surface on the processing process. The processing modes that formed the experimental base selected take into account the matrix of the practical plan described above. The force, cutting temperature, and current wear on the flank surface were recorded. These data from field experiments formed the basis for constructing the predictive evaluation function.

Table 1. Values of the studied processing parameters were obtained as a result of experiments for the original and textured tool.

No.	V, m/min	S, mm/rev	t mm	F_t H	Θ C	$F_{t(texture)}$ H	$\Theta_{(texture)}$ C	$h_{f(texture)} - h_f$
1	17	0.1	0.3	193.3	287	170.7	200	−0.013
2	17	0.1	0.4	234.9	295	249.3	240	−0.012
3	17	0.1	0.5	353	310	372.7	270	−0.016
4	17	0.125	0.3	226.6	305	221.4	240	−0.013
5	17	0.125	0.4	297.7	315	269.4	290	−0.014
6	17	0.125	0.5	422.1	326	409.6	310	−0.02
7	17	0.15	0.3	281.1	315	253	272	−0.024
8	17	0.15	0.4	327	320	315.5	300	−0.023
9	17	0.15	0.5	476.4	332	484.8	325	0.017
10	27	0.1	0.3	181	360	193.3	342	0.018
11	27	0.1	0.4	279.3	371	258.5	358	−0.032
12	27	0.1	0.5	450	387	356	387	−0.03
13	27	0.125	0.3	200	376	218.7	347	−0.036
14	27	0.125	0.4	335.6	385	273.8	365	−0.033
15	27	0.125	0.5	520	395	404.1	391	−0.033
16	27	0.15	0.3	223.9	391	250	368	0.014
17	27	0.15	0.4	380	395	328.4	385	−0.035
18	27	0.15	0.5	600	400	489.7	396	−0.032

3.2. Predictive Assessment of the Microtexturing Effect on Wear Resistance

The established practice of constructing empirical models for cutting parameters suggests using the multiplicative dependence of the studied parameter for modeling through processing modes of the form:

$$F_t = CV^{a_1} S^{a_2} t^{a_3} \tag{4}$$

where t—is the cutting depth, S—is the feed per revolution, V—is the cutting speed, and C is the coefficient that takes into account the features of this pair of materials [27].

A similar expression is given for the cutting temperature θ.

At the same time, it assumed that the nature of the change in cutting forces and temperatures is monotonous. These values either grow monotonically or decrease monotonically. This type of change is usually well described by parabolic or hyperbolic curves, which are conveniently approximated by power functions. The values of the exponents and express the dependence of the change in the parameter under study on the processing modes can be obtained having formed an experimental plan for this type of model with the alternating iterations of processing modes.

However, it is difficult to assess the nature of the change in cutting parameters when there is a discrete microtexture on the rake surface. Changes in feed modes and cutting depth can have an unpredictable effect on the parameters under study. Changing the width and depth of cutting entails changes in the contact point. The entry into the cutting process of new discrete microtexture elements or the exit of individual components from the contact spot is likely to complicate the dependence (4) seriously when there is a microtexture on the rake surface, which consist of individual elements characterized by directivity, the density of layout, etc.

Based on this assumption, consider a variant of the empirical form of the model, where exponents of degrees are functions of the feed and depth of cut. Then, the model for the cutting force takes the following form:

$$F_t = CV^{a_1(S,t)} S^{a_2(S,t)} t^{a_3(S,t)} \tag{5}$$

After the logarithm of this expression, a function comes to the following form:

$$\ln(F_t) = \ln(C) + a_1(S,t)\ln(V) + a_2(S,t)\ln(S) + a_3(S,t)\ln(t) \tag{6}$$

In the resulting expression, we have the dependence of the cutting force logarithm on the cutting modes and coefficients logarithms, which are a function of the reliance on the feed and cutting depth. Let us rewrite (6), bringing all dependencies to a single form of logarithmic argument functions:

$$\ln(F_t) = \ln(C) + a_1'(\ln(S), \ln(t))\ln(V) + a_2'(\ln(S), \ln(t))\ln(S) \\ + a_3'(\ln(S), \ln(t))\ln(t) \tag{7}$$

The expression (4) has the same type of parameters, reduced to the dependence of the cutting force logarithm on the processing modes logarithms:

$$\ln(F_t) = f(\ln(V), \ln(S), \ln(t)) \tag{8}$$

This function of dependence on the logarithms of the processing mode is represented as a polynomial for regression analysis, limited to the second order of nonlinearity:

$$\ln(F_t) = \ln(C) + a_1\ln(V) + a_2\ln(S) + a_3\ln(t) + a_4\ln^2(S) + a_5\ln^2(t) \\ + a_6\ln(V)\ln(S) + a_7\ln(V)\ln(S) + a_8\ln(S)\ln(t) \tag{9}$$

Thus, we can obtain an analytical dependence of the logarithm of the cutting force on the logarithms of the processing modes having further carried out the process of determining the coefficients of this expression using regression analysis, and by potentiating, we obtain a model of the cutting force of the following type:

$$F_t = CV^{a_1}S^{(a_2+a_4\ln(S)+a_6\ln(V)+a_8\ln(t))}t^{(a_3+a_5\ln(t)+a_7\ln(V))} \tag{10}$$

Conducting similar reasoning for the cutting temperature, we obtain its model in the form:

$$\theta = CV^{b_1}S^{(b_2+b_4\ln(S)+b_6\ln(V)+b_8\ln(t))}t^{(b_3+b_5\ln(t)+b_7\ln(V))} \tag{11}$$

As a result, we obtain models of thermal force factors that allow the possibility of mutual changes in processing modes and power coefficients when determining temperatures and cutting forces. When considering the effective use of a microtextured tool under specified cutting conditions, as a result of comparing it with a tool without texture, we introduce a criterion for evaluating relative efficiency based on the following functionality:

$$I_\% = \frac{(F_t\theta)_t}{F_t\theta} \tag{12}$$

where $(F_t\theta)_t$ is the product of the tangential component of the cutting force and the cutting temperature for the textured tool, and $F_t\theta$ is the product of the force tangential component and temperature for the original tool. At the same time, the criterion itself (the area of effectiveness of the microtextured tool) is determined by the inequality:

$$I_\%(S, t, V) < 1 \tag{13}$$

This hypothetical assumption will be tested during experimental studies and becomes confidential. The value of $I_\%$ is a predictive estimation function. To construct this value, we based it on the empirical dependencies of the cutting force and temperature in expressions (10) and (11). Let us consider the source of determining the numerical values of the coefficients for the models of cutting force and temperature before proceeding to determine the final type of expression (12). We will find these coefficients by conducting a factor experiment with a complete search of the feed and cutting depth, interval selection of speed, and calculation of values using the least squares method in the form:

$$A = \left(G^T G\right)^{-1} G^T Y \tag{14}$$

where A is the vector of the desired coefficients of the models of the form (a0, a1, a2 ...), G is the formed matrix of the experiment plan with encoded values of variables and their mutual enumeration, Y is the vector of the results of the investigated quantity importance (cutting force or temperature) for each combination of processing modes under the experiment plan.

First, let us consider the procedure for calculating the coefficients for the cutting force. We reduce the initial form of the chosen multiplicative model Equation (10) to a polynomial form using a logarithm. As a result, we obtain an expression of the following form:

$$\ln(F_t) = \ln(C) + a_1 \ln(V) + a_2 \ln(S) + a_3 \ln(t) + a_4 \ln^2(S) + a_5 \ln^2(t) \\ + a_6 \ln(V) \ln(S) + a_7 \ln(V) \ln(S) + a_8 \ln(S) \ln(t) \tag{15}$$

As a result, we obtain a polynomial whose coefficients can found using regression analysis. At the same time, the logarithm of the correction factor $\ln(C)$ will be the zero coefficient of the regression model. Let us rewrite the logarithm expression in the following form:

$$\ln(F_t) = a_0 + a_1 \ln(X_1) + a_2 \ln(X_2) + a_3 \ln(X_3) + a_4 \ln^2(X_2) + a_5 \ln^2(X_3) \\ + a_6 \ln(X_1) \ln(X_2) + a_7 \ln(X_1) \ln(X_3) + a_8 \ln(X_2) \ln(X_3) \tag{16}$$

where $\ln(C)$ is replaced by the coefficient a_0, and the original designations of the processing modes V, S, and t by X_1, X_2, and X_3, respectively.

For the next stage of finding coefficients by the least square's method, it will be necessary to encode variables, reducing the ranges of values of each variable to the interval $(-1, 1)$. It is possible to perform this operation using the following expression:

$$\ln(x_i) = \frac{\ln(X_i) - \frac{\ln(X_{i\,min}) + \ln(X_{i\,max})}{2}}{\frac{\ln(X_{i\,max}) - \ln(X_{i\,min})}{2}}, \quad i = 1,\, 2,\, 3 \tag{17}$$

where X_i is the initial value of the variable models (speed, feed, and cutting depth) used in the experiment, and $\ln(X_i)$ is the result of coding from -1 to 1.

Next, it is necessary to make a matrix of the experiment plan G with encoded variables and their mutual enumeration of values using the values of the minimum value $(\ln(X_i) = -1, X_i - min)$, the maximum value $(\ln(X_i) = 1, X_i - max)$ and the average value of each variable in various combinations. In this experimental plan, a complete search was performed for the feed parameters X_2 and the cutting depth X_3, and the interval selection of the cutting speed X_1.

During the experiment, it is necessary to follow the obtained matrix, setting the appropriate modes from each set of variables and measuring the resulting value of the cutting force tangential component. As a result, a vector of the resulting Y values is obtained, and it is possible to apply expression (14) to calculate the desired coefficients of the model for the cutting force.

The type of model for the cutting temperature is similar to the kind of model for the cutting force. After performing a logarithm to bring it to a polynomial form and rewriting the designations of the model parameters, we have the following type of model:

$$\ln(\Theta) = b_0 + b_1 \ln(X_1) + b_2 \ln(X_2) + b_3 \ln(X_3) + b_4 \ln^2(X_2) + b_5 \ln^2(X_3) \\ + b_6 \ln(X_1) \ln(X_2) + b_7 \ln(X_1) \ln(X_3) + b_8 \ln(X_2) \ln(X_3), \tag{18}$$

Accordingly, the coding of variables and the experimental plan be similar for the cutting temperature, which can measure simultaneously with the cutting force in the same matrix of the practical plan G. As a result, using the logarithms of the values of the observed temperature data, we will be able to find the importance of the coefficients of the model using the least squares method. As a result, we obtain an expression for finding the vectors of the model's coefficients of cutting force **A** [9 × 1] and cutting temperature **B** [9 × 1] in the following form:

$$[A, B] = \left(G^T G\right)^{-1} G^T [Y_P, Y_\theta] \tag{19}$$

where Y_P [18 × 1], Y_θ [18 × 1] are the vectors of the results of the cutting force values and temperature, respectively, obtained in the experiment, the matrix of the experiment plan G, therefore, has a dimension of 18 × 9.

Let us consider a way to calculate the relative efficiency criterion using calculated cutting forces and temperature models. First, write down once again the expressions for the logarithms of the cutting force and temperature in the polynomial form:

$$\ln(F_t) = a_0 + a_1 \ln(X_1) + a_2 \ln(X_2) + a_3 \ln(X_3) + a_4 \ln^2(X_2) + a_5 \ln^2(X_3) \\ + a_6 \ln(X_1)\ln(X_2) + a_7 \ln(X_1)\ln(X_3) + a_8 \ln(X_2)\ln(X_3) \tag{20}$$

$$\ln(\Theta) = b_0 + b_1 \ln(X_1) + b_2 \ln(X_2) + b_3 \ln(X_3) + b_4 \ln^2(X_2) + b_5 \ln^2(X_3) \\ + b_6 \ln(X_1)\ln(X_2) + b_7 \ln(X_1)\ln(X_3) + b_8 \ln(X_2)\ln(X_3), \tag{21}$$

Having made the potentiation of these expressions and given transformations by grouping such multipliers, we obtain terms that will take the following form:

$$F_t = e^{a_0} X_1^{a_1} X_2^{(a_2 + a_4 \ln(X_2) + a_6 \ln(X_1) + a_8 \ln(X_3))} X_3^{(a_3 + a_5 \ln(X_3) + a_7 \ln(X_1))} \tag{22}$$

$$\theta = e^{b_0} X_1^{b_1} X_2^{(b_2 + b_4 \ln(X_2) + b_6 \ln(X_1) + b_8 \ln(x_3))} X_3^{(b_3 + b_5 \ln(x_3) + b_7 \ln(x_1))} \tag{23}$$

Here, the selected model variant for the studied parameters is quite convenient for calculating the value of the relative efficiency criterion. We obtain the same multiplicative form of expression in which it is simply necessary to add up the corresponding coefficients found for each of the models when the force model is multiplied by the temperature model:

$$F_t\theta = e^{a_0 + b_0} X_1^{a_1 + b_1} X_2^{(a_2 + b_2 + (a_4 + b_4)\ln(X_2) + (a_6 + b_6)\ln(X_1) + (a_8 + b_8)\ln(X_3))} \\ \times X_3^{(a_3 + b_3 + (a_5 + b_5)\ln(X_3) + (a_7 + b_7)\ln(X_1))} \tag{24}$$

In dividing one product of force by temperature for a textured tool by the product for the original instrument, it is sufficient to subtract the corresponding values of the divisor from the exponents of the multipliers of the divisible. Based on this, the final expression of the predictive evaluation functional for the criterion of relative efficiency will take the form:

$$\frac{(F_t\theta)_t}{F_t\theta} = e^{c_0} X_1^{c_1} X_2^{(c_2 + c_4 \ln(X_2) + c_6 \ln(X_1) + c_8 \ln(X_3))} \times X_3^{(c_3 + c_5 \ln(X_3) + c_7 \ln(X_1))} \\ c_i = (a_i + b_i)_t - (a_i + b_i), \quad i = 0, 1, \ldots, 8 \tag{25}$$

Thus, we immediately calculate the function of the predictive evaluation of the criterion of relative efficiency, which eventually has the same form as the original models of cutting forces and temperatures. It turns out that we can compare the textured and the original tool by determining the coefficients.

3.3. Mathematical Modeling of Thermal Force Parameters When Comparing Conventional and Textured Tools

It is possible to perform mathematical modeling to create a predictive assessment of the success of the modified cutting tool after making an experimental base during practical tests as a result of turning an iron–nickel alloy with carbide plates with the presence of microtextures and in the initial state under various modes and recording the results of the force measuring, cutting temperature and wear on the flank surface.

First, we analyze the comparison of the classical type with constant coefficients of degree indicators empirical models and the extended model proposed version, when the degree indicators are a function of the feed modes and cutting depth themselves. Let us make this comparison using the example of the cutting force model calculation. To begin with, we will encode the values of the cutting modes used in the experiment using the expression (17). We obtain the following encoded values of variables for building models (Tables 2–4):

Table 2. Encoded values of cutting speed V.

$X_1 = V$, m/min	$\ln(X_1)$	$\ln(x_1)$
27	3.296	1
17	2.833	−1

Table 3. Encoded values of feed S.

$X_2 = S$, mm/rev	$\ln(X_2)$	$\ln(x_2)$
0, 15	−1.897	1
0, 125	−2.079	0.1
0, 1	−2.303	−1

Table 4. Encoded values of cutting depth t.

$X_3 = t$, mm	$\ln(X_3)$	$\ln(x_3)$
0, 5	−0.693	1
0, 4	−0.916	0.126
0, 3	−1.204	−1

The view of the cutting force tangential component model in coded variables has the following form:

$$\ln(F_t) = a_0 + a_1 \ln(x_1) + a_2 \ln(x_2) + a_3 \ln(x_3) + a_4 \ln^2(x_2) + a_5 \ln^2(x_3) \\ + a_6 \ln(x_1) \ln(x_2) + a_7 \ln(x_1) \ln(x_3) + a_8 \ln(x_2) \ln(x_3) \tag{26}$$

In this case, the matrix of the experiment plan (G), taking into account the coding of the modes used in the experiments, have the following form Table 5:

Table 5. Experiment plan (matrix G).

No.	K	$\ln(x_1)$	$\ln(x_2)$	$\ln(x_3)$	$\ln^2(x_2)$	$\ln^2(x_3)$	$\ln(x_1)\ln(x_2)$	$\ln(x_1)\ln(x_3)$	$\ln(x_2)\ln(x_3)$
1	1	−1	−1	−1	1	1	1	1	1
2	1	−1	−1	0.126	1	0.016	1	−0.126	−0.126
3	1	−1	−1	1	1	1	1	−1	−1
4	1	−1	0.1	−1	0.01	1	−0.1	1	−0.1
5	1	−1	0.1	0.126	0.01	0.016	−0.1	−0.126	0.013
6	1	−1	0.1	1	0.01	1	−0.1	−1	0.1
7	1	−1	1	−1	1	1	−1	1	−1
8	1	−1	1	0.126	1	0.016	−1	−0.126	0.126
9	1	−1	1	1	1	1	−1	−1	1
10	1	1	−1	−1	1	1	−1	−1	1
11	1	1	−1	0.126	1	0.016	−1	0.126	−0.126
12	1	1	−1	1	1	1	−1	1	−1
13	1	1	0.1	−1	0.01	1	0.1	−1	−0.1
14	1	1	0.1	0.126	0.01	0.016	0.1	0.126	0.013
15	1	1	0.1	1	0.01	1	0.1	1	0.1
16	1	1	1	−1	1	1	1	−1	−1
17	1	1	1	0.126	1	0.016	1	0.126	0.126
18	1	1	1	1	1	1	1	1	1

We can calculate the coefficient vector A_G for an extended model using the following formula

$$A_{\mathbf{G}} = \left(G^T G\right)^{-1} G^T Y, \text{ where } A_{\mathbf{G}} = (a_0, a_1, a_2, a_3, \dots, a_8)^T \tag{27}$$

The vector of coefficients A_G calculated for the extended matrix G have the form (Table 6):

Table 6. The vector A_G coefficients for an extended model.

Coefficient	A_G
a_0	5.6679
a_1	0.0359
a_2	0.151
a_3	0.3836
a_4	−0.00079
a_5	0.0835
a_6	−0.0166
a_7	0.0933
a_8	0.0012

We can write a variant of the extended model with encoded variables in the following form by these coefficients:

$$\begin{aligned} \ln(F_t) = {}& 5.6679 + 0.0359 \ln(x_1) + 0.151 \ln(x_2) + 0.3836 \ln(x_3) - 0.00079 \ln^2(x_2) \\ & + 0.0835 \ln^2(x_3) - 0.0166 \ln(x_1)\ln(x_2) + 0.0933 \ln(x_1)\ln(x_3) + 0.0012 \ln(x_2)\ln(x_3) \end{aligned} \tag{28}$$

Next, we perform similar calculations to calculate the model of the cutting temperature dependence on the cutting speed, feed, and depth. View of the extended model in coded variables:

$$\begin{aligned} \ln(\theta) = {}& b_0 + b_1 \ln(x_1) + b_2 \ln(x_2) + b_3 \ln(x_3) + b_4 \ln^2(x_2) + b_5 \ln^2(x_3) + b_6 \ln(x_1)\ln(x_2) \\ & + b_7 \ln(x_1)\ln(x_3) + b_8 \ln(x_2)\ln(x_3) \end{aligned} \tag{29}$$

The matrices of experimental data are similar to those of cutting forces calculation. The vector of the resulting values of the investigated quantity is $Y_\Theta = \ln(\theta)$, where θ are the values recorded in the experiments. Taking into account these values, we can calculate the coefficient vector B_G for an extended model using the following formula:

$$B_{\mathbf{G}} = \left(G^T G\right)^{-1} G^T Y_\theta \tag{30}$$

We have for the components of the vector B_G (Table 7):

Table 7. The components of the vector B_G.

Coefficient	B_G
b_0	5.8429
b_1	0.1055
b_2	0.0354
b_3	0.0286
b_4	−0.0077
b_5	0.0088
b_6	−0.00561
b_7	−0.0042
b_8	−0.00909

We will calculate the coefficients of the models for the case with a microtextured surface according to a similar scenario considered when calculating models of cutting force and temperature. The only difference here is in the vectors of the resulting values obtained in the experiment. The values of the coefficients for the extended cutting force and temperature models are equal (Table 8):

Table 8. Models coefficients for the case with a microtextured surface.

Coefficient	$A_{G(texture)}$	Coefficient	$B_{G(texture)}$
a_0	5.5678	b_0	5.7586
a_1	0.0086	b_1	0.1628
a_2	0.1427	b_2	0.0746
a_3	0.3289	b_3	0.0882
a_4	0.0336	b_4	−0.0112
a_5	0.1078	b_5	−0.0032
a_6	−0.0068	b_6	−0.0465
a_7	−0.0119	b_7	−0.0354
a_8	−0.0111	b_8	−0.0210

Thus, the models of cutting forces and temperatures are calculated based on the accumulated experimental data during the turning of an iron–nickel alloy, which can be used to calculate the predictive function.

3.4. Using the Predictive Evaluation Functionality

Let us determine the resulting values of the relative efficiency criterion and consider its relationship with the difference in wear on the flank surface (h_f *(texture)* − h_f) in each case. Let us multiply the corresponding importance of force and temperature to determine the values of the criterion, find their ratio, and compare this ratio with the difference in the wear for the two considered options. The studied processing modes and the corresponding values of the cutting force and temperature, and the results of the calculations are presented in Table 9.

Table 9. Results of calculating the values of the relative efficiency criterion.

No.	V m/min	S mm/rev	t mm	$F_t \Theta$	$(F_t \Theta)_{(texture)}$	$(F_t \Theta)_{(texture)}/F_t \Theta$	$h_{f(texture)} - h_f$
1	17	0.1	0.3	55.477	34.140	0.615389	−0.013
2	17	0.1	0.4	69.295	59.832	0.863433	−0.012
3	17	0.1	0.5	109.430	100.629	0.919574	−0.016
4	17	0.125	0.3	69.113	53.136	0.768828	−0.013
5	17	0.125	0.4	93.775	78.126	0.833117	−0.014
6	17	0.125	0.5	137.604	126.976	0.92276	−0.02
7	17	0.15	0.3	88.546	68.816	0.777174	−0.024
8	17	0.15	0.4	104.640	94.650	0.90453	−0.023
9	17	0.15	0.5	158.164	157.560	0.996176	0.017
10	27	0.1	0.3	65.160	66.108	1.014558	0.018
11	27	0.1	0.4	103.620	92.543	0.893097	−0.032
12	27	0.1	0.5	174.150	137.772	0.791111	−0.03
13	27	0.125	0.3	75.200	75.888	1.009161	−0.036
14	27	0.125	0.4	129.206	99.937	0.77347	−0.033
15	27	0.125	0.5	205.400	158.003	0.769246	−0.033
16	27	0.15	0.3	87.544	92.000	1.050889	0.014
17	27	0.15	0.4	150.100	126.434	0.842332	−0.035
18	27	0.15	0.5	240.000	193.921	0.808005	−0.032

According to the assumption, the value of the ratio greater than one indicates the worst efficiency of the cutting process and the expectation of more intensive tool wear. In the considered cases, the value of the proposed criterion exceeds the value of one in three instances: under the numbers 10, 13, and 16. For these cutting modes, it is expected that the difference in the importance of the wear value on the back surface between the tool with microtexture and the tool without microtexture would be positive. The significance of this difference is equal, respectively: 0.017, −0.036, and 0.014. In this case, in two out of three points, a higher wear value was obtained for a tool with a microtexture. Additionally, among other cases, when the ratio value is less than one and a lower wear value is expected for a tool with a microtexture, a positive difference was obtained in one set of modes. The expectations about the final amount of wear on the rear surface based on the proposed criterion coincide in 16 out of 18 cases based on the 18 cases considered.

Let us now calculate the model for the available forecast according to the proposed algorithm above using the coefficient values of previously computed models. The expression of the predictive evaluation available for the criterion of relative efficiency has the form:

$$\frac{(F_t\theta)_t}{F_t\theta} = e^{c_0} X_1^{c_1} X_2^{(c_2+c_4\ln(X_2)+c_6\ln(X_1)+c_8\ln(X_3))} \times X_3^{(c_3+c_5\ln(X_3)+c_7\ln(X_1))} \quad c_i = (a_i+b_i)_t - (a_i+b_i), \ i = 0, 1, \ldots, 8 \quad (31)$$

Thus, it is necessary to sum and subtract the corresponding coefficients of the degrees of the models for the textured and original tools. The vector of the functional magnitude coefficients calculated by the formula $(A_G + B_G)_{texture} - (A_G + B_G)$ is presented below (Table 10):

Table 10. The vector of the functional magnitude coefficients.

Coefficient	$(A_G + B_G)_{texture}$	$A_G + B_G$	Criterion Coefficient
c_0	11.32653	11.51088	−0.18435
c_1	0.171558	0.141421	0.030137
c_2	0.217384	0.186554	0.030831
c_3	0.417228	0.412301	0.004927
c_4	0.022411	−0.00852	0.030931
c_5	0.104585	0.092425	0.01216
c_6	−0.05334	−0.02223	−0.03112
c_7	−0.04738	0.089169	−0.13655
c_8	−0.03215	−0.0079	−0.02424

Thus, the calculated form of the predictive evaluation functional model for the relative efficiency criterion will be as follows:

$$\frac{(F_t\theta)_t}{F_t\theta} = e^{-0.184} X_1^{0.0301} X_2^{(0.0308+0.0309\ln(X_2)-0.0311\ln(X_1)-0.0242\ln(X_3))} \times X_3^{(0.0049+0.0121\ln(X_3)-0.1365\ln(X_1))} \quad (32)$$

Consider the results of the values calculated using this model (Table 11):

Table 11. The predictive evaluation functional model coefficients.

Experiment Number	Coefficient Value
1	0.670949
2	0.782311
3	0.934612
4	0.709069
5	0.806954
6	0.940961
7	0.797175
8	0.885492
9	1.007808
10	0.996542
11	0.884264
12	0.803953
13	0.98962
14	0.857088
15	0.760579
16	1.04546
17	0.883761
18	0.765463

The results of these calculations show values above 1 in two cases. There are experiments number 9 when a combination of minimum speed and maximum feed and cutting depth was used, and number 16 when a combination of top speed and feed and minimum cutting depth was used. These cases should correspond to the inefficiency of using microtexture on the tool surface according to the accepted assumption. Let us consider in more detail the model results on the plane of values.

For the convenience of further analysis, we reduce the model to a polynomial form by performing a logarithm of the expression:

$$\ln\left(\frac{(F_t\theta)_{texture}}{F_t\theta}\right) = -0.184 + 0.0301\ln(x_1) + 0.0308\ln(x_2) + 0.0049\ln(x_3) + 0.0309\ln^2(x_2) \\ +0.012\ln^2(x_3) - 0.031\ln(x_1)\ln(x_2) - 0.136\ln(x_1)\ln(x_3) - 0.02424\ln(x_2)\ln(x_3) \tag{33}$$

In this form, the values above 0 indicate areas of inefficient use of microtexture on the tool surface, and values below 0 indicate areas of effective use of microtexture.

Let us construct the planes of the values of the criterion logarithm depending on the processing parameters in encoded values (Figures 14 and 15). It noted that the feed value does not significantly affect the results of the values. In turn, changes in the importance of the speed and depth of cutting cause transitions from the area of effective use of microtexture to the area of its inefficient use.

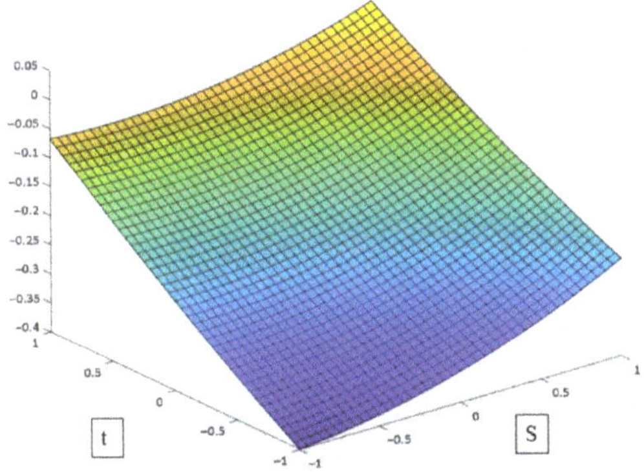

Figure 14. The surface of the change in criterion logarithm value from the cutting depth and feed encoded values at a speed of $V = 17$ m/min.

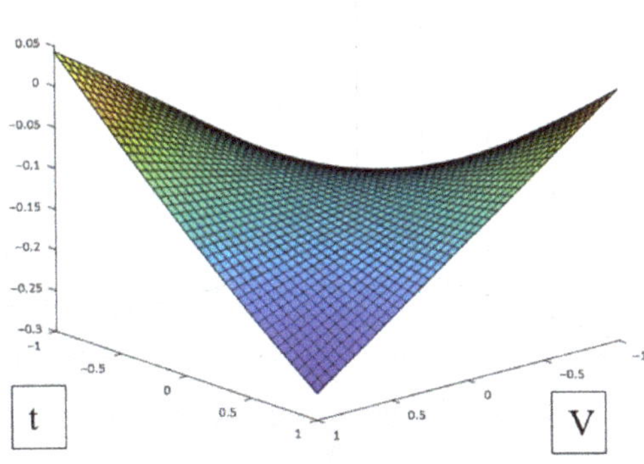

Figure 15. The surface of the change in the criterion logarithm value from the cutting depth and cutting speed encoded values at $S = 0.15$ mm/rev.

Let us consider the case of the criterion values transition from the size of practical application of microtexture to the size of ineffective values. Let us take the chance of top feed and processing depth and consider the range of practical values depending on the speed.

We set the efficiency condition for the expression of the criterion in the polynomial form:

$$-0.184 + 0.0301 \ln(x_1) + 0.0308 \ln(x_2) + 0.0049 \ln(x_3) + 0.0309 \ln^2(x_2) + 0.012 \ln^2(x_3)$$
$$-0.031 \ln(x_1) \ln(x_2) - 0.136 \ln(x_1) \ln(x_3) - 0.02424 \ln(x_2) \ln(x_3) < 0 \tag{34}$$

Let us substitute into this expression the maximum values of feed and processing depth in coded values that correspond to one:

$$-0.184 + 0.0301 \ln(x_1) + 0.0308 + 0.0049 + 0.0309 + 0.012 - 0.031 \ln(x_1) - 0.136 \ln(x_1) - 0.02424 < 0 \tag{35}$$

We give similar terms and express the condition for the speed values in coded variables when the use of microtexture is effective:

$$\ln(x_1) > 0.95 \tag{36}$$

As a result, we obtain the condition of the encoded speed value, under which the processing conditions with a microtextured tool will be more effective. We reverse the conversion of the encoded velocity value to the values used in the experiment:

$$V > 17.2 \text{ m/min} \tag{37}$$

We obtain the condition that with the top feed and cutting depth used in the experiments, which are, respectively, $S = 0.15$ mm/rev and $t = 0.5$ mm, for the effective use of microtextures, the speed must be higher than 17.2 m/min. We examine this condition in more detail in the next section.

3.5. Verification of the Theoretical Forecast of the Effectiveness of the Tool with Microtexturing of the Working Surface

The criterion of relative efficiency establishes the area of effective use of a cutting tool with a microtexture. In particular, a condition for the efficiency of using a microtextured tool obtained when the speed was higher than 17.2 m/min. This value is close to the boundary of the minimum velocity value used in the experiment at 17 m/min. The result of processing at the minimum speed showed that the efficiency criterion takes a value higher than one, and the relative wear of the tool with microtexture at the same time turns out higher than the wear of the original instrument.

Although based on the modeling requirements, we can judge the results of the model's operation only under the conditions of the selected range of processing modes. Let us consider a further decrease in the cutting speed at the same maximum feed and cutting depth. For the experiment, we choose a cutting speed of 12.5 m/min, a feed of 0.15 mm/rev, and a cutting depth of 0.5 mm. The results of the experiment are shown in the Table 12 below:

Table 12. Results of additional experiment processing parameters.

Tool	V m/min	S mm/rev	t mm	F_t H	Θ C	h_f mm
Textured	12.5	0.15	0.5	497.4	302	0.199
Untextured	12.5	0.15	0.5	488.4	296	0.189

The value of the relative efficiency criterion for these processing modes is 1.039, which indicates a lower efficiency of using a tool with a microtexture and assumes a large amount of wear. Under these processing conditions, the wear difference on the back surface between the microtextured tool and the original one is equal to $h_{r(textures)} - h_r = 0.01$. Based on these values, it assumed that at a speed of $V = 17.2$ m/min, there is a transition point from the effective use of microstructure on the surface of the tool to the size of its inefficient use.

The method of predictive evaluation used in work suggests evaluating the durability properties of the tool without conducting labor-intensive durability tests. However, it has a hypothetical character, since the wear is a random process, and the nature of the wear curves can be quite variable. Therefore, to confirm the method, resistance tests were carried out in the zones of processing modes, where the criterion of relative efficiency shows values below one.

In these tests, the maximum values of the cutting speed and depth and the minimum feed were used ($V = 25$ m/min, $S = 0.1$ mm/rev, $t = 0.5$ mm). The tests carried out until the wear on the flank surface was equal to 0.4 mm. The cutting time until the critical wear criterion for initial plate was reached was approximately 30 min. The results of the resistance tests are shown in Figure 16.

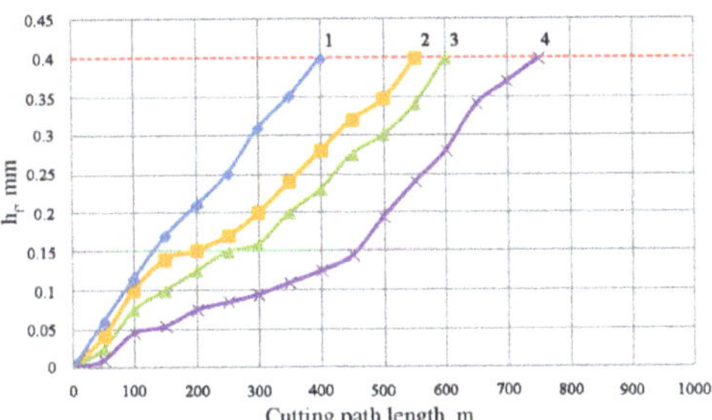

Figure 16. Comparison of wear on the flank surface for textured and original tools at $V = 25$ m/min, $S = 0.1$ mm/rpm, $t = 0.5$ mm: 1—initial, 2—initial textured, 3—with coating, 4—coated and textured.

In the course of resistance tests, a tool with a microtexture on the rake surface showed increased resistance values under operation conditions with lubrication in the zone of the practical importance of processing modes according to the criterion of relative efficiency. At the same time, the durability of the modified tool turned out to be about 30% higher compared to the initial instrument, and about 25% compared to the coated tool.

4. Conclusions

1. The article presents new technological techniques aimed at solving the problem of increasing the efficiency of processing iron–nickel alloys with a carbide tool.
2. A method is proposed and implemented for the one-parameter setting of the structure of the print when microtexturing the rake surface of a carbide plate in the form of strips, focused on their automated application. Tests of a tool with a microtexture revealed a significant effect of a microtexture filled with a MoS2-based lubricant on the processed material adhesion to the tool. The build-up zone has characteristic boundaries that repeat the geometry of the microtexture.
3. The comprehensive tests of the cutting tool for durability under various processing modes with the thermal power parameters measurement during the turning of an iron–nickel alloy made it possible to create an experimental basis for further mathematical modeling
4. Mathematical models of thermal power parameters of the cutting process based on multiplicative power functions with indicators in the form of feed and cutting depth linear functions have been developed. The function of predictive evaluation is proposed and justified using the criterion of relative efficiency. An analytical expression of this function is obtained and makes it possible to predict the rational use of a cutting tool with a microtextured working surface based on numerical calculations.
5. Predictive estimates of the turning efficiency have been experimentally tested on the example of alloy C0.3Cr15Ni35Mo7Mn7Fe. It is established that the values calculated on the basis of the developed forecasting method have a high convergence with the results of experimental tests. It was possible to obtain an increase in the tool durability by 1.3–1.5 times due to the microtexturing operation.

Author Contributions: Conceptualization, M.S.; Methodology, M.M.; Software, E.O.; Formal analysis, M.S.; Data curation, E.O. and M.M.; Writing—original draft, S.F.; Project administration, S.F. All authors have read and agreed to the published version of the manuscript.

Funding: The research was carried out at the expense of the grant of the Russian Science Foundation No. 22-19-00694. The study was carried out on the equipment of the Center of collective use of MSUT "STANKIN" supported by the Ministry of Higher Education of the Russian Federation (project No. 075-15-2021-695 from 26 July 2021, unique identifier RF 2296.61321 $\times$ 0013).

Institutional Review Board Statement: Not applicable.

Informed Consent Statement: Not applicable.

Data Availability Statement: Not applicable.

Conflicts of Interest: The authors declare no conflict of interest.

References

1. Fedorov, S.V.; Swe, M.H.; Kapitanov, A.V.; Egorov, S.B. Wear of carbide inserts with complex surface treatment when milling nickel alloy. *Mech. Ind.* **2017**, *18*, 710. [CrossRef]
2. Fedorov, S.V.; Swe, M.H. Refractory phases synthesis at the surface microalloying using a wide aperture electron beam. *IOP Conf. Ser. J. Phys. Conf. Ser.* **2017**, *830*, 012076. [CrossRef]
3. Sousa, V.F.C.; Silva, F.J.G. Recent advances on coated milling tool technology—A Comprehensive review. *Coatings* **2020**, *10*, 235. [CrossRef]
4. Grigoriev, S.N.; Vereschaka, A.A.; Fyodorov, S.V.; Sitnikov, N.N.; Batako, A.D. Comparative analysis of cutting properties and nature of wear of carbide cutting tools with multi-layered nano-structured and gradient coatings produced by using of various deposition methods. *Int. J. Adv. Manuf. Technol.* **2017**, *90*, 3421–3435. [CrossRef]
5. Vereschaka, A.S.; Grigoriev, S.N.; Sotova, E.S.; Vereschaka, A.A. Improving the efficiency of the cutting tools made of mixed ceramics by applying modifying nano-scale multilayered coatings. *Adv. Mater. Res.* **2013**, *712–715*, 391–394. [CrossRef]
6. Siwawut, S.; Saikaew, C.; Wisitsoraat, A.; Surinphong, S. Cutting performances and wear characteristics of WC inserts coated with TiAlSiN and CrTiAlSiN by filtered cathodic arc in dry face milling of cast iron. *Int. J. Adv. Manuf. Technol.* **2018**, *97*, 3883–3892. [CrossRef]
7. Grigoriev, S.N.; Volosova, M.A.; Vereschaka, A.A.; Fyodorov, S.V.; Seleznev, A.E. Properties of (Cr,Al,Si)N-(DLC-Si) composite coatings deposited on a cutting ceramic substrate. *Ceram. Int.* **2020**, *46*, 18241–18255. [CrossRef]
8. Fedorov, S.V.; Ostrikov, E.A.; Mustafaev, E.S.; Hamdy, K. The formation of the cutting tool microgeometry by pulsed laser ablation. *Mech. Ind.* **2018**, *19*, 703. [CrossRef]
9. Grigoriev, S.; Melnik, Y.; Metel, A. Broad fast neutral molecule beam sources for industrial scale beam-assisted deposition. *Surf. Coat. Technol.* **2002**, *156*, 44–49. [CrossRef]
10. Metel, A.; Bolbukov, V.; Volosova, M.; Grigoriev, S.; Melnik, Y. Source of metal atoms and fast gas molecules for coating deposition on complex shaped dielectric products. *Surf. Coat. Technol* **2013**, *225*, 34–39. [CrossRef]
11. Ozel, T.; Biermann, D.; Enomoto, T.; Mativenga, P. Structured and textured cutting tool surfaces for machining applications. *CIRP Ann. Manuf. Technol.* **2021**, *70*, 495–518. [CrossRef]
12. Sharma, V.; Pandey, P.M. Recent advances in turning with textured cutting tools: A review. *J. Clean. Prod.* **2016**, *137*, 701–715. [CrossRef]
13. Lei, S.; Devarajan, S.; Chang, Z. A study of micropool lubricated cutting tool in machining of mild steel. *J. Mater. Process. Technol.* **2009**, *209*, 1612–1620. [CrossRef]
14. Deng, J.; Song, W.; Zhang, H.; Yan, P.; Liu, A. Friction and wear behaviors of the carbide tools embedded with solid lubricants in sliding wear tests and in dry cutting processes. *Wear* **2011**, *270*, 666–674. [CrossRef]
15. Wu, Z.; Deng, J.; Su, C.; Luo, C.; Xia, D. Performance of the micro-texture self-lubricating and pulsating heat pipe self-cooling tools in dry cutting process. *Int. J. Refract. Met. Hard Mater.* **2014**, *45*, 238–248. [CrossRef]
16. Kawasegi, N.; Sugimori, H.; Morimoto, H.; Morita, N.; Hori, I. Development of cutting tools with microscale and nanoscale textures to improve frictional behavior. *Precis. Eng.* **2009**, *33*, 248–254. [CrossRef]
17. Deng, J.; Wu, Z.; Lian, Y.; Qi, T.; Cheng, J. Performance of carbide tools with textured rake-face filled with solid lubricants in dry cutting processes. *Int. J. Refract. Met. Hard Mater.* **2012**, *30*, 164–172. [CrossRef]
18. Xing, Y.; Deng, J.; Zhao, J.; Zhang, G.; Zhang, K. Cutting performance and wear mechanism of nanoscale and microscale textured Al_2O_3/TiC ceramic tools in dry cutting of hardened steel. *Int. J. Refract. Met. Hard Mater.* **2014**, *43*, 46–58. [CrossRef]
19. Fatima, A.; Whitehead, D.J.; Mativenga, P.T. Femtosecond laser surface structuring of carbide tooling for modifying contact phenomena. *Proc. Inst. Mech. Eng. Part B J. Eng. Manuf.* **2014**, *228*, 1325–1337. [CrossRef]
20. Obikawa, T.; Kamio, A.; Takaoka, H.; Osada, A. Micro-texture at the coated tool face for high performance cutting. *Int. J. Mach. Tools Manuf.* **2011**, *51*, 966–972. [CrossRef]

21. Xing, Y.; Deng, J.; Li, S.; Yue, H.; Meng, R.; Gao, P. Cutting performance and wear characteristics of Al_2O_3/TiC ceramic cutting tools with WS_2/Zr soft-coatings and nano-textures in dry cutting. *Wear* **2014**, *318*, 12–26. [CrossRef]
22. Fatima, A.; Mativenga, P.T. A comparative study on cutting performance of rake-flank face structured cutting tool in orthogonal cutting of AISI/SAE 4140. *Int. J. Adv. Manuf. Technol.* **2014**, *78*, 2097–2106. [CrossRef]
23. Grigoriev, S.N.; Kozochkin, M.P.; Sabirov, F.S.; Kutin, A.A. Diagnostic Systems as Basis for Technological Improvement. *Proc. CIRP* **2012**, *1*, 599–604. [CrossRef]
24. Grigoriev, S.N.; Sinopalnikov, V.A.; Tereshin, M.V.; Gurin, V.D. Control of parameters of the cutting process on the basis of diagnostics of the machine tool and workpiece. *Meas. Tech.* **2012**, *55*, 555–558. [CrossRef]
25. Vereschaka, A.; Grigoriev, S.; Tabakov, V.; Migranov, M.; Sitnikov, N.; Milovich, F.; Andreev, N. Influence of the Nanostructure of Ti-TiN-(Ti,Al,Cr)N Multilayer Composite Coating on Tribological Properties and Cutting Tool Life. *Tribol. Int.* **2020**, *150*, 106388. [CrossRef]
26. Grigoriev, S.; Vereschaka, A.; Milovich, F.; Tabakov, V.; Sitnikov, N.; Andreev, N.; Sviridova, T.; Bublikov, J. Investigation of multicomponent nanolayer coatings based on nitrides of Cr, Mo, Zr, Nb, and Al. *Surf. Coat. Technol.* **2020**, *401*, 126258. [CrossRef]
27. Grigoriev, S.N.; Gurin, V.D.; Volosova, M.A.; Cherkasova, N.Y. Development of residual cutting tool life prediction algorithm by processing on CNC machine tool. *Materwiss. Werksttech.* **2013**, *44*, 790–796. [CrossRef]

Communication

Reliability Enhancement of 14 nm HPC ASIC Using Al$_2$O$_3$ Thin Film Coated with Room-Temperature Atomic Layer Deposition

Po-Chou Chen [1], Shu-Mei Chang [1], Hao-Chung Kuo [2], Fu-Cheng Chang [2], Yu-An Li [3] and Chao-Cheng Ting [2,*]

[1] Institute of Organic and Polymeric Materials, National Taipei University of Technology, Taipei 10608, Taiwan
[2] Department of Photonics & Institute of Electro-Optical Engineering, College of Electrical and Computer Engineering, National Chiao Tung University, Hsinchu City 30010, Taiwan
[3] Pure Metallica Co., Ltd., Taipei City 11493, Taiwan
* Correspondence: chaochengting.mse00g@g2.nctu.edu.tw; Tel.: +886-918528040; Fax: +886-3-5716631

Abstract: In this research, a 14 nm high-performance computing application-specific integrated circuit was coated with a 5–20 nm Al$_2$O$_3$ thin film by atomic layer deposition in room-temperature conditions to study its performance in terms of reliability with different thicknesses. An open/short test, standby current measurement, interface input/output performance test, and phase-locked loops functional test were used to verify chip performance. Furthermore, an unbiased highly accelerated temperature and humidity stress test and a 72 h wear-out test were used to study the effects of the atomic layer deposition coating. The results showed that the coating thickness of 15 nm provided the best performance in the wear-out test, as well as the unbiased highly accelerated temperature humidity stress. This study demonstrates that room-temperature atomic layer deposition is a promising technique for enhancing the reliability of advanced node semiconductor chips.

Keywords: room temperature-ALD; high-performance computing; ASIC; wear-out test; uHAST

Citation: Chen, P.-C.; Chang, S.-M.; Kuo, H.-C.; Chang, F.-C.; Li, Y.-A.; Ting, C.-C. Reliability Enhancement of 14 nm HPC ASIC Using Al$_2$O$_3$ Thin Film Coated with Room-Temperature Atomic Layer Deposition. *Coatings* **2022**, *12*, 1308. https://doi.org/10.3390/coatings12091308

Academic Editor: Sergey N. Grigoriev

Received: 1 August 2022
Accepted: 30 August 2022
Published: 7 September 2022

Publisher's Note: MDPI stays neutral with regard to jurisdictional claims in published maps and institutional affiliations.

1. Introduction

Atomic layer deposition (ALD) process technology was first developed in the 1970s. In 1977, Dr. Tuomo Suntola of Finland officially applied for the first patent related to ALD technology [1–3]. Between 1983 and 1998, ALD technology was applied to the production of electronic displays at Helsinki Airport, Finland. In the late 1990s, as the semiconductor industry began to introduce ALD processes, a large amount of research and development funding and manpower was invested into the rapid growth of ALD process technology. In the year 2007, Intel used ALD process technology to grow a hafnium oxide (HfO$_2$) gate-oxide layer for a metal-oxide-semiconductor field-effect transistor on a 45 nm microprocessor, further establishing the importance of ALD process technology in the semiconductor industry [4].

ALD is a surface-chemical reaction-based technology featuring excellent atomic thickness accuracy, large area uniformity, and conformity of the film on high-aspect-ratio structures [5].

The coating of thermally fragile substrates by ALD may provide exciting new applications in diverse areas ranging from packaging to microelectronics. For example, thin ALD films may serve as an efficient hole-blocking layer in a complementary metal-oxide-semiconductor image sensor to enhance the signal to noise ratio [6,7]. ALD coatings on polymers may also be important as gas diffusion barriers for flexible electronic devices or organic light-emitting diodes [8,9]. Moreover, applications of ALD processes have been demonstrated at low temperatures of <100 °C [10–18]. Room-temperature catalytic SiO$_2$ ALD has been realized using SiCl$_4$ and H$_2$O plus pyridine or NH$_3$ as the catalyst [19,20]. The ALD of ZnSe and CdS has also been demonstrated at room temperature [21]. The ALD technique is widely used in the electronic industry [22]. However, few studies have addressed the advantages of deposition on a packaged chip. Most ALD processes occur at temperatures >100 °C that would degrade organic, polymeric, or biological material. The material normally used for the packaging of semiconductor chips is organic polyimide. As the technology advances, the performance in terms of reliability becomes critical, and sensitivity to moisture in harsh environments becomes a factor when using the chip. In this study, a 14 nm high-performance computing application specific integrated circuit (ASIC) was selected to verify the effect of changing the thickness of the ALD coating on performance in terms of reliability, which was proposed for the first time. An open/short test, standby current measurement, interface input/output performance test, and phase-locked loops (PLL) function test were used to verify the chip yield. The lowest operating voltage (LOV) was used as the performance index for reliability after wear-out tests and unbiased highly accelerated temperature and humidity stress test (uHAST). The ASIC with a 15 nm Al_2O_3 ALD coating showed the best performance in both the uHAST and the 72 h wear-out test. The reasons for failure of the 25 nm thick coating remain unclear and are worthy of further investigation. The results of this study provide an interesting evaluation of the relationship between ALD coating thickness and performance in terms of reliability, which can pave the way for a new package scheme in advanced semiconductor processes.

The aim of this study is to establish a protective nanoscale thin film using the ALD technology for electronic components in a prepackaged high-performance computing chip, which can improve the electrical performance and reliability after high-temperature and high-humidity stress tests.

2. Materials and Methods

The ALD technique was applied using the Lunamia model LA-1T from Pure Metallica Co., Ltd., Taipei City, Taiwan. The Lunamia is a type of capacitive plasma-enhanced ALD equipment. To apply the plasma ALD coating of the Al_2O_3 layer, the reaction conditions were trimethylaluminum exposure (from PentaPro Materials Inc., Hsinchu County, Taiwan, purity: 6N), argon purge (from Jing De Gases Co., Ltd., Kaohsiung City, Taiwan, purity: 5N), and oxygen (from Jing De Gases Co., Ltd., purity: 6N) with a plasma power of 100 W. Each step was controlled by the automatic processes described below.

First, the inert gas was introduced into the deionized water tank, as well as water vapor, using a plasma power of 100 W, which covered the substrate with hydroxyl groups, thereby forming a reactive layer (i.e., prereaction seasoning). The precursor was introduced into the chamber, along with the carrier gas Ar, at a flow rate of 500 sccm to form a surface-adsorbed monolayer for later reaction.

Second, the precursor and other byproducts were removed by introducing the inert gas Ar, followed by pumping the chamber to a base pressure of 10^{-6} Torr. Third, the reaction gas O_2 was then introduced, supplemented by plasma. This O_2/plasma mixture allowed oxygen radicals to react with the surface-adsorbed Al precursor, thereby forming an alumina monolayer, i.e., the desired material for the ALD process.

The fourth and final step of the ALD reaction cycle involved removal of the excess reaction gas and byproducts by again introducing the inert gas Ar, followed by pumping the chamber to the base pressure. Table 1 lists the ALD process parameters. The O_2 process was to ensure the complete oxidation of the trimethylaluminum. The radio frequency (RF) power of 100 W was selected by considering the degree of oxidation with the minimum of substrate reflected power. The soak time was considered for the full diffusion of the precursor for better process uniformity. The Ar purge time was to ensure no aluminum precursor remained forming gas-phase reaction-generating particles.

Table 1. Parameters of the ALD process.

Step Process	Time (s)	Pressure (mTorr)	RF (W)	Gas Flow (sccm)	Soak Time (s)
O_2 treatment	1	5	100	300	5
1st Ar purge	3	5	0	500	0
TMA exposure	1	5	0	500	0
2nd Ar purge	3	5	0	500	0

The thickness and the quality of the ALD coatings were determined by ellipsometry. The desired thickness could be controlled by the cycle number, whereby one ALD cycle deposited 0.24 nm of Al_2O_3 thin film. The ALD thicknesses of 5–20 nm were selected to study their effect on performance in terms of reliability.

The uHAST was carried out at a temperature of 130 °C and 85% relative humidity (RH) for 96 h at a relative pressure of 2.3 atm. The stressed duration was 96 h under the JESD22-A118 regulation. After an ambient temperature test, various electrical measurements were performed, including an open/short test, power short test, ISB (power standby current test), direct current (DC) characterization test, and lowest operation voltage functional test (0.62–0.78 V) using an Agilent 9300 SOC series (Santa Clara, CA, USA). The uHAST and automatic testing equipment is displayed in Figure 1.

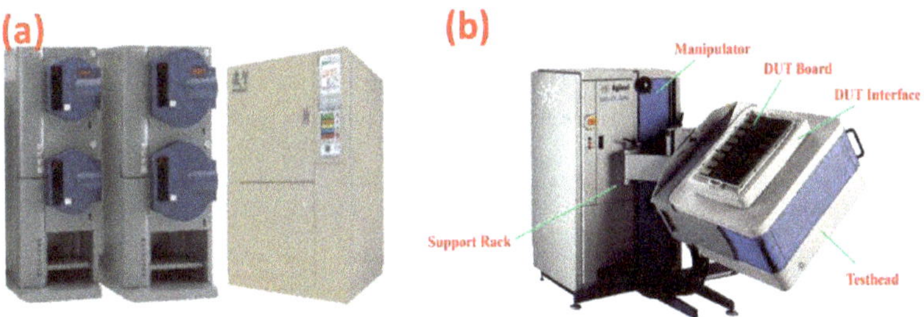

Figure 1. The uHAST (**a**) and testing equipment (Agilent 9300 SOC series, (**b**)). Reprinted/adapted with permission from Ref. [23]. 2022, Integrated Service Technology Inc.

3. Results

3.1. Background of the 14 nm HPC ASIC

The cryptocurrency mining chip is a 14 nm ASIC. This chip operates in harsh conditions when used in high-performance computing. Thus, its reliability is crucial, as cryptocurrency mining machines require reliable computing performance to ensure their profitability. A 14 nm ASIC can provide lower power consumption, along with high computing performance, making it suitable for mining applications.

Figure 2 shows a block diagram illustrating the functions of the 14 nm ASIC. The ASIC consists of five main functional circuits, including the PLL which fundamentally provides the clock source used for controlling the calculation via an automatic feedback design. The PLL clock serves as a gateway to the calculation hash core, which features translated algorithms used to directly operate the software to generate cryptocurrency. The control unit rechecks the calculation result and outputs the data through the IO_OUT pin.

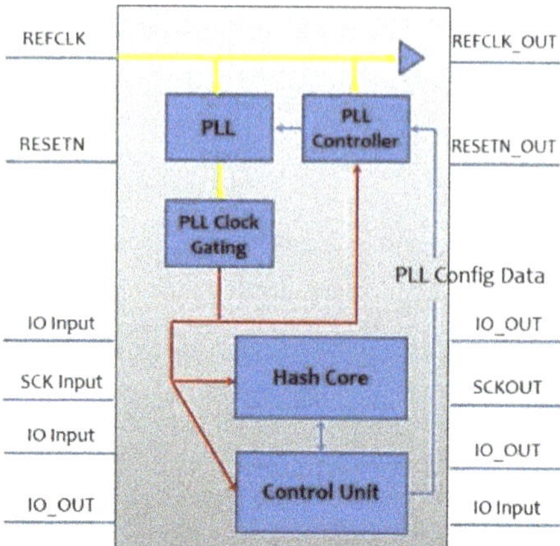

Figure 2. The function block diagram illustrating the functions of the 14 nm ASIC.

3.2. Yield and Reliability of the ALD-Coated ASIC Chip

Figure 3 illustrates the conditions of the coating structure and the thermal dispatch pathway of the flip-chip land grid array (FCLGA). Two exposed heat sinks were used as the main heat dispatch material. Inside the package, the chip was connected via a copper pillar bump. During high-speed operation, the chip generates tremendous heat and becomes vulnerable to moisture in the environment. In order to prevent its penetration into the chip, different ALD Al_2O_3 coatings thicknesses were applied, and we studied their effect on performance in terms of reliability.

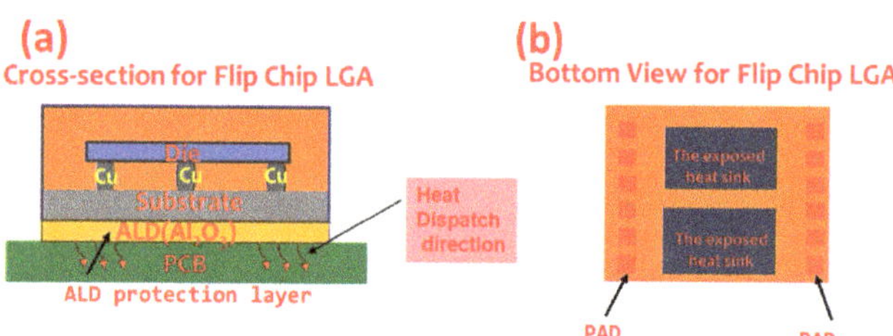

Figure 3. Cross section diagram illustrating the conditions of the ALD coating structure (**a**) and the bottom view of the flip-chip land grid array (**b**).

Figure 4 shows the cross-section FIB image of the 50 nm ALD-coated thin film. The ALD-coated thin film established a uniform and continuous protection layer of the integrated chip. The diagram in Figure 5 (top panel) schematically demonstrates the ALD coating on the 14 nm ASIC. The results showed that the 15 nm thick ALD alumina coating exhibited the lowest operation voltage performance after a 1000 h uHAST, whereby the operation voltage was reduced to 0.62 V, thus indicating its ability to resist harsh conditions. The similar results of a 72 h wear-out test were also performed (not shown here). On the other hand, the chip with a 20 nm thick coating failed the DC 72 h wear-out stress test. The reason for failure was not clear and is worthy of further investigation. It is suspected that the chlorine ion may have reacted with the alumina coating, thereby causing interconnect path damage. The previous work in corrosion research indicated that the existence of Al_2O_3 accelerated the penetration of the chlorine ion (Cl^-) and thus corrosion occurred [24]. The mechanism of how the corrosion induced by moisture and Cl^- occurred was also proposed by Fu et al. [25].

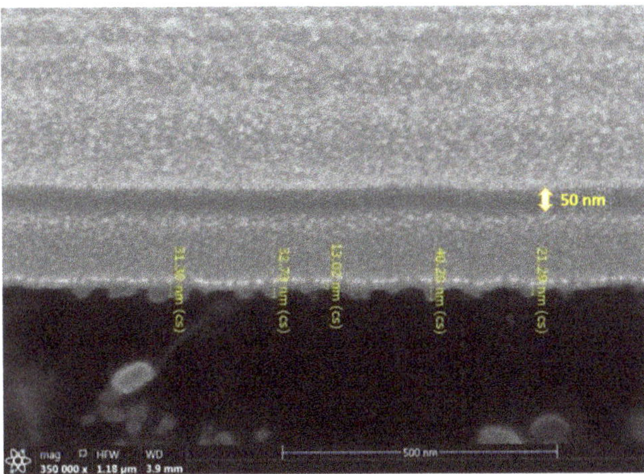

Figure 4. Schematic diagram of the ALD-coated ASIC (top); cross-section FIB image of 50 nm ALD-coated thin film.

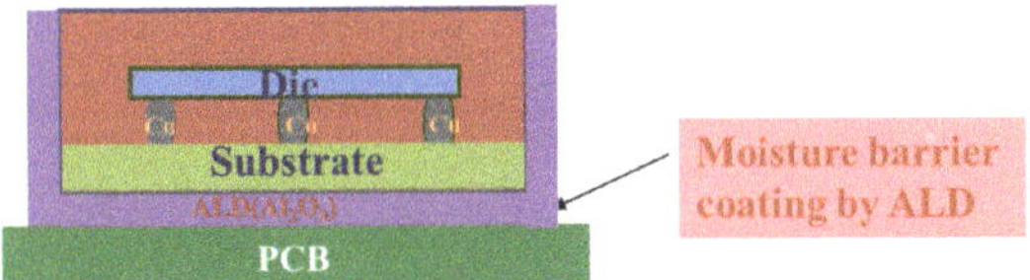

Figure 5. Schematic diagram showing the packaged printed circuit board structure and the coated alumina protection layer.

Figure 5 shows how the ALD coating protected the printed circuit board (PCB) structure; the alumina layer prevented moisture from penetrating into the main chip, while also facilitating good heat dispatch. The ALD layer established a capped circuit structure, allowing the thermal stress of the surface and the chip to be alleviated. Otherwise, moisture could penetrate through the land grid metal pad and the adjacent plastic substrate due to the difference in thermal expansion coefficient between the metal pad and epoxy polymer molding, thus leading to thermal stress accumulation and thermal strain distribution problems. In our study, the ALD coating alleviated the stress-induced electronic performance decay. The effect of an epoxy molding compound on thermal stresses in IC packages during the manufacturing process were carefully discussed [26,27]. In the study, the failure of the chip was caused by the thermal deformations of the die/EMC bimaterial. The experimental results also showed that the ability to resist the harsh environment varied with the tested thicknesses of 0, 5, 10, 15, and 20 nm, according to 1000 h of an uHAST and 72 h of a wear-out salt-spray test.

Figure 6 illustrates the open/short testing method. A current of 100 mA was collected from the input/output (IO) pad, and the voltage was measured. The ALD coating with a thickness of 20 nm failed the open/short test, according to the criteria listed in the figure (-0.2 to 0.7 V). This may have been due to the thickness of the Al_2O_3 coating, which caused poor pad connection after the 72 h wear-out test. The sheet resistance test was performed with different ALD thicknesses on the conducting substrate. The sheet resistance was 1.6×10^7 ohm/sq with an Al_2O_3 thickness of 16 nm.

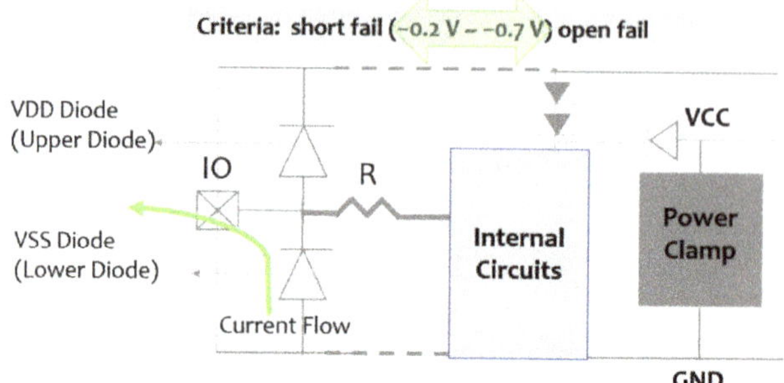

Figure 6. The test criteria and methodology of the open/short test.

Table 2 lists the uHAST results with different Al_2O_3 thicknesses. The chip with a 15 nm thick protection layer demonstrated the lowest operation voltage (0.66 V). It was also discovered that a thinner protection layer led to a higher operation voltage shift. The yield was only 50% with an operation voltage of 0.78 V when the ALD coating thickness was below 10 nm. Figure 7 shows the operation voltage results as a function of ALD thickness before and after the uHAST. Each sample was evaluated under the same stress conditions, thus demonstrating the correlation of voltage shift with thickness. The samples with ALD thicknesses of 15 and 20 nm both surpassed the lowest operation voltage shift after the stress test.

Table 2. The uHAST results as a function of Al_2O_3 thickness.

Test Item	Non-ALD	Thickness of ALD-Coated Al_2O_3 Thin Film			
		5 nm	10 nm	15 nm	20 nm
Open test	100%	100%	100%	100%	100%
Short test	100%	100%	100%	100%	100%
Stand by current	100%	100%	100%	100%	100%
Input voltage	100%	100%	100%	100%	100%
Output voltage	100%	100%	100%	100%	100%
Pull down/pull up	100%	100%	100%	100%	100%
PLL function test	0%	0%	50%	100%	100%

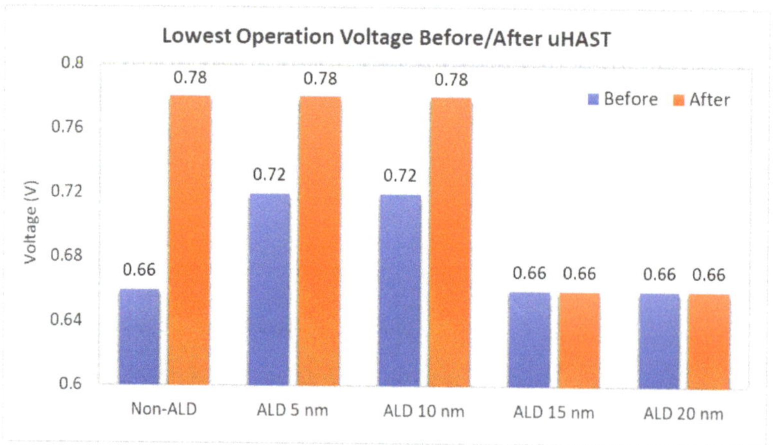

Figure 7. The Al_2O_3 thicknesses versus the lowest operating voltage results before and after the 1000 h uHAST.

4. Future Aspects

In this study, an ALD fabricated Al_2O_3 thin film was used to establish a protection layer on the high-performance computing chip, which can also be applied to a bonding wire to form an insulating and heat dispatch layer. Detailed diagrams of these applications are shown in Figure 8. The figure shows the patent design. The original bonding wire is marked "31", while the functional layers are marked "411" and "412", i.e., the electromagnetic interference shielding layer or heat dispatch layer. The outer insulating layer is marked "42". The methodology can also be used in 3D inkjet flexible printed touch sensors to enhance the sensor lifetime [28].

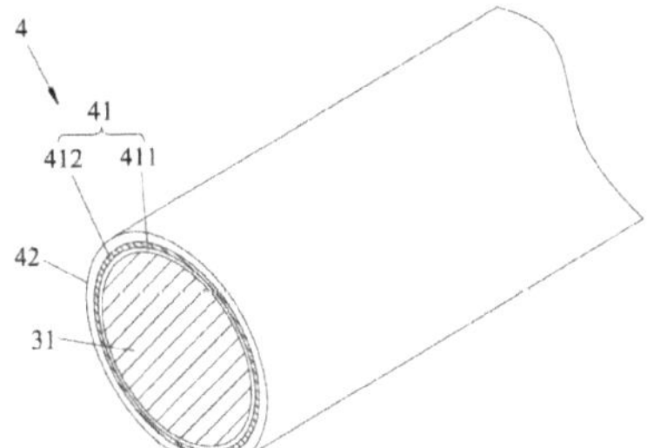

Figure 8. Schematic diagram of the bonding wire ALD-coated with Al_2O_3 thin film.

5. Conclusions

In this study, the overall protective layer using the atomic layer deposition technology established after encapsulation with electronic components obviously improved the lifetime in harsh environments. The excavator integrated circuit tested in this paper is prone to generate high heat during operation as a packaging technology in a flip-chip land grid array package type, whereby the stress caused by high temperature can easily destroy the interface between plastic substrates, solder, and metal mats, thereby reducing its service lifetime. The creation of an Al_2O_3 protective thin film layer with a thickness of 15 nm through atomic layer deposition technology significantly improved the component damage caused by thermal stress. The protective film with a thickness of 15 nm showed the same lowest operation voltage at 0.66 V before and after the 1000 h unbiased highly accelerated temperature and humidity stress test. This research showed that atomic layer deposition can be effectively used to block moisture and oxygen from entering electronic devices, to avoid their rapid deterioration. The protection layer can also improve the mechanical protection and reduce the damage caused by moving collisions, which has potential applications in organic light-emitting diodes, micro light-emitting diodes, and semiconductor packages for resisting moisture. In addition, the atomic layer deposition can be used to fabricate luminescent material layers, biomedicine sensor components, and water-resistant thin films for use in energy engineering and other applications related to coating technology.

6. Patents

The related patents described in this research are pending.

Author Contributions: Methodology, P.-C.C. and S.-M.C.; validation, H.-C.K. and F.-C.C.; writing—original draft preparation, Y.-A.L.; writing—review and editing, C.-C.T. All authors have read and agreed to the published version of the manuscript.

Funding: This research received no external funding.

Institutional Review Board Statement: The study did not involve humans or animals.

Informed Consent Statement: Not applicable.

Data Availability Statement: Data are available from the corresponding author.

Acknowledgments: The authors acknowledge the reliability evaluation support of the IST group.

Conflicts of Interest: The authors declare no conflict of interest.

References

1. Suntola, T.; Antson, J. Method for Producing Compound Thin Films. U.S. Patent 4,058,430, 15 November 1977.
2. Abegunde, O.O.; Akinlabi, E.T.; Oladijo, O.P.; Akinlabi, S.; Ude, A.U. Overview of thin film deposition techniques. *AIMS Mater. Sci.* **2019**, *6*, 174–199. [CrossRef]
3. Naghdi, S.; Rhee, K.Y.; Hui, D.; Park, S.J. A review of conductive metal nanomaterials as conductive, transparent, and flexible coatings, thin films, and conductive fillers: Different deposition methods and applications. *Coatings* **2018**, *8*, 278. [CrossRef]
4. Profijt, H.; Potts, S.; Van de Sanden, M.; Kessels, W. Plasma-assisted atomic layer deposition: Basics opportunities, and challenges. *J. Vac. Sci. Technol. A Vac. Surf. Films* **2011**, *29*, 050801. [CrossRef]
5. George, S.M. Atomic layer deposition: An overview. *Chem. Rev.* **2010**, *110*, 111–131. [CrossRef]
6. Chang, C.Y.; Pan, F.M.; Lin, J.S.; Yu, T.Y.; Li, Y.M.; Chen, C.Y. Lateral amorphous selenium metal-insulator-semiconductor-insulator-metal photodetectors using ultrathin dielectric blocking layers for dark current suppression. *J. Appl. Phys.* **2016**, *120*, 234501. [CrossRef]
7. Hu, H.; Dong, B.; Hu, H.; Chen, F.; Kong, M.; Zhang, Q.; Wan, L. Atomic layer deposition of TiO_2 for a high-efficiency hole-blocking layer in hole-conductor-free perovskite solar cells processed in ambient air. *ACS Appl. Mater. Interfaces* **2016**, *8*, 17999–18007. [CrossRef] [PubMed]
8. Chen, G.; Weng, Y.; Sun, F.; Zhou, X.; Wu, C.; Yan, Q.; Guo, T.; Zhang, Y. Low-temperature atomic layer deposition of Al_2O_3/alucone nanolaminates for OLED encapsulation. *RSC Adv.* **2019**, *9*, 20884. [CrossRef]
9. Weaver, M.; Michalski, L.; Rajan, K.; Rothman, M.; Silvernail, J.; Brown, J.J.; Burrows, P.E.; Graff, G.L.; Gross, M.E.; Martin, P.M. Organic light-emitting devices with extended operating lifetimes on plastic substrates. *Appl. Phys. Lett.* **2002**, *81*, 2929–2931. [CrossRef]
10. Di Mauro, A.; Cantarella, M.; Nicotra, G.; Privitera, V.; Impellizzeri, G. Low temperature atomic layer deposition of ZnO: Applications in photocatalysis. *Appl. Catal. B Environ.* **2016**, *196*, 68–76. [CrossRef]
11. Vayrynen, K.; Mizohata, K.; Ra isa nen, J.; Peeters, D.; Devi, A.; Ritala, M.; Leskela, M. Low-temperature atomic layer deposition of low-resistivity copper thin films using Cu(dmap)$_2$ and tertiary butyl hydrazine. *Chem. Mater.* **2017**, *29*, 6502–6510. [CrossRef]
12. Vangelista, S.; Mantovan, R.; Lamperti, A.; Tallarida, G.; Kutrzeba-Kotowska, B.; Spiga, S.; Fanciulli, M. Low-temperature atomic layer deposition of MgO thin films on Si. *J. Phys. D Appl. Phys.* **2013**, *46*, 485304. [CrossRef]
13. Jurca, T.; Moody, M.J.; Henning, A.; Emery, J.D.; Wang, B.; Tan, J.M.; Lohr, T.L.; Lauhon, L.J.; Marks, T.J. Low-temperature atomic layer deposition of MoS$_2$ films. *Angew. Chem. Int. Ed.* **2017**, *56*, 4991–4995. [CrossRef] [PubMed]
14. Dendooven, J.; Ramachandran, R.K.; Devloo-Casier, K.; Rampelberg, G.; Filez, M.; Poelman, H.; Marin, G.B.; Fonda, E.; Detavernier, C. Low-temperature atomic layer deposition of platinum using (methylcyclopentadienyl) trimethylplatinum and ozone. *J. Phys. Chem. C* **2013**, *117*, 20557–20561. [CrossRef]
15. Putkonen, M.; Sippola, P.; Svärd, L.; Sajavaara, T.; Vartiainen, J.; Buchanan, I.; Forsström, U.; Simell, P.; Tammelin, T. Low-temperature atomic layer deposition of SiO_2/Al_2O_3 multilayer structures constructed on self-standing films of cellulose nanofibrils. *Philos. Trans. R. Soc. A* **2018**, *376*, 20170037. [CrossRef]
16. Nam, T.; Kim, J.-M.; Kim, M.-K.; Kim, H.; Kim, W.-H. Low-temperature atomic layer deposition of TiO_2, Al_2O_3, and ZnO thin films. *J. Korean Phys. Soc.* **2011**, *59*, 452–457. [CrossRef]
17. Kwak, I.; Park, J.H.; Grissom, L.; Fruhberger, B.; Kummel, A. Mechanism of low temperature ALD of Al_2O_3 on graphene terraces. *ECS Trans.* **2016**, *75*, 143. [CrossRef]
18. Vandalon, V.; Kessels, W. What is limiting low-temperature atomic layer deposition of Al_2O_3? A vibrational sum-frequency generation study. *Appl. Phys. Lett.* **2016**, *108*, 011607. [CrossRef]
19. Klaus, J.W.; Sneh, O.; George, S.M. Growth of SiO_2 at room temperature with the use of catalyzed sequential half-reactions. *Science* **1997**, *278*, 1934–1936. [CrossRef]
20. Klaus, J.; George, S. Atomic layer deposition of SiO_2 at room temperature using NH$_3$-catalyzed sequential surface reactions. *Surf. Sci.* **2000**, *447*, 81–90. [CrossRef]
21. Knez, M.; Nielsch, K.; Niinistö, L. Synthesis and surface engineering of complex nanostructures by atomic layer deposition. *Adv. Mater.* **2007**, *19*, 3425–3438. [CrossRef]
22. Biercuk, M.; Monsma, D.; Marcus, C.; Becker, J.; Gordon, R. Low-temperature atomic-layer-deposition lift-off method for microelectronic and nanoelectronic applications. *Appl. Phys. Lett.* **2003**, *83*, 2405–2407. [CrossRef]
23. Temperature and Humidity Test—iST-Integrated Service Technology. Available online: https://www.istgroup.com/en/service/temperature-humidity/ (accessed on 1 June 2021).
24. Chiu, Y.T.; Chiang, T.H.; Yang, P.F.; Huang, L.; Hung, C.P.; Uegaki, S.; Lin, K.L. The corrosion behavior of Ag alloy wire bond on Al pad in molding compounds of various chlorine contents under biased-HAST. In Proceedings of the 2016 International Conference on Electronics Packaging (ICEP), Hokkaido, Japan, 20–22 April 2016; pp. 497–501.
25. Fu, S.W.; Lee, C.C. A corrosion study of Ag–Al intermetallic compounds in chlorine-containing epoxy molding compounds. *J. Mater. Sci. Mater. Electron.* **2017**, *28*, 15739–15747. [CrossRef]
26. Tsai, M.Y.; Wang, C.T.; Hsu, C.H. The effect of epoxy molding compound on thermal/residual deformations and stresses in IC packages during manufacturing process. *IEEE Trans. Compon. Packag. Manuf. Technol.* **2006**, *29*, 625–635. [CrossRef]

27. Lancaster, A.; Keswani, M. Integrated circuit packaging review with an emphasis on 3D packaging. *Integration* **2018**, *60*, 204–212. [CrossRef]
28. Palanisamy, S.; Thangaraj, M.; Moiduddin, K.; Al-Ahmari, A.M. Fabrication and performance analysis of 3D inkjet flexible printed touch sensor based on AgNP electrode for infotainment display. *Coatings* **2022**, *12*, 416. [CrossRef]

Article

Diagnostic Techniques for Electrical Discharge Plasma Used in PVD Coating Processes

Sergey Grigoriev [1], Sergej Dosko [2], Alexey Vereschaka [2,*], Vsevolod Zelenkov [2] and Catherine Sotova [1]

[1] VTO Department, Moscow State University of Technology STANKIN, Vadkovsky per.1, 127055 Moscow, Russia
[2] Institute of Design and Technological Informatics of the Russian Academy of Sciences (IDTI RAS), 127055 Moscow, Russia
* Correspondence: dr.a.veres@yandex.ru

Abstract: This article discusses the possibilities of two methods for monitoring Physical Vapor Deposition (PVD) process parameters: multi-grid probe, which makes it possible, in particular, to determine the energy distribution of ions of one- or two-component plasma and spectrum analyzer of the glow discharge plasma electromagnetic radiation signal based on the Prony–Fourier multichannel inductive spectral analysis sensor. The energy distribution curves of argon ions in the low-voltage operation mode of ion sources with closed electron current have been analyzed. With a decline in the discharge current, the average ion energy decreases, and the source efficiency (the ratio of the average ion energy W to the discharge voltage U) remains approximately at the same level of $W/U \approx 0.68, \ldots, 0.71$ in the operating voltage range of the source. The spectrum analyzer system can obtain not only the spectra at the output of the sensor, but also the deconvolution of the spectrum of the electromagnetic radiation signal of the glow discharge plasma. The scheme of a spectrum analyzer is considered, which can be used both for monitoring and for controlling the processing process, including in automated PVD installations.

Keywords: electrical discharge plasma; analyzing methods; Physical Vapor Deposition (PVD); energy distribution

Citation: Grigoriev, S.; Dosko, S.; Vereschaka, A.; Zelenkov, V.; Sotova, C. Diagnostic Techniques for Electrical Discharge Plasma Used in PVD Coating Processes. *Coatings* **2023**, *13*, 147. https://doi.org/10.3390/coatings13010147

Academic Editor: Engang Fu

Received: 19 November 2022
Revised: 5 January 2023
Accepted: 9 January 2023
Published: 11 January 2023

1. Introduction

1.1. Method of Physical Vapor Deposition and Issues of Process Diagnostics

The processes of ion-plasma treatment and the deposition of coatings through the Physical Vapor Deposition (PVD) technology are actively used in various industries with the aim to increase surface hardness, resistance to wear and corrosion, and control of tribological parameters and optical properties of the products [1–5].

The PVD methods are an effective way to improve the working efficiency and prolong service life of metal-cutting tools and various parts of tool, chemical, machine-building, automotive, aviation, and medical industries. The methods of ion-plasma treatment of surfaces have a number of advantages over other coating deposition methods, including Chemical Vapor Deposition (CVD), spraying with electricity, thermal spraying, as well as electrolytic and galvanic methods [4,5]. In particular, the PVD method provides perfect adhesion to a wide range of various materials, both conductive and dielectric ones. The advantages of the PVD method also include the possibility of depositing high-quality coatings at considerably low temperatures (compared to the CVD methods), the possibility of using a wide range of coating compositions and structures, and the possibility of effectively controlling the parameters of a coating during its deposition [4,5].

When any method of surface modification and coating deposition is used, the key task is to ensure the quality and reproducibility of the results of the technology process carried out in accordance with the selected method. The PVD methods are based on the impact of high-energy flows of gas or metal plasma on the products. This exposure changes the structure and properties of the coating surface and its energy state, the degree of which depends on the type and mode of the treatment. The result of such exposure depends on many parameters that are difficult to control, including the chemical and phase composition and the structural state of the substrate material, which change during the exposure to the plasma flow [6–8]. The listed parameters directly depend on the energy characteristics of the plasma flow. When developing new technological devices for the deposition of PVD coatings and ion-plasma surface treatment, as well as for the control of the technology ion-plasma processes, it is necessary to use the methods of diagnostics and dynamic measurement of various physical properties of the surface and the parameters of plasma flows. Some of these methods are considered below.

1.2. Key Methods of Operational Control of the Deposition Process Parameters

The development of the PVD technologies and the modification of surfaces of solid bodies requires the study methods providing for the quick control of the state of the process and the results obtained at various stages of technology cycles.

Although the process of ion-plasma treatment of surfaces by the PVD method applies such standard control techniques as measurement of the arc current and bias voltage, the composition and pressure of the reaction gas, and the turntable rotation speed [9,10]. Due to the complexity of the ion-plasma treatment process and the variety of influencing factors, the mentioned control parameters are often not enough to ensure unambiguous reproducibility of the process results. The results of ion-plasma treatment are significantly affected by such factors as the spatial arrangement of samples, sizes and weight of technological equipment, the cathode wear degree, the condition of inner surface of the chamber, and a number of other parameters, the direct control of which is difficult or impossible [6,11–17]. Thus, there is a need for dynamic control of the plasma parameters and the state of the modified surface.

With regard to the operational control over the parameters of the ion-plasma treatment and the state of modified surfaces, the following considerably effective methods may be considered [18–30]:

- Langmuir probe methods [19–22];
- application of a charged particle energy analyzer (multigrid probe) [23–25];
- and use of a broadband spectrum analyzer based on a multichannel inductive sensor and Prony–Fourier (PF) spectral analysis.

Various technological ion-plasma processes of surface treatment and coating deposition in vacuum are determined mainly by the parameters of the plasma acting on the surface and the temperature of the very workpieces. Therefore, due to a change in the parameters of the environment, one technological cycle can cover various processes. an increase of the ion energy makes it possible to conduct ion cleaning, and a decrease of the ion energy ensures the deposition of coatings. The energy of ions bombarding the surface of the substrate is one of the most important parameters to be measured and controlled during the development of cleaning and etching technologies for various materials, as well as the PVD technologies [31–36]. The plasma density (concentration) and the temperature of electrons and ions determine the equilibrium temperature of the treated workpieces and, consequently, the diffusion processes and the processes of crystal structure growth on the surface of the treated workpieces.

The main parameters of the low temperature plasma generated by various electric discharges used in the technological processes of treating the surfaces of the workpieces in vacuum (glow discharge, magnetron discharge, vacuum arc discharge with a sacrificial cathode, etc.) include plasma density, electron temperature, ion energy, and ion current density on the surface of the workpieces [37,38]. The non-equilibrium nature and high-power density make plasma a considerably complex object to control and study.

There is a large number of diagnostic methods for determining various parameters of both high temperature and low temperature plasma. Of them, the most widely used are electromagnetic, spectral, microwave, and probe methods [39–43].

The spectral methods give accurate information for optically thin plasma, while for optically dense plasma, the information content of the method is low. Some alternative diagnostic methods are discussed in [44–46].

One of the most common methods used in scientific research and in the industrial practice are microwave methods [38–40]. Scattering of external high-frequency electromagnetic radiation on free electrons leads to a change in the frequency of scattering wave, the measurements of which determine the distribution function of particles according to their speeds, density, temperature, and directed velocity. Probe methods of diagnostics are the most common and easy to implement, allowing to determine the main parameters of ion-plasma flows necessary for the control of the PVD technological processes.

Physical processes in technological installations are often oscillatory in nature, so the universal model proposed by Bulgakov [47] can be used to describe them, and the Prony method [48,49] is the most acceptable for model identification in terms of physical adequacy and accuracy of approximation.

Estimating the parameters of an exponential signal is an important task, since the response of a linear system to an impulse action is the sum of just such signals. Thus, by evaluating the parameters of the signals at the output of the system, it is possible to solve the problem of identifying the system and its state. The use of the Fourier transform for this purpose does not always give acceptable results. This is due to the fact that the Fourier transform is intended to estimate the signal spectrum, not the frequency, and, moreover, in the classical version it is not statistically stable [50].

For non-stationary oscillations, spectral analysis is required, with which you can:

1. perform spectral estimation of segments of time series without side effects in time windows of limited duration;
2. use a non-stationary time series model (for example, increasing or decreasing in the time window);
3. determine own frequency spectrum of a segment of the time series;
4. and determine damping at natural frequencies.

Considering the above requirements, it seems promising to use the Prony method to determine the time-dependent damping spectra of non-stationary oscillations. When identifying the similarity of segments of a time series, there are not a number of restrictions inherent in the Fourier transform of time series [50].

The Langmuir probe measurements [19–22,41,51–59] are widely used low temperature plasma diagnostic methods. This method makes it possible to determine the main parameters of the low temperature plasma in the local region, including electron temperature, electron density, and ion current density, and, in some cases, to approximately estimate the ion temperature. Probe measurements are reliable, provided that the lengths of the particle path exceed the dimensions of the probe. Interpretation of data in the presence of considerably large magnetic fields and collisions is complicated. In accordance with the method Langmuir probe measurements, a small electrode—a probe—is placed in the plasma (to reduce disturbances caused in the plasma, the probe should be small) and its current-voltage curve is taken (the probe current value depending on its voltage), which is called the plasma probe characteristics. Probes can be in the form of a flat disk, a cylinder, or a sphere. A typical diameter is from 10^{-3} to 10^{-1} cm for a cylindrical probe and from 10^{-2} to 10^{-1} cm for a spherical probe. To take the plasma probe characteristics, it is important that the electrons have the energy distribution close to Maxwellian distribution, which is provided by the plasma of vacuum stationary discharges used in technological processes. The scope of the method for gas pressure covers the range from 10^{-2} to 10^{2} Pa and from 10^{7} to 10^{5} cm^3—for the concentration of charged particles.

This article discusses the possibilities of two methods for monitoring PVD process parameters:

- multi-grid probe, which makes it possible, in particular, to determine the energy distribution of ions of one- or two-component plasma generated by a vacuum arc evaporator;

- and spectrum analyzer of the glow discharge plasma electromagnetic radiation signal based on the Prony–Fourier multichannel inductive spectral analysis sensor.

2. Materials and Methods

2.1. The Langmuir Probe Measurements

To determine the parameters of the ionic component, a charged particle energy analyzer (multigrid probe) was used [60–65]. Figure 1 exhibits the method for measuring the energy of charged particles using a retarding potential energy analyzer [66].

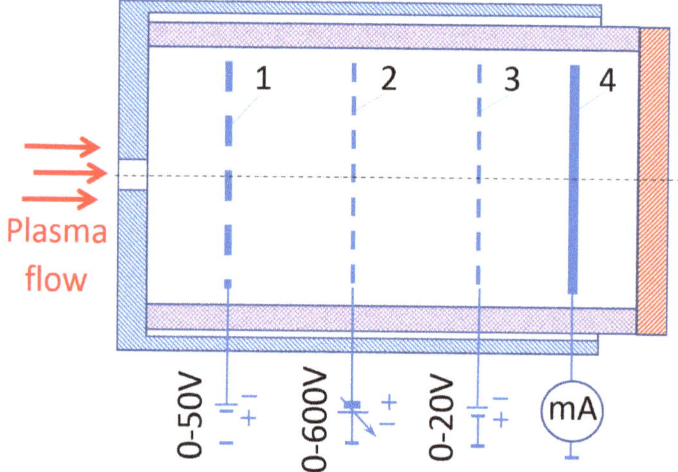

Figure 1. Schematic diagram of multigrid analyzer probe [59]. 1—electron-intercepting grid, 2—analyzing (retarding) grid, 3—antidynatron grid, 4—collector.

In the course of measurements, the plasma enters the analyzer through the hole, and the increasing retarding potential, which decelerates the ions, is applied to the analyzing grid 2, while the electrons are intercepted by the grid 1. The ions with energy higher than the retarding potential barrier of the grid 2 form the current of the collector 4. Based on the measurement results, a retarding parameter—a curve of the collector current depending on the retarding potential—is being plotted. To determine the energy distribution function of particles (differential energy spectrum), it is necessary to differentiate the resulting retarding curve. To improve the accuracy of measurement, an antidynatron grid is used in the probe to suppress the secondary emission of electrons that occurs under the action of ion bombardment of the collector surface.

The disadvantages of the described method include:

- high sensitivity of the probe to its contamination;
- and when working with metal plasma, probes quickly fail, or their readings change due to the dust deposited on the input grid; the application of probes is complicated in the presence of considerably large magnetic fields in the plasma.

At high plasma densities, when the free paths of electrons and ions are short, the probe method is practically not applicable.

From the point of view of vacuum technologies for modifying surfaces of solid bodies, the advantage of the probe methods of analysis is in the possibility of integrating the probes, due to their small sizes, into the technological volumes and the possibility of carrying out operational control and adjusting the technological process.

A multigrid probe was used to determine the energy distributions of ions in the two-component plasma, generated by a vacuum arc evaporator with two evaporating electrodes similar to those described by Zelenkov et al. [67]. The device consists of an evaporable cathode, an evaporable anode, and a passive (non-evaporable) anode (see Figure 2).

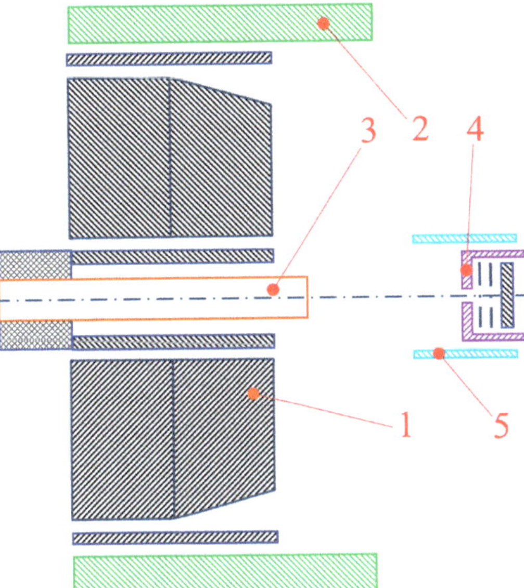

Figure 2. Scheme of multigrid probe [67]. 1—cathode, 2—passive anode, 3—evaporable anode, 4—probe (energy analyzer), 5—screen.

2.2. Broadband Spectrum Analyzer

An effective method for analyzing the plasma electromagnetic radiation signal (in particular, the glow discharge plasma) is to use a broadband spectrum analyzer based on a multichannel inductive sensor, analog-to-digital converter (ADC), and Prony–Fourier (PF) spectral analysis. The mentioned spectrum analyzer may be used both for monitoring and for controlling the ion-plasma process of surface treatment of solid bodies, including in automated technological environments, with high reproducibility of results. Changes in the shape, amplitudes, and frequencies of the recorded electromagnetic pulses during the treatment process are closely related to the pulsations of the current in the discharge during the technological process of treatment. In particular, such method may be used to determine the moment when the PVD unit reaches the operating mode, to control the stages of surface thermal activation and coating deposition, as well as to promptly respond to deviations from the specified parameters of the technological process.

The wide use in recent decades of the resonance methods for the study of substances in gaseous, liquid, and solid states is justified by their universality [68–73]. The concept of resonance means an increase in the response of an oscillating system to a periodic external action when its frequency approaches one of the natural frequencies of the system. All oscillating systems are able to resonate and may have a very different nature. In the substances, such systems may be electrons, electron shells of atoms, magnetic and electric moments of atoms and molecules, impurity centers in crystals and individual crystals, as well as their groups. However, in all cases, the general picture of the resonance is preserved, that is, near the resonance, the amplitude of the oscillations and the energy, transferred to the oscillatory system from the outside, increase. This increase stops when the energy losses compensate for its gain.

Resonance methods can be attributed to the most sensitive and accurate methods for studying the substances (and, consequently, affecting them) [68–70]. The resonance methods allow obtaining various types of information about the chemical composition, structure, symmetry, and internal interactions between the structural units of a substance. A substance, depending on its internal structure, has its own unique set of natural oscillation frequencies (frequency or energy spectrum). Natural frequencies f_k can be in a wide range from 10^2 to 10^{22} Hz. Such sets of frequencies is a kind of visiting card of the substance. Electromagnetic radiation is a common and effective type of periodic external action. The frequencies of electromagnetic waves are in the ranges of $10^2 \dots 10^8$ Hz (radio waves), $10^9, \dots, 10^{11}$ Hz (radio microwaves), $10^{13}, \dots, 10^{14}$ Hz (infrared light), 10^{15} Hz (visible light), $10^{15}, \dots, 10^{16}$ Hz (ultraviolet light), $10^{17}, \dots, 10^{20}$ Hz (X-ray emission), and $10^{20}, \dots, 10^{22}$ Hz (γ-radiation).

Many researchers have proved [74–77] that the energy processes in plasma are of broadband nature with the frequency spectrum of $10^4, \dots, 10^{20}$ Hz, as a result of which it is almost impossible to cover it with one sensitive element. To meet this challenge, it is proposed to use a set of sensitive elements with some overlap of the probable frequency range of the process. Each of the elements may be considered as a linear filter tuned to the specific frequency subrange. In mathematical terms, this can be described as follows:

$$\Omega_n \cong \bigcup_{i=1}^{M} \Omega_i \tag{1}$$

where Ω_n is the frequency spectrum of the process; Ω_i is a potential fragment of the process spectrum contained in the signal at the output of the ith sensitive element (of the coil).

$$y(t) = \sum_{r=1}^{\infty} a_r e^{\delta_r t} + \sum_{k=1}^{\infty} \left[A_k e^{(\delta_k + i2\pi f_k)t} + A_k^* e^{(\delta_k + i2\pi f_k)t} \right] \tag{2}$$

where a_r, is the amplitude; f_k is the frequency; t is the time (sec); δ_k is the attenuation factor.

A computational experiment was carried out in order to check the efficiency of the spectral analysis process. A model signal, which is the sum of cosine waves and noise, was used as a test signal:

$$y(t) = \sum_{k=1}^{8} A_k \cos(2\pi f_k) + \varepsilon(t) \tag{3}$$

The values of natural frequencies and their corresponding amplitudes, which are the parameters of the model, are contained in Table 1.

Table 1. Values of natural frequencies and their corresponding amplitudes, which are the parameters of the model.

A_k	0.7	1.0	1.5	0.5	0.6	1.0	1.0	1.0
f_k, Hz	10,000	12,000	15,000	20,000	25,000	50,000	75,000	100,000

To carry out experimental studies, a broadband spectrum analyzer of electromagnetic radiation from glow discharge plasma with time-frequency signal separation was developed. The spectrum analyzer (Figure 3) consists of a multichannel measuring unit, an analog-to-digital converter (ADC), and a software module, carrying out the Prony–Fourier (PF) spectral analysis [78–84].

The measuring unit is made in the form of a set of 18 inductive sensors, which are coils of copper wire, evenly spaced along the periphery of a disk made of dielectric material. During the investigation, the measuring unit was installed at the viewing window of the vacuum chamber of the PVD unit [85–87]. To reduce the influence of external noise on the sensor, a screen in the form of a Faraday cup with grounding is installed on the unit body [84–86].

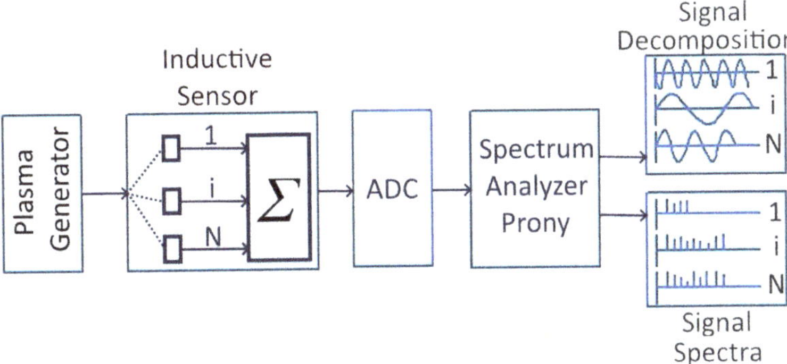

Figure 3. Structural diagram of the broadband Prony spectrum analyzer of electromagnetic radiation from glow discharge plasma.

The measuring unit is a distributed system with multiple inputs which number is equal to the number of coils and one output. To identify the transfer function from each input to the output, a special procedure was developed, embedded in the software module. Each coil of the measuring unit possesses a set of natural frequencies and perceives (resonates) a certain spectrum of electromagnetic radiation of the glow discharge plasma, thus acting as a physical bandpass filter. Further, the signals coming from all the sensor coils are summed up and fed through one channel to the ADC and then to the Prony spectrum analyzer, which performs the frequency-time separation of the signal. At the output of the Prony spectrum analyzer, signal frequency estimates, an analytical Fourier spectrum, an analytical energy spectrum of a signal, and its analytical time decomposition over specified frequency ranges can be obtained.

3. Results and Discussion

3.1. Analysis of the Energy Distribution of Ions in Two-Component Plasma Using a Multigrid Probe

The retarding characteristics of ions were obtained at the pressure in the vacuum chamber of ~6 × 10^{-3} Pa and titanium cathode currents reaching 180 A and 300 A. The magnitude of the magnetic field near the cathode surface did not exceed 8 × 10^{-4} T. The probe was installed at a distance of 150 mm from the cathode cutoff. Figure 4 exhibits the energy distribution of titanium ions.

At arc currents of 150, . . . , 180 A, the average energy of titanium ions was ~35, . . . , 40 eV, and at current of 300 A, the ion energy decreased to ~30 eV. The average ion energy decreases, as with an increase in the arc discharge current, the number of high-energy ions in the plasma flow decreases, and the energy of titanium ions shifts to lower values. This occurs due to an increase in the evaporation rate and plasma concentration and, consequently, the degree of plasma flow randomness in the near-cathode region due to an increase in the number of particle collisions.

In case of sublimating materials with high saturated vapor pressure (C, Cr), an evaporable anode was made in the form of a rod. An anode in the form of a crucible made of refractory materials was used in case of substances with low vapor pressure. In the describe case, the anode was a rod 6 mm in diameter made of spectrally pure carbon. To isolate graphite plasma ions from the two-component plasma, a screen was installed in front of the probe diaphragm, at a distance of 50 mm from the cathode surface, under a floating potential, which prevented titanium ions generated by the cathode spot from entering the analyzer. The probe characteristics were taken at two values of the electron current to the evaporable anode: 70 A and 120 A. In both cases, the cathode current was 150 A. The burning voltage of the arc discharge with the evaporable graphite anode was 40–43 V when the anode current varied from 50 to 150 A, respectively. The energy distribution curves of carbon ions generated by the evaporable anode are exhibited in Figure 3.

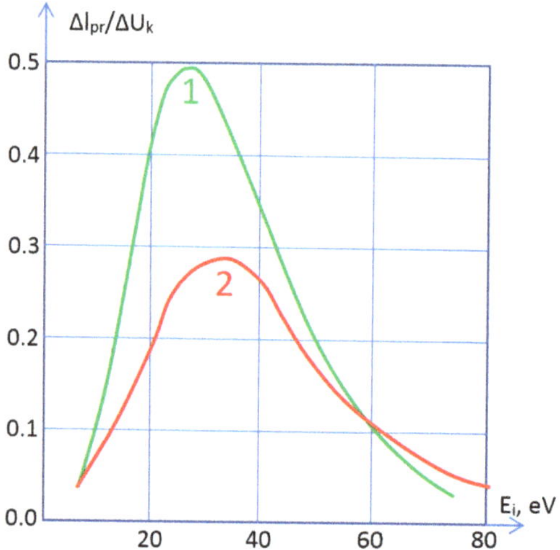

Figure 4. Energy distribution of titanium ions. 1—discharge current of 180 A, 2—discharge current of 300 A.

The energy spectrum of the ions of the cathode material almost coincides with the spectrum of ions during the operation of the arc evaporator in the normal mode without a sacrificial anode.

The estimative analysis of the spectrum of carbon ions generated by the anode (Figure 5) finds that with an increase in the anode current, the average energy of carbon ions decreases noticeably. Therefore, when the anode current is 70 A, the average ion energy reaches ~15, ... , 18 eV, and as the current increases to 120 A, it decreases to ~9, ... , 10 eV.

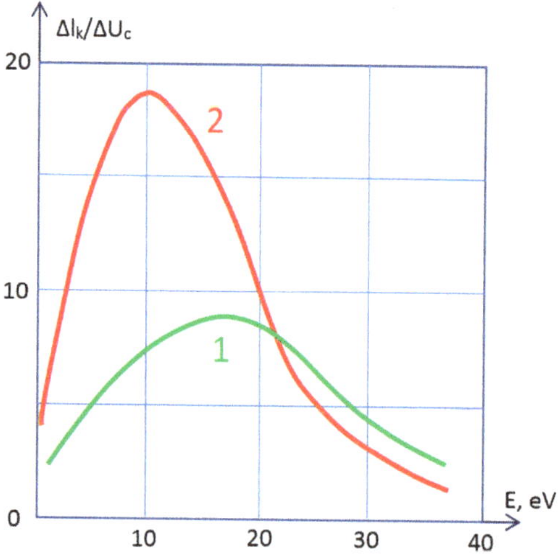

Figure 5. Energy distribution of carbon ions. 1—anode current of 70 A, 2—anode current of 120 A.

The effect of the decrease in the energy of graphite ions may occur, as with an increase in the anode current, the temperature of the anode surface, the rate of evaporation of graphite, and, consequently, the density of plasma and vapor increase. The growth of the plasma density and, importantly, vapor density leads to a decrease in the degree of plasma ionization and an increase in the randomness of particle motion in the plasma flow due to an increase in the number of collisions of ions with particles having thermal velocities. This is confirmed by the fact that at low anode currents, not exceeding 70 A, the average energy of graphite ions correlates well with the anode potential.

3.2. Plotting Curves of the Distribution of Argon Ions Using a Source of Gas Ions with Closed Electron Current

Gas ion sources with a closed electron current are an effective tool for the implementation of ionic surface treatment processes [88–92]. Technological ion sources are used for cleaning and activating the surface of products before applying thin-film coatings, ion assistance in the processes of coating deposition, ion-beam etching, and shaping.

There is a high-voltage mode of operation of ion sources with the discharge voltage from 600 to 3000 V and higher and a low-voltage mode with the voltage usually up to 300–500 V. The first mode is usually applied in the processes of depositing thin layers by target sputtering and for cleaning and etching surfaces. The second mode is used in the processes of coating deposition with etching and in the processes of ion cleaning. The energy of accelerated ions generated by a source is determined by the operating voltage of the discharge. Figure 6 depicts the energy distribution curves of argon ions depending on the discharge voltage for the low-voltage mode of operation of the ion source, obtained using a multigrid probe [93–95].

It can be noted from the curves that the average ion energy decreases with a decline in the discharge current, and the source efficiency (the ratio of the average ion energy W to the discharge voltage U) remains approximately at the same level in the operating voltage range of the source ions.

For carrying out the processes of coating deposition with etching, an energy value from 100 to 200 eV is sufficient. A built-in multigrid probe analyzer of ion energy may be used to adjust the technological processes of ion processing. For some ion treatment processes (for example, surface cleaning), the above ion energy is not enough. To increase the energy of the bombarding ions and improve the efficiency of etching and cleaning, a negative potential (bias potential) can be applied to the substrate.

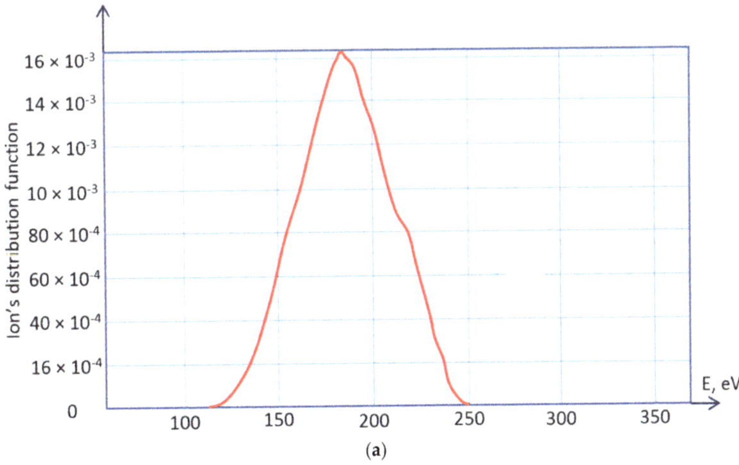

Figure 6. *Cont.*

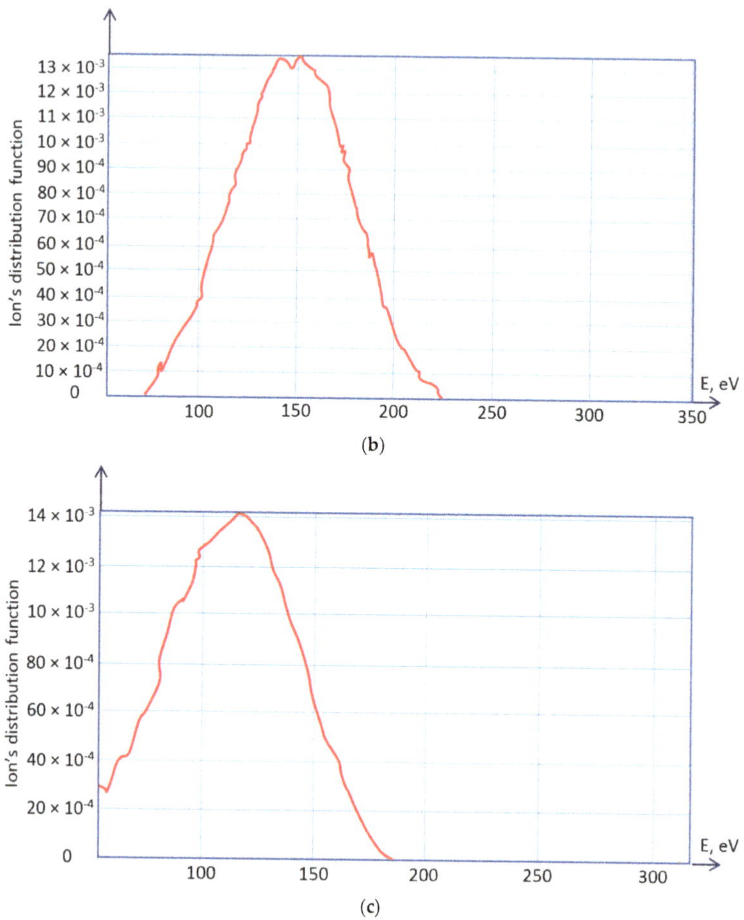

(b)

(c)

Figure 6. Distribution curves of argon ions depending on the energy of the low-voltage ion source. At discharge voltage of: (**a**) −250 V, (**b**) −230 V, (**c**) −200 V.

3.3. Spectrum Analyzer of the Signal of Electromagnetic Radiation of Glow Discharge Plasma Based on Multichannel Inductive Sensor

Figure 7 exhibits a graphical relationship between the oscillatory process synthesized in accordance with relationship (3) and Table 1.

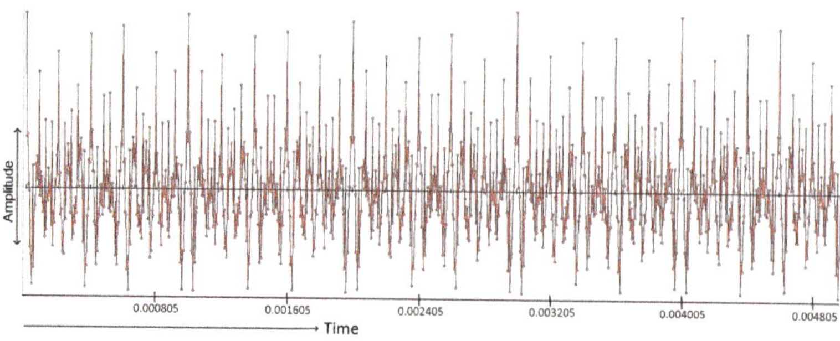

Figure 7. Relationship between the synthesized test signal of the oscillatory process and the time.

Since the linear bandpass filters in the model signal are tuned to the following frequency subranges: (5, … , 18), (18, … , 40), (40, … , 120) kHz, then its analytical decomposition is tuned to the corresponding subranges. The results of the analytical modal decomposition are presented in Figure 8. The left side of the figure shows graphs displaying all modes from the corresponding specified frequency, and the right side of the spectrum is A^2, squared Prony amplitudes [85,87].

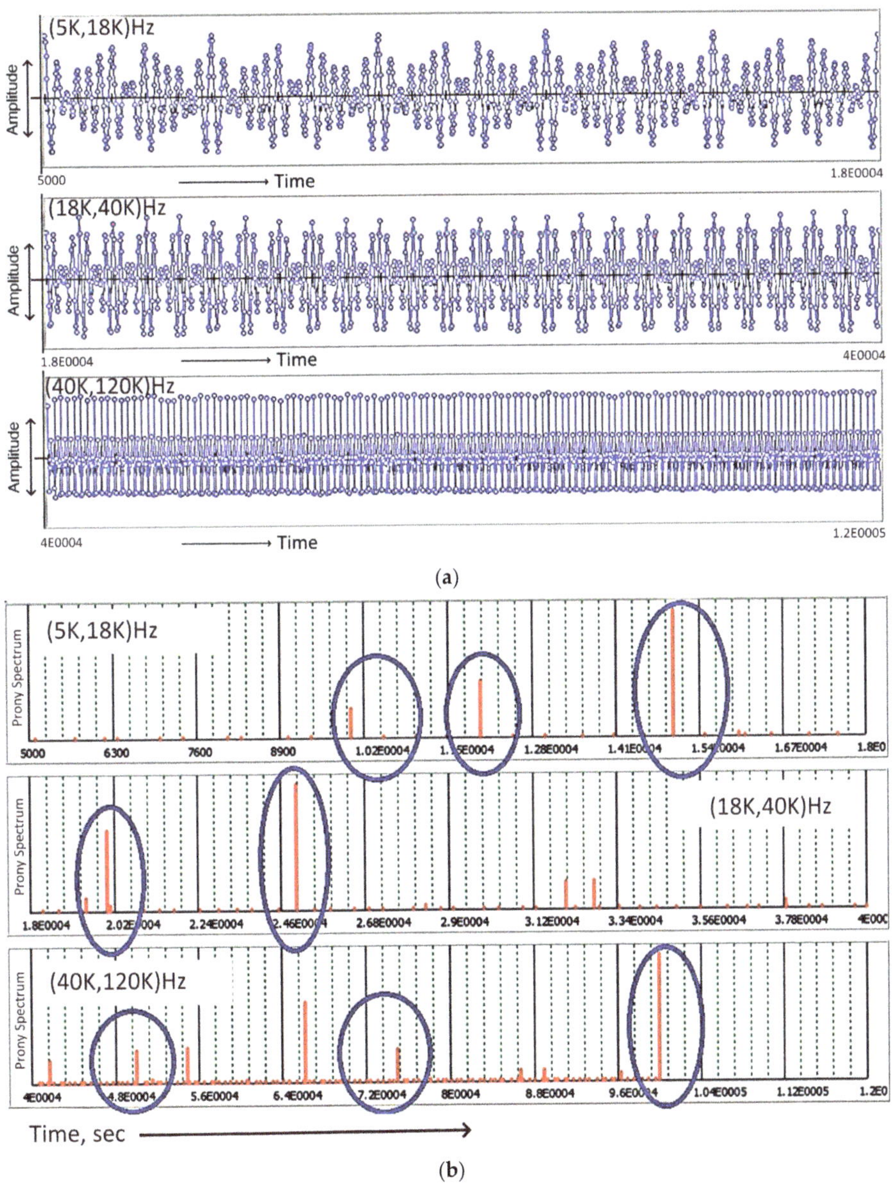

Figure 8. Decomposition of the signal by the frequency subbands of the sensors. (**a**) time signal, (**b**) its A^2—the Prony spectrum. Model signal modes are circled with blue ovals [79,80].

Modes of the model signal are selected on the A^2 Prony spectrum (circled with blue ovals), and the remaining methods correspond to the random component or are mathematical models, that is, they are not actually present in the signal, but appear as a result of signal approximation.

Figure 9 shows a stabilization diagram for the frequency estimates of the model signal. The stabilization diagram in this case of estimates of natural frequencies shows the stability of the appearance of a mode with the corresponding natural frequency for a given order (maximum number of modes) of the approximation model (1). Physical modes i.e., actually present in the signal are characterized by vertical lines already after the number of modes equal to 15. The stabilization diagram clearly showed modes with model frequencies, namely: 10, 12, 15, 20, 25, 50, 75, and 100 kHz.

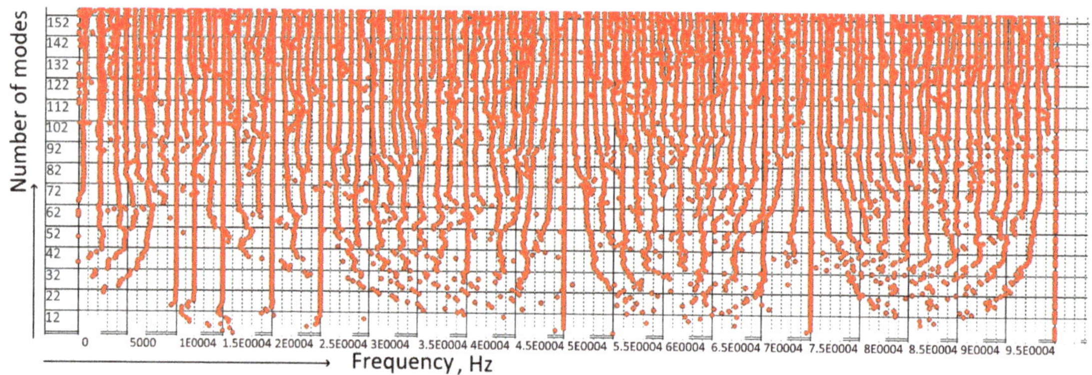

Figure 9. Stabilization diagram for frequency estimates of the model process. The horizontal axis exhibits frequency, and the vertical axis—the number of modes (model complexity), i.e., the number of components in the approximation model, which may be much larger than in the original test model.

Figure 10 shows the analytical Fourier spectrum of the total model time signal, the spectral lines of which correspond with high accuracy to the original natural frequencies of the model. The analytical Fourier spectrum (Prony–Fourier spectrum) is the result of the analytical integration of the signal model (1) obtained as a result of the Prony approximation procedure [86,87]. The analytical Prony–Fourier spectrum has an increased resolution compared to conventional discrete Fourier transforms.

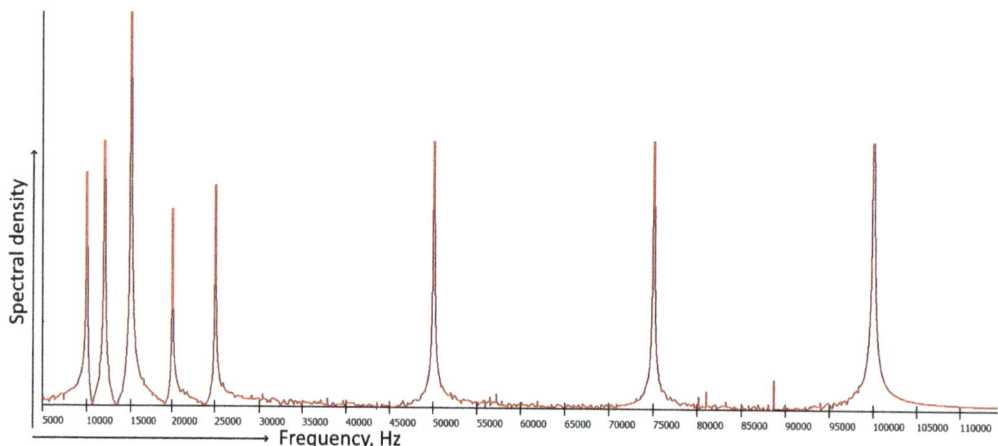

Figure 10. Spectrum of the total model signal.

Figure 11 shows a graphical dependence of the recorded changes in the electromagnetic radiation of the glow discharge plasma, obtained at the output of the ADC during the processing in the PVD installation, and in Figure 12a,b, respectively, the analytical Fourier spectrum (spectral density) and the analytical energy spectrum of the signal after processing in the Prony spectrum analyzer.

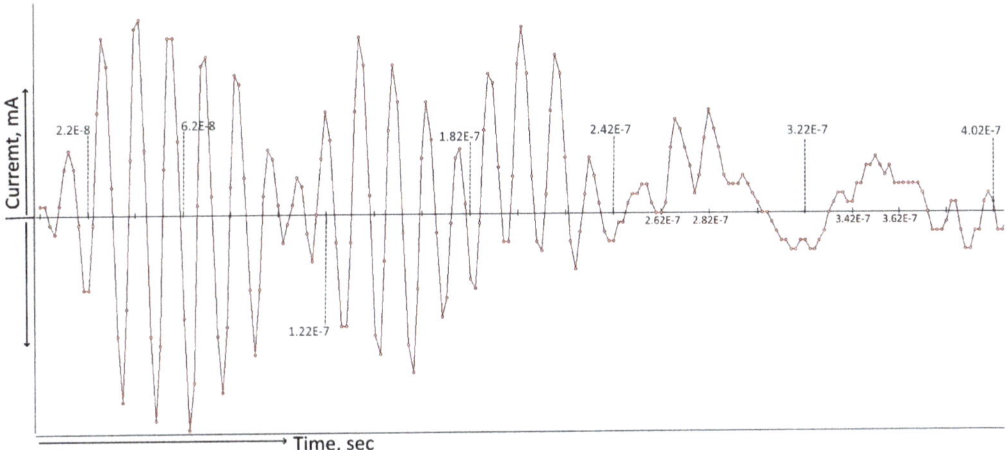

Figure 11. Graphical relationship between the changes in the electromagnetic radiation from the glow discharge plasma obtained at the output of the ADC.

An analysis of the spectra presented in Figure 12 shows that their structure is saturated and contains several clearly distinguishable frequencies, i.e., at a qualitative level, it is quite consistent with the ongoing physical processes. For final conclusions about the effectiveness of the proposed scheme of the spectrum analyzer, additional experimental studies are planned.

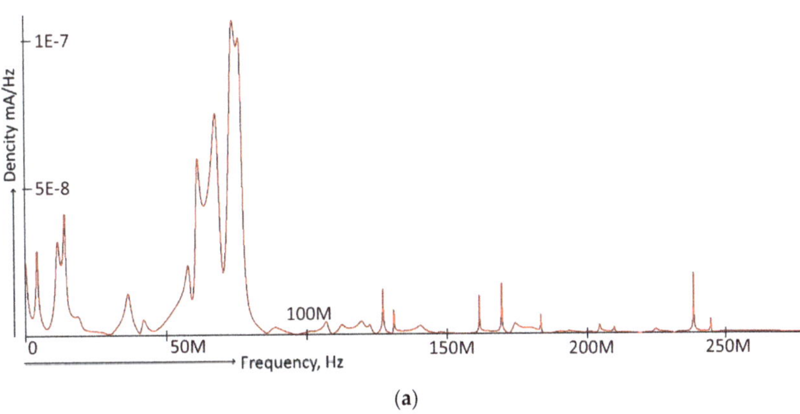

(a)

Figure 12. *Cont.*

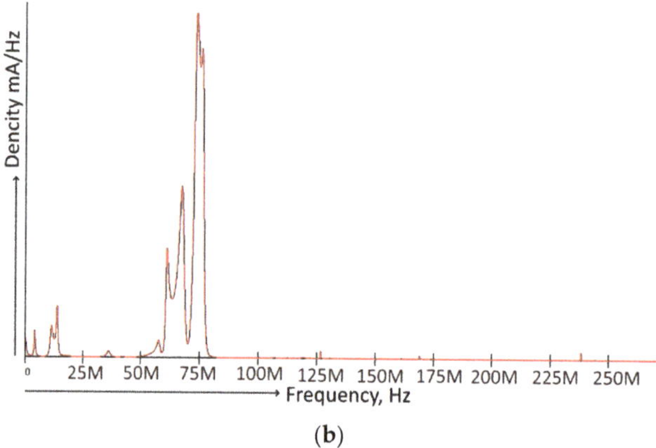

(**b**)

Figure 12. (**a**) Analytical Fourier spectrum of the signal at the output of the Prony spectrum analyzer (relationship between the amplitude spectrum of the signal (density mA/Hz) and frequency, Hz) and (**b**) its energy spectrum.

4. Conclusions

The article describes the main advantages and disadvantages of the methods in relation to the continuity of vacuum technological processes directed to modification of surfaces of solid bodies.

1. The advantage of the probe methods for plasma diagnostics is the information content and ease of their adaptation to technological equipment.
2. The possibility of determining the energy distribution of ions in the two-component plasma has been described. The energy distribution curves of ions in the two-component plasma generated by a vacuum arc evaporator with sacrificial cathode and anode have been plotted.
3. Energy distribution curves of argon ions in the low-voltage mode of operation of ion sources with a closed electron current have been analyzed. The average ion energy decreases with a decline in the discharge current, and the source efficiency (the ratio of the average ion energy W to the discharge voltage U) remains approximately at the same level of $W/U \approx 0.68, \ldots, 0.71$ in the operating voltage range of the source.
4. A scheme of a spectrum analyzer is proposed, which can be used both for monitoring and for controlling the processing process, including in automated PVD installations.
5. The proposed magnetic induction method for monitoring the glow discharge plasma parameters can be used to determine the moment the PVD facility enters the operating mode and completes the process steps, as well as to perform operational intervention in the process, taking into account changes in electromagnetic pulses.
6. The proposed electromagnetic radiation signal spectrum analyzer based on a multichannel inductive sensor makes it possible to analyze broadband plasma electromagnetic radiation signals with the possibility of obtaining not only spectra at the sensor output, but also deconvolution of the glow discharge plasma electromagnetic radiation signal spectrum.

Thus, it can be argued that the proposed scheme of the spectrum analyzer is fundamentally operable, however, its verification requires specially planned experiments.

Author Contributions: Conceptualization, S.D. and A.V.; methodology, V.Z.; investigation, S.D., C.S. and V.Z.; resources, S.G.; data curation, A.V.; writing—original draft preparation, A.V., S.D. and V.Z.; writing—review and editing, A.V.; project administration, S.G.; funding acquisition, S.G. All authors have read and agreed to the published version of the manuscript.

Funding: This work was supported financially by the Ministry of Science and Higher Education of the Russian Federation (project No FSFS-2021-0003).

Institutional Review Board Statement: Not applicable.

Informed Consent Statement: Not applicable.

Data Availability Statement: Not applicable.

Conflicts of Interest: The authors declare no conflict of interest.

References

1. Helmersson, U.; Lattemann, M.; Bohlmark, J.; Ehiasarian, A.P.; Gudmundsson, J.T. Ionized physical vapor deposition (IPVD): A review of technology and applications. *Thin Solid Film.* **2006**, *513*, 1–24. [CrossRef]
2. Mehran, Q.M.; Fazal, M.A.; Bushroa, A.R.; Rubaiee, S. A Critical Review on Physical Vapor Deposition Coatings Applied on Different Engine Components. *Crit. Rev. Solid State Mater. Sci.* **2018**, *43*, 158–175. [CrossRef]
3. Baptista, A.; Silva, F.; Porteiro, J.; Míguez, J.; Pinto, G. Sputtering physical vapour deposition (PVD) coatings: A critical review on process improvement andmarket trend demands. *Coatings* **2018**, *8*, 402. [CrossRef]
4. Choy, K.L. Chemical vapour deposition of coatings. *Prog. Mater. Sci.* **2003**, *48*, 57–170. [CrossRef]
5. Grigoriev, S.; Vereschaka, A.; Milovich, F.; Tabakov, V.; Sitnikov, N.; Andreev, N.; Sviridova, T.; Bublikov, J. Investigation of multicomponent nanolayer coatings based on nitrides of Cr, Mo, Zr, Nb, and Al. *Surf. Coat. Technol.* **2020**, *401*, 126258. [CrossRef]
6. Lewis, A.S.; Brown, S.D.; Tutwiler, R.L. Unified control approach for an electron beam physical vapour deposition process. *Int J. Robot. Autom.* **2000**, *15*, 145–151.
7. Metel, A.S.; Grigoriev, S.N.; Melnik, Y.A.; Bolbukov, V.P. Broad beam sources of fast molecules with segmented cold cathodes and emissive grids. *Instrum. Exp. Tech.* **2012**, *55*, 122–130. [CrossRef]
8. Metel, A.; Grigoriev, S.; Melnik, Y.; Panin, V.; Prudnikov, V. Cutting Tools Nitriding in Plasma Produced by a Fast Neutral Molecule Beam. *Jpn. J. Appl. Phys.* **2011**, *50*, 08JG04. [CrossRef]
9. Volosova, M.A.; Grigor'ev, S.N.; Kuzin, V.V. Effect of Titanium Nitride Coating on Stress Structural Inhomogeneity in Oxide-Carbide Ceramic. Part 4. Action of Heat Flow. *Refract. Ind. Ceram.* **2015**, *56*, 91–96. [CrossRef]
10. Vereschaka, A.; Tabakov, V.; Grigoriev, S.; Sitnikov, N.; Milovich, F.; Andreev, N.; Sotova, C.; Kutina, N. Investigation of the influence of the thickness of nanolayers in wear-resistant layers of Ti-TiN-(Ti,Cr,Al)N coating on destruction in the cutting and wear of carbide cutting tools. *Surf. Coat. Technol.* **2020**, *385*, 125402. [CrossRef]
11. Colombo, D.A.; Echeverría, M.D.; Moncada, O.J.; Massone, J.M. PVD TiN and CrN coated austempered ductile iron: Analysis of processing parameters influence on coating characteristics and substrate microstructure. *ISIJ Int.* **2012**, *52*, 121–126. [CrossRef]
12. Grigoriev, S.; Vereschaka, A.; Zelenkov, V.; Sitnikov, N.; Bublikov, J.; Milovich, F.; Andreev, N.; Mustafaev, E. Specific features of the structure and properties of arc-PVD coatings depending on the spatial arrangement of the sample in the chamber. *Vacuum* **2022**, *200*, 111047. [CrossRef]
13. Grigoriev, S.; Vereschaka, A.; Zelenkov, V.; Sitnikov, N.; Bublikov, J.; Milovich, F.; Andreev, N.; Sotova, C. Investigation of the influence of the features of the deposition process on the structural features of microparticles in PVD coatings. *Vacuum* **2022**, *202*, 111144. [CrossRef]
14. Khoroshikh, V.M.; Leonov, S.A.; Belous, V.A. Influence of substrate geometry on ion-plasma coating deposition process. In Proceedings of the International Symposium on Discharges and Electrical Insulation in Vacuum, ISDEIV 2, Bucharest, Romania, 15–19 September 2008; Volume 4676863, pp. 591–594.
15. Grigoriev, S.N.; Melnik, Y.A.; Metel, A.S.; Panin, V.V.; Prudnikov, V.V. Broad beam source of fast atoms produced as a result of charge exchange collisions of ions accelerated between two plasmas. *Instrum. Exp. Tech.* **2009**, *52*, 602–608. [CrossRef]
16. Grigoriev, S.N.; Teleshevskii, V.I. Measurement problems in technological shaping processes. *Meas. Tech.* **2011**, *54*, 744–749. [CrossRef]
17. Grigoriev, S.; Melnik, Y.; Metel, A. Broad fast neutral molecule beam sources for industrial scale beam-assisted deposition. *Surf. Coat. Technol.* **2002**, *156*, 44–49. [CrossRef]
18. Bagotsky, V.S. *Fundamentals of Electrochemistry*; Willey Interscience: Hoboken, NJ, USA, 2006.
19. Popov, T.K.; Dimitrova, M.; Ivanova, P.; Kovačič, J.; Gyergyek, T.; Dejarnac, R.; Stöckel, J.; Pedrosa, M.A.; López-Bruna, D.; Hidalgo, C. Advances in Langmuir probe diagnostics of the plasma potential and electron-energy distribution function in magnetized plasma. *Plasma Sources Sci. Technol.* **2016**, *25*, 033001. [CrossRef]
20. Bilik, N.; Anthony, R.; Merritt, B.A.; Aydil, E.S.; Kortshagen, U.R. Langmuir probe measurements of electron energy probability functions in dusty plasmas. *J. Phys. D* **2015**, *48*, 105204. [CrossRef]
21. Tanışlı, M.; Şahin, N.; Demir, S. An investigation on optical properties of capacitive coupled radio-frequency mixture plasma with Langmuir probe. *Optik* **2017**, *142*, 153–162. [CrossRef]
22. Seo, D.-C.; Chung, T.-H. Observation of the transition of operating regions in a low-pressure inductively coupled oxygen plasma by Langmuir probe measurement and optical emission spectroscopy. *J. Phys. D* **2001**, *34*, 2854–2861. [CrossRef]
23. Yuan, Y.; Bansky, J.; Engemann, J.; Brockhaus, A. Ion energy determination for r.f. plasma and ion beam using a multigrid retarding field analyser. *Surf. Coat. Technol.* **1995**, *74–75 Pt 1*, 534–538. [CrossRef]

24. Mišina, M.; Musil, J. Plasma diagnostics of low pressure microwave-enhanced d.c. sputtering discharge. *Surf. Coat. Technol.* **1995**, *74–75 Pt 1*, 450–454. [CrossRef]

25. Kitami, H.; Miyashita, M.; Sakemi, T.; Aoki, Y.; Kato, T. Quantitative analysis of ionization rates of depositing particles in reactive plasma deposition using mass-energy analyzer and Langmuir probe. *Jpn. J. Appl. Phys.* **2015**, *54* (Suppl. S1), 01AB05. [CrossRef]

26. Powers, E.J.; Caldwell, G.S. Fast-fourier-transform spectral-analysis techniques as a plasma fluctuation diagnostic tool. *IEEE Trans. Plasma Sci. IEEE Nucl. Plasma Sci. Soc.* **1975**, *2*, 261–272.

27. Krebbers, R.; Liu, N.; Jahromi, K.E.; Nematollahi, M.; Harren, F.J.M.; Cristescu, S.M.; Khodabakhsh, A. Fourier transform spectroscopy based on a mid-infrared supercontinuum source for plasma study. *Opt. InfoBase Conf. Pap.* **2022**, MF3C.4.

28. Xie, W.; Jiang, W.; Gao, Y. Spectral analysis of Ar plasma-arc under different experimental parameters. *Optik* **2013**, *124*, 420–424. [CrossRef]

29. Grigoriev, S.N.; Kozochkin, M.P.; Sabirov, F.S.; Kutin, A.A. Diagnostic Systems as Basis for Technological Improvement. *Proc. CIRP* **2012**, *1*, 599–604. [CrossRef]

30. Grigoriev, S.; Metel, A. Plasma- and beam- assisted deposition methods. In *Nanostructured Thin Films and Nanodispersion Strengthened Coatings*; Nato Science Series, Series II: Mathematics, Physics and Chemistry; Springer: Dordrecht, The Netherland, 2004; Volume 155, pp. 147–154.

31. Metel, A.S.; Grigoriev, S.N.; Melnik, Y.A.; Panin, V.V. Filling the vacuum chamber of a technological system with homogeneous plasma using a stationary glow discharge. *Plasma Phys. Rep.* **2009**, *35*, 1058–1067. [CrossRef]

32. Abramov, A.S.; Vinogradov, A.Y.; Kosarev, A.I.; Smirnov, A.S.; Orlov, K.E.; Shutov, M.V. Study of ion bombardment of amorphous silicon films during plasma-chemical deposition in a high-frequency discharge. *JTF* **1998**, *68*, 52.

33. Hamers, E.A.G.; Fontcuberta, I.; Morral, A.; Niikura, C.; Brenot, R.; Roca, I.; Cabarrocas, P. Contribution of ions to the growth of amorphous, polymorphous, and microcrystalline silicon thin films. *J. Appl. Phys.* **2000**, *88*, 3674–3688. [CrossRef]

34. Pelhos, K.; Donnelly, V.M.; Kornblit, A.; Green, M.L.; van Dover, R.B.; Manchanda, L.; Hu, Y.; Morris, M.; Bower, E. Etching of high-k dielectric $Zr_{1-x}Al_xO_y$ films in chlorine-containing plasmas. *J. Vac. Sci. Technol. A* **2001**, *19*, 1361–1366. [CrossRef]

35. Sungauer, E.; Pargon, E.; Mellhaoui, X.; Ramos, R.; Cunge, G.; Vallier, L.; Joubert, O.; Lill, T. Etching mechanisms of HfO_2, SiO_2, and poly-Si substrates in BCl3 plasmas. *J. Vac. Sci. Technol. B* **2007**, *25*, 1640–1646. [CrossRef]

36. Pearton, S.J.; Abernathy, C.R.; Ren, F.; Lothian, J.R.; Wisk, P.W.; Katz, A. Dry and wet Etching Characteristics of InN, AIN, and GaN Deposited by Electron Cyclotron Resonance Metalorganic Molecular Beam Epitaxy. *J. Vac. Sci. Technol. A* **1993**, *11*, 1772. [CrossRef]

37. Metel, A.; Bolbukov, V.; Volosova, M.; Grigoriev, S.; Melnik, Y. Equipment for deposition of thin metallic films bombarded by fast argon atoms. *Instrum. Exp. Tech.* **2014**, *57*, 345–351. [CrossRef]

38. Hartfuss, H.J.; Geist, T.; Hirsch, M. Heterodyne methods in millimetre wave plasma diagnostics with applications to ECE, interferometry and reflectometry. *Plasma Phys. Control Fusion* **1997**, *39*, 1693–1769. [CrossRef]

39. Thiry, D.; Konstantinidis, S.; Cornil, J.; Snyders, R. Plasma diagnostics for the low-pressure plasma polymerization process: A critical review. *Thin Solid Film.* **2016**, *606*, 19–44. [CrossRef]

40. Williams, C.B.; Amais, R.S.; Fontoura, B.M.; Jones, B.T.; Nóbrega, J.A.; Donati, G.L. Recent developments in microwave-induced plasma optical emission spectrometry and applications of a commercial Hammer-cavity instrument. *TrAC Trends Anal. Chem.* **2019**, *116*, 151–157. [CrossRef]

41. Schwabedissen, A.; Benck, E.C.; Roberts, J.R. Comparison of electron density measurements in planar inductively coupled plasmas by means of the plasma oscillation method and Langmuir probes. *Plasma Sources Sci. Technol.* **1998**, *7*, 119–129. [CrossRef]

42. Welzel, S.; Hempel, F.; Hübner, M.; Lang, N.; Davies, P.B.; Röpcke, J. Quantum cascade laser absorption spectroscopy as a plasma diagnostic tool: An overview. *Sensors* **2010**, *10*, 6861–6900. [CrossRef]

43. Benedikt, J.; Hecimovic, A.; Ellerweg, D.; Von Keudell, A. Quadrupole mass spectrometry of reactive plasmas. *J. Phys. D* **2012**, *45*, 403001. [CrossRef]

44. Liu, C.; Ying, P. Theoretical study of novel B-C-O compounds with non-diamond isoelectronic. *Chin. Phys. B* **2022**, *31*, 026201. [CrossRef]

45. Cheng, M.; Yan, X.; Cui, Y.; Han, M.; Wang, X.; Wang, J.; Zhang, R. An eco-friendly film of pH-responsive indicators for smart packaging. *J. Food Eng.* **2022**, *321*, 110943. [CrossRef]

46. Wu, Y.; Zhao, Y.; Han, X.; Jiang, G.; Shi, J.; Liu, P.; Khan, M.Z.; Huhtinen, H.; Zhu, J.; Jin, Z.; et al. Ultra-fast growth of cuprate superconducting films: Dual-phase liquid assisted epitaxy and strong flux pinning. *Mater. Today Phys.* **2021**, *18*, 100400. [CrossRef]

47. Bulgakov, B.V. *Fluctuations*; Publishing House of Technical and Theoretical Literature: Moscow, Russia, 1954; p. 891.

48. Hauer, J.F.; Scharf, L.L. Initial results in prony analysis of power system response signals. *IEEE Trans. Power Syst.* **1990**, *5*, 80–89. [CrossRef]

49. Hauer, J.F. Application of prony analysis to the determination of modal content and equivalent models for measured power system response. *IEEE Trans. Power Syst.* **1991**, *6*, 1062–1068. [CrossRef]

50. Lobos, T.; Rezmer, J. Real-time determination of power system frequency. *IEEE Trans. Instrum. Meas.* **1997**, *46*, 877–881. [CrossRef]

51. Lebedev, Y.A. *Electrical Probes in Reduced Pressure Plasma*; Institute of Petrochemical Synthesis, A.V. Topchiev RAS: Moscow, Russia, 2003.

52. Raiser, Y.P. *Physics of the Gas Discharge*; Intellect Publishing House: Dolgoprudny, Russia, 2009.

53. Demidov, V.I.; Kolokolov, N.B.; Kudryavtsev, A.A. *Probe Methods for Studying Low-Temperature Plasma*; Energoatomizdat: Moscow, Russia, 1996.

54. Mott-Smith, H.; Langmuir, I. The theory of collectors in gaseous discharges. *Phys. Rev.* **1926**, *28*, 727–763. [CrossRef]

55. Godyak, V.A.; Piejak, R.B.; Alexandrovich, B.M. Measurement of electron energy distribution in low-pressure RF discharges. *Plasma Sources Sci. Technol.* **1992**, *1*, 36–58. [CrossRef]

56. Godyak, V.A.; Demidov, V.I. Probe measurements of electron-energy distributions in plasmas: What can we measure and how can we achieve reliable results? *J. Phys. D* **2011**, *44*, 233001. [CrossRef]

57. Demidov, V.I.; Ratynskaia, S.V.; Rypdal, K. Electric probes for plasmas: The link between theory and instrument. *Rev. Sci. Instrum.* **2002**, *73*, 3409. [CrossRef]

58. Irfan, Z.; Bashir, S.; Butt, S.H.; Hayat, A.; Ayub, R.; Mahmood, K.; Akram, M.; Batool, A. Evaluation of electron temperature and electron density of laser-Ablated Zr plasma by Langmuir probe characterization and its correlation with surface modifications. *Laser Part. Beams* **2020**, *38*, 84–93. [CrossRef]

59. Li, P.; Hershkowitz, N.; Wackerbarth, E.; Severn, G. Experimental studies of the difference between plasma potentials measured by Langmuir probes and emissive probes in presheaths. *Plasma Sources Sci. Technol.* **2020**, *29*, 025015. [CrossRef]

60. Kimura, H.; Odajima, K.; Sugie, T.; Maeda, H. Application of multigrid energy analyzer to the scrape-off layer plasma in DIVA. *Jpn. J. Appl. Phys.* **1979**, *18*, 2275–2281. [CrossRef]

61. Ingram, S.G.; Braithwaite, N.S.J. Ion and electron energy analysis at a surface in an RF discharge. *J. Phys. D* **1988**, *21*, 1496. [CrossRef]

62. Bohm, C.; Perrin, J. Retarding-field analyzer for measurements of ion energy distributions and secondary electron emission coefficients in low-pressure radio frequency discharges. *Rev. Sci. Instrum.* **1993**, *64*, 31. [CrossRef]

63. Tonks, L.; Mott-Smith, H.M., Jr.; Langmuir, I. Flow of ions through a small orifice in a charged plate. *Phys. Rev.* **1926**, *28*, 104. [CrossRef]

64. Cubric, D.; Kholine, N.; Konishi, I. Electron optics of spheroid charged particle energy analyzers. *Nucl. Instrum. Methods Phys. Res.* **2011**, *645*, 234–240. [CrossRef]

65. Walker, C.G.H.; Zha, X.; El Gomati, M.M. A parallel acquisition charged particle energy analyser using a magnetic field. *Surf. Interface Anal.* **2017**, *49*, 34–46. [CrossRef]

66. Veselovzorov, A.N.; Dlugach, E.D.; Pogorelov, A.A.; Svirsky, E.B.; Smirnov, V.A. Study of the formation of ion flows in variable electric fields of a stationary plasma engine. *Tech. Phys.* **2013**, *83*, 37–42.

67. Dorodnov, A.M.; Zelenkov, V.V.; Kuznetsov, A.N. Stationary vacuum arc with two evaporated electrodes. *Ther. Phys. High Temp.* **1997**, *35*, 983–1008.

68. Mohd Bahar, A.A.; Zakaria, Z.; Md Isa, A.A.; Ruslan, E.; Alahnomi, R.A. A review of characterization techniques for materials' properties measurement using microwave resonant sensor. *J. Telecommun. Electron. Comput. Eng.* **2015**, *7*, 1–6.

69. Itoh, T.; Yamamoto, Y.S. Between plasmonics and surface-enhanced resonant Raman spectroscopy: Toward single-molecule strong coupling at a hotspot. *Nanoscale* **2021**, *13*, 1566–1580. [CrossRef] [PubMed]

70. Hafez, M.G.; Iqbal, S.A.; Asaduzzaman, N.; Hammouch, Z. Dynamical behaviors and oblique resonant nonlinear waves with dual-power law nonlinearity and conformable temporal evolution. *Discrete Cont. Dyn. Syst.-S* **2021**, *14*, 2245–2260. [CrossRef]

71. Lewandowski, W.; Vaupotič, N.; Pociecha, D.; Górecka, E.; Liz-Marzán, L.M. Chirality of Liquid Crystals Formed from Achiral Molecules Revealed by Resonant X-Ray Scattering. *Adv. Mater.* **2020**, *32*, 1905591. [CrossRef]

72. Mitrano, M.; Wang, Y. Probing light-driven quantum materials with ultrafast resonant inelastic X-ray scattering. *Commun. Phys.* **2020**, *3*, 184. [CrossRef]

73. Cao, Y.; Feng, C.; Jakli, A.; Zhu, C.; Liu, F. Deciphering chiral structures in soft materials via resonant soft and tender X-ray scattering. *Giant* **2020**, *2*, 100018. [CrossRef]

74. Pace, D.C.; Shi, M.; Maggs, J.E.; Morales, G.J.; Carter, T.A. Exponential frequency spectrum in magnetized plasmas. *Phys. Rev. Lett.* **2008**, *101*, 085001. [CrossRef]

75. Pace, D.C.; Shi, M.; Maggs, J.E.; Morales, G.J.; Carter, T.A. Exponential frequency spectrum and Lorentzian pulses in magnetized plasmas. *Phys. Plasmas* **2008**, *15*, 122304. [CrossRef]

76. Milaniak, N.; Audet, P.; Griffiths, P.R.; Massines, F.; Laroche, G. Interpretation of artifacts in Fourier transform infrared spectra of atmospheric pressure dielectric barrier discharges: Relationship with the plasma frequency between 300 Hz and 15 kHz. *J. Phys. D* **2020**, *53*, 015201. [CrossRef]

77. Viktorov, M.E.; Ilichev, S.D. Method for Studying the Dynamics of Fast Frequency Sweeping Events in the Spectra of Non-Thermal Electromagnetic Plasma Emission. *Radiophys. Quant. Electron.* **2019**, *62*, 286–292. [CrossRef]

78. Vladimirov, S.N.; Maydanovskiy, A.S. Comparative Analysis of Natural Fluctuations in Two Classes of Self-Excited Oscillatory Systems. *Sov. J. Commun. Technol. Electron.* **1987**, *32*, 69–75.

79. Lv, H.; Liu, H.; Tan, Y.; Sun, Z. Improved methodology for identifying Prony series coefficients based on continuous relaxation spectrum method. *Mater. Struct.* **2019**, *52*, 86. [CrossRef]

80. Xia, X.; Luo, A.; Luo, S.; Xu, J. Harmonic current detection based on improved extended Prony spectrum estimation. *Dianli Zidonghua Shebei/Electr. Power Autom. Equip.* **2010**, *30*, 6–11, 21.

81. Marple Jr, S.L. Digital Spectral Analysis with Applications. *J. Acoust. Soc. Am.* **1989**, *86*, 2043. [CrossRef]

82. Kirenkov, V.V.; Gusarov, S.V.; Dosko, S.I.; Volkov, N.V. Method for diagnosing the state of mechanical systems based on modal analysis in the time domain. *Bull. Mstu Stank.* **2012**, *1*, 90.
83. Kukharenko, B.G. Spectral analysis technology based on fast Prony transform. *Inform. Technol.* **2008**, *4*, 38–42.
84. Dosko, S.I.; Logvin, V.A.; Sheptunov, S.A.; Yuganov, E.V. Identification of the induction sensor model and deconvolution of the input signal spectrum. *Sci.-Intensive Technol. Mech. Eng.* **2018**, *12*, 32–38.
85. Grigoriev, S.N.; Gurin, V.D.; Volosova, M.A.; Cherkasova, N.Y. Development of residual cutting tool life prediction algorithm by processing on CNC machine tool. *Materwiss. Werksttech.* **2013**, *44*, 790–796. [CrossRef]
86. Grigoriev, S.N.; Sinopalnikov, V.A.; Tereshin, M.V.; Gurin, V.D. Control of parameters of the cutting process on the basis of diagnostics of the machine tool and workpiece. *Meas. Tech.* **2012**, *55*, 555–558. [CrossRef]
87. Grigoriev, S.N.; Vereschaka, A.A.; Fyodorov, S.V.; Sitnikov, N.N.; Batako, A.D. Comparative analysis of cutting properties and nature of wear of carbide cutting tools with multi-layered nano-structured and gradient coatings produced by using of various deposition methods. *Int. J. Adv. Manuf. Technol.* **2017**, *90*, 3421–3435. [CrossRef]
88. Alekseev, V.V.; Zelevkov, V.V.; Krivoruchko, M.M.; Keem, J.E. Ion Sorces. U.S. Patent 2004/0195521 A1, 7 October 2004.
89. Han, C.; Rok Ahn, J.; Jung Ahn, S.; Joon Park, C. Optimization of closed ion source for a high-sensitivity residual gas analyzer. *J. Vac. Sci. Technol. A* **2014**, *32*, 021603. [CrossRef]
90. Zhurin, V.V. *Industrial Ion Sources: Broadbeam Gridless ion Source Technology*; Wiley: Hoboken, NJ, USA, 2011; p. 326.
91. Arnold, T.; Pietag, F. Ion beam figuring machine for ultra-precision silicon spheres correction. *Precis Eng.* **2015**, *41*, 119–125. [CrossRef]
92. Shvydkiy, G.V.; Zadiriev, I.I.; Kralkina, E.A.; Vavilin, K.V. Acceleration of ions in a plasma accelerator with closed electron drift based on a capacitive radio-frequency discharge. *Vacuum* **2020**, *180*, 109588. [CrossRef]
93. Hufner, S. *Photoelectron Spectroscopy: Principles and Application*; Springer: New York, NY, USA, 2003; p. 684.
94. Roberts, A.J.; Macak, K.; Takahashi, K. Test of the Consistency of Angle Resolved XPS Data for Depth Profile Reconstruction using the Maximum Entropy Method. *J. Surf. Anal.* **2009**, *15*, 291–294. [CrossRef]
95. Lu, S.; Yin, Z.; Liao, S.; Yang, B.; Liu, S.; Liu, M.; Yin, L.; Zheng, W. An asymmetric encoder–decoder model for Zn-ion battery lifetime prediction. *Energy Rep.* **2022**, *8*, 33–50. [CrossRef]

MDPI
St. Alban-Anlage 66
4052 Basel
Switzerland
www.mdpi.com

Coatings Editorial Office
E-mail: coatings@mdpi.com
www.mdpi.com/journal/coatings